Horse Power to Nuclear Power

Horse Power to Nuclear Power

Memoir of an Energy Pioneer

DONALD B. TRAUGER

Hillsboro Press
PROVIDENCE PUBLISHING CORPORATION
FRANKLIN, TENNESSEE

TENNESSEE HERITAGE LIBRARY

Printed in the United States of America

06   05   04   03   02   1   2   3   4   5

Library of Congress Catalog Card Number: 2001099117

ISBN: 1-57736-253-5

Cover design by Gary Bozeman.

Cover photos courtesy of www.arttoday.com, Donald Trauger, Ethel Trauger, and Tennessee Valley Authority.

Author photo courtesy of Carolyn Jourdan.

Hillsboro Press
PROVIDENCE PUBLISHING CORPORATION

238 Seaboard Lane • Franklin, Tennessee 37067
800-321-5692
www.providencepubcorp.com

For
Elaine, Byron, and Tom

Contents

Prologue

B orn into a pioneering farm family with modest expectations for the future, I began my life's work on the frontier of a new energy source. More than a century after my grandmother homesteaded on the Great Plains, I closed a career of research in energy systems—a time span from horse power to nuclear power. From my first experience, with the Manhattan District Project, then through a long tenure at the Oak Ridge National Laboratory, I came to see adequate energy as the life blood of modern civilization, vital to human progress.

In this memoir I review the development of energy systems as they relate to the needs and well-being of the United States. The viewpoint is that of the pioneers who made energy an ample commodity. Adequate energy sources have been a major factor in the progress of the United States from a fledgling nation in 1776 to today's super power. In succession, we have burned wood, coal, oil, and natural gas to heat our homes, power our factories, and provide transportation.

My vision of nuclear power shortly after its revelation in 1939 was as the next, giant step in humanity's long development of energy sources. Those early ponderings envisioned an organization not unlike the United Nations, but with greater authority as a necessary accompaniment to developing nuclear energy for power or weapons. The naiveté of my vision was shattered by the realities of World War II and by the invitation to participate in the highly secret weapons-related project at Columbia University, which was called the Manhattan District Project. Nuclear weapons arrived before international measures were in place for their control.

Today, the fossil energy sources upon which we depend have become increasingly uncertain for the long term. Use of coal, our largest fossil energy reserve, may become constrained because of its contribution to global warming; natural gas reserves are projected as adequate for only half a century; oil reserves have been depleted, and to maintain our way of life we rely on high levels of imported oil; and the United States has little experience with renewable energy systems. In this context, decades—even a century—become short-term planning time frames.

From this perspective, I find crowded, rushing, interstate highway traffic and the immensity of its energy consumption to be more than mere nuisance: I find them threatening. Seemingly, there is little public concern for the energy supply so long as the auto service stations and the airport fuel trucks are functioning. How remarkable that our national economy enables such levels of abandon while it provides the means to supply the energy to comfort-condition homes and offices at the travelers' destinations. I revel in the luxury of life in a society supported by an energy system that makes this possible. But is it sustainable for all people? We know that most societies on earth do not have such comfort. It may surprise many that the U.S. energy reserves may not support such abandon through the lifetimes of our youth. This book explores the options available and efforts required for long-term energy sustainability.

A vivid boyhood recollection from farm life is the energy required to turn the black soil of Nebraska with a horse-drawn plow. The horses perspired and heaved for breath at each break as they were turned to reverse the course through the field. Farm life became easier when we transferred the labor from horses to a tractor. The source of energy then shifted from the oats and hay we produced to feed the horses to gasoline that was extracted and refined from oil stored by nature deep in the earth. The chapters describing farm life of the drought and depression period of the 1930s also depict hardships that may be prophetically illustrative of problems confronting a society deprived of adequate energy.

Pioneer and *Pioneering* are recurring terms in the book. They refer to my maternal grandmother, an English immigrant, who as a child arrived at a remote home site in Nebraska by covered wagon, to the pioneers of the Manhattan District Project, and to many new developments in energy that appear necessary for the long-term future of the United States. Those later pioneers with whom I worked include Hans Bethe, Enrico Fermi, Willard Libby, James Rainwater, Edward Teller, Eugene Wigner, and Harold Urey. Those men who held or later earned Nobel Prizes were rivaled by others such as Manson Benedict, Eugene Booth, John Dunning, and Francis Slack. It is a privilege to describe accomplishments led by those scientists and implemented by persons, such as myself, who also can claim roles as pioneers in the new and epoch-making nuclear field.

This tracing of U.S. energy development through two centuries and into the future is written as a memoir that includes projections related to energy-systems development. It includes many stories that relate to the energy theme or life in general. A brief bibliography at the end of the book provides a number of more scholarly writings that present the science, engineering, and history of the subject for the interested reader. My assignment within the effort at Columbia University to separate uranium isotopes was equally applicable to weapons and nuclear power. Thus, my career in developing energy systems began with intense interest and dedication but also with a great tension between the benefits of nuclear energy and its potential use for destruction. My account of nuclear-weapons development during World War II is written from a "worm's eye view" within one segment of the Manhattan District Project. To my knowledge these specific events have no prior documentation. It is a saga of intense effort driven by the belief that Germany was pursuing a nuclear weapon from a position of scientific prominence, a belief that intensified when their effort became cloaked in secrecy.

I also hold it a privilege to have worked at the Oak Ridge National Laboratory and therefore describe its programs for new and innovative energy technologies. The experience ranged from tasks within a project to develop a nuclear powered aircraft, to programs for improving the safety and efficiency of nuclear-electric generating plants. Systems for the safe handling and disposal of radioactive waste were developed but have not yet been publicly accepted or fully deployed. Innovative programs, such as that for the High Temperature Gas-Cooled Reactor, involved several international organizations. Indeed, a satisfying aspect of my experience in nuclear energy development derived from the effective cooperation and friendships that evolved among the engineers of many countries.

My tenure at the Oak Ridge National Laboratory (ORNL) occurred in a time when its programs expanded from the development of nuclear systems to those of a broadly based energy laboratory. Nuclear energy is prominent in several chapters because my work centered in that field and because of its role and promise as a major resource. However, the laboratory also accomplished pioneering work by initiating energy programs for conservation, renewable sources, efficiency in use, safety, and environmental impact before these needs were apparent to either the public or the government. ORNL provided early recognition of the many energy systems that must be developed and deployed appropriately for our lifestyle to be maintained. Those that are renewable by growing plants or make more direct use of the sun's radiation need increased emphasis for evaluation and utilization. Greater efficiency in energy use for manufacturing processes offers attractive returns and lasting value. I was fortunate to become familiar with most of these fields during fourteen years as an ORNL associate director.

From these perspectives I attempt to suggest pathways for future long-term technological and social stability. I have chosen a projection time frame of one hundred years because that is short enough to seem relevant to my grandchildren and sufficient to provide room for the foresight of longer-term problems. I am immodest enough to hope that this book will be helpful to political leaders and others responsible for planning the efficient use and effective conservation of the marvelous energy resources with which the earth is blessed. Needed changes will not come easily. Nuclear energy, for example, is a difficult technology compared with burning natural gas from limited reserves, but nuclear power, now the source of 20 percent of U.S. electric power, can be abundant and is environmentally friendly.

We must recognize that ample energy systems are essential to transportation, industry, commercial enterprises, and the creature comforts we enjoy. Improved energy availability can also lessen international tensions over resources that may cause conflict and war. My hope and prayer is that more effective cooperation can be reached among peoples and among governments, which seems imperative for survival in an age of nuclear and other weapons of fearsome capability. We must find new pioneers who will lead the world to a lasting stability that assures peace.

PART ONE

1838~1938

Family Pioneering
and Perseverance

The Downey Sod House, 1874–1923.

ONE

Pioneering Family Heritage

G randmother's description of sharing her parents' excitement on the first train to cross that prairie land made an indelible impression on me. Anna Coates Downey, my maternal grandmother, made the trip to Nebraska in 1871 as a child of six years with her parents, Thomas and Sarah Coates, and four siblings. They had traveled from England in a sailing vessel to make their home on a pristine prairie and, after visiting briefly in Michigan with English friends, continued to Lincoln, the state capitol of five years where the rail service ended. Three days after their arrival, they were excited to take the very first passenger train on the newly completed Burlington & Missouri Railroad extension to Crete, about twenty-five miles to the southwest. In Crete they purchased a plow, a covered wagon, a team of horses, a cow, and other supplies to supplement their English possessions and set forth into the frontier.

Their first American home site was located on the upper north bank of Turkey Creek, about twenty miles from Crete. During that first spring and summer, the two older sons excavated the deep loam soil of the upper creek bank to form two underground structures, called "dugouts." They felled trees that grew only along the creek in the otherwise level, treeless plain and split the logs to span the excavations. They covered the logs with layers of tree bark and dense prairie sod to form a nearly waterproof, contoured roof. One dugout was to be their home and the other a barn for the horses and cow. In addition to breaking and transporting sod for the roof, my great-grandfather and the boys plowed the land and planted crops to produce grain for sale and to feed the family and livestock through the winter. Mary, the oldest child, Anna, and even little Harry helped their mother prepare and plant a garden in the floodplain across the stream. Thus began the Nebraska pioneering of the seven-member Thomas Coates family.

Standing recently near the doorway location of that home, my granddaughter, Hallie, and I were fascinated by the meandering

Granddaughter Hallie with Don at the dugout site, 1999. Photograph by Robert Trauger.

stream carelessly caressing its floodplain, the same stream so often sparkling in the eyes and thoughts of her great-great-grandmother, Anna. Grandmother Anna, as a young girl, saw it as crystal clear and sparkling from sunlight beams filtered by foliage. Looking upward, she viewed massive oak, elm, ash, and towering cottonwood trees, some laced with wild grape vines. Adjacent to these trees were native plums that produce delicious fruit in September. Looking to the north from the roof level, her vista was of the seemingly endless prairie with waving, tall grass in summer and wind-swept snow in much of winter. That land was to provide the cash crops to achieve their goal of a more prosperous life. The garden was lifesaving during the severe drought of 1873. Spring-fed Turkey Creek provided water for its plants as well as for the animals and household. Earlier droughts had been reported by explorers of the 1820s and 1850s; the cycle continued until I experienced the drought of 1934.

Apparently, the creek was properly named, for it supported an abundance of wild turkeys. Other game included catfish, antelope, rabbit, prairie chicken, and even buffalo. Grandmother remembered deer frequently passing the dugout's door. Snakes and rodents also were visitors, even inside the dwelling. Hunting expeditions by the men were not difficult because the animals had not become gun shy. However, the hunters sometimes encountered Native Americans with whom they were expected to share the trophies. From today's perspective, that seems only fair compensation from intruders on Indian soil, but that was not the attitude in 1871. A book entitled *A History of Nebraska*, published in 1882, states, "The area continues to be infested with Indians."

Grandmother remembered the family as initially living with fear that these native neighbors would resent to the point of violence appropriation of their land and game. Her brother, Joe, once encountered an Indian who took his gun. Frightened and expecting the worst, Joe readily relinquished the firearm; however, after a brief inspection, the weapon was returned to him with a friendly gesture. Grandmother also told of Indians coming to the door asking for food; it was never refused. She remembered only good relationships with the friendly natives.

This account of the Coates family journey and early days in Nebraska is largely from my recollection of stories that Grandmother related to me or to others in my presence. Grandmother's descriptions of her experiences as a child were vivid and perceptive. This was particularly true when we visited the site of their dugout home, still identifiable as a depression in the creek bank. Unfortunately, I did not know to ask more questions or to take notes; it might have enabled me to better understand, for example, how Grandmother's experiences have affected me through her teaching of my mother.

Imagine the contrast between Coates family life in the stone house of an English community with centuries of recorded history and life in a home excavated into the bank of Turkey Creek. The transition to pioneering life must have been most difficult for the women as they established housekeeping in a dwelling having partially lined earthen walls, packed-dirt floors, and a rustic roof. For the men, accustomed to the worn and rocky land of former farms, tilling the rich, rock-free, fertile soil was a pleasure, once the tough prairie sod had been broken.

Even so, the effort was demanding for both men and women who were driven by visions of future abundance.

That enterprise at Turkey Creek was typical of U.S. farm energy systems up to that time. Two horses pulled a plow that was carefully and firmly guided by the farmer who walked behind. The plowed land was leveled for planting and the seed covered by dragging brush across the surface. The basic energy source was the abundant native grass that provided food for the horses. But much would change as steam engines, riding plows, and other improvements in machinery were widely introduced. The Coates family, in pioneer fashion, adopted the new patterns for farming as their resources permitted.

One nearby neighbor of the Coates's dugout site had obtained one square mile of prairie land adjoining Turkey Creek to breed and raise draft horses. This enterprise provided horse power to meet a rapidly growing demand, as more and more settlers arrived to convert the rich prairie land into farms. Even though steam-powered equipment had become available, horses continued to be a major energy source in moving soil for railroads, in road and street construction, and for transportation. Horse power would decline but nevertheless remain important to meet energy needs for more than sixty years.

My maternal grandfather Downey's family history is not as well known to me, but the genealogy has been traced to a marriage in New York State. William Downey and Nancy Gossard married on January 15, 1829, and were the parents of William Jr., born in Yates County, New York, on May 13, 1835. William Jr. married Nancy Stephens of Ohio on January 8, 1860, and came to live near Exeter, Nebraska, early in 1872. Their son, Robert Herman Downey, born in Kalamazoo County, Michigan, on June 26, 1865, married my Grandmother Annie (as I knew her) on December 28, 1886. I must have known great-grandmother Nancy who died in 1923, but I recall her only through family stories of a lively lady who crossed the Atlantic Ocean seven times in sailing ships, suffering seasickness on each occasion.

The Coates and Downey families have not researched their English roots, but memories of long-lost letters suggest that both families had resided in or near Worksop, Nottinghamshire. After living in the Nebraska dugout for two years, the Coates family moved to a sod house nearer the Exeter community, presumably to be neighbors of the Downeys. Both the Coates and Downey families then lived in sod houses made of turf carefully cut as slabs from the prairie grass soil and piled to form thick walls. Farm buildings, mostly built later, were of wood. They often shared labor and equipment to manage the farms efficiently. Each of their eighty-acre farms could be tilled with the energy of four horses; wheeled machines on which the men could ride soon shifted the work further to the horses.

The Exeter farming community was established in the 1870s during that great period of westward migration and railroad building. It soon developed an international character. The town's population was predominantly English, but the trade area included Scandinavians to the north, Germans to the east and northeast, Czechs to the south, with Russians and others intermingled throughout. Most of these newcomers fiercely struggled to become Americanized, whether they had come from abroad or from families already rooted here.

The towns along the Burlington & Missouri River Railroad were sited and named by the railroad. (Later it was known as the Burlington, and now is the Burlington, Northern & Santa Fe. A branch line of the Northwestern Railroad also served the town from the late 1800s until after World War II.) The naming and location of Exeter interested me through descriptions by early pioneers; the following account derives both from those stories and from historical records.

Railroad construction was supported by the federal government through grants of thousands of square miles of land by providing title to odd-numbered sections (one square mile each) for forty miles on either side of the route chosen for the rails. In turn, the railroad sold that land to settlers who could best be attracted to purchase if the amenities of a village were available. The Burlington & Missouri River Railroad chose to locate towns at eight-mile intervals. A logic attached to that spacing was that a farmstead located midway between towns was less than an hour away on horseback, an hour with a fast team and buggy. Named alphabetically and starting at Lincoln, the towns now are Crete, Dorchester, Friend, Exeter, Fairmont, Grafton, Harvard, and so on. (The first two towns, one name beginning with an A, the other with a B, no longer exist.) A pioneer, George Friend, had staked and filed papers for a town he called Friendsville and had recently built a house on the site. The railroad honored the claim when he agreed to the name Friend as more consistent with their choices. Town names and locations west of Friend were then allocated using the alphabetical order previously planned. This naming deviated from a plan developed by a committee formed by the railroad. The misplacement of Exeter at nine miles from Friend and seven from

Fairmont and the alphabetical disorder apparently derived in part or whole from the efforts of an entrepreneur, whose story follows.

A physician, H. G. Smith, having tired of his medical practice in Michigan, chose to go west in 1870, presumably to seek his fortune in Nebraska. He walked the fifty miles from the Lincoln railroad terminal to the open area that became Exeter. Upon observing the line of stakes defining the future railroad route, he hurried back to Lincoln and placed a homestead claim on 160 acres of government land that he thought would be the center of the new town. Enterprising Dr. Smith apparently was not an expert in measuring distances; his claim was west of the planned site. Undaunted, Smith then persuaded the railroad to locate Exeter on his property instead of one mile east on an area better suited by terrain and recommended by the committee. Exeter grew rapidly and by 1879 had become incorporated and boasted several stores and churches, a school, and even a newspaper.

The Downey farm, when inherited by my grandparents, had been increased to 160 acres, ample to support their four children, of whom my mother, Ethel, was the third. She was preceded in order by Aunt Pearl and Uncle Chester and followed by Uncle Harold. Mother was born in the Downey "soddy" on September 10, 1891. She described a comfortable life in the four-room sod house that stood on a slight elevation of the plain. The thick, sod walls provided insulation from hot winds of summer and blizzards of winter.

It was a luxurious "soddy" with plastered interior walls and a surfaced floor. Even so, wildlife visitors kept life interesting and sometimes frightening. Both Mother and Grandmother told stories of rodents and snakes, including rattlers, that had dug or pushed their way through the sod and cracks in the plaster to visit the family, and birds found nesting sites in the outer walls. A photograph shows this sod house as a simple but attractive structure, with sash windows and a wood-shingled roof. My mother lived in that home until she was twelve. When after fifty years the walls of the soddy became badly weathered by rain and the persistent winds of the Midwest, the old home yielded to the elements and fell into disrepair. It was replaced by an attractive two-story frame house, but the soddy stood as a storeroom for another decade before it was eventually destroyed.

The farm where I grew up was established by my grandfather, Manuel Edmund Trauger, known in the town as "M. E." During an initial visit to Exeter in 1877, Grandfather bought a tract of land that he soon sold. He returned in 1878 to purchase 160 acres from the railroad to establish a home site. The farm was expanded to 240 acres by my parents, Charley and Ethel Downey Trauger, at the time of their marriage in 1919. Their adjoining "eighty," as land areas there are known, included a fine house of modest size and other farm buildings. Wet-weather Johnson Creek traverses both tracts to provide slight relief from the region's flatland character. Most of the creek's flood plain and slopes were retained for pasture and hay in an early demonstration of soil conservation, when most farms were developed for row crops of wheat, corn, and oats.

Although Grandpa Trauger was a farmer, he was not a dedicated tiller of the land. His loves were reading, writing, and the Exeter Methodist Episcopal Church, which he helped found. We have learned little of his boyhood life in the tiny community of Flatbrookville, New Jersey, or of his education. We know that he taught school in New Jersey as a young man and that he visited rural communities in Illinois and Iowa while progressing west to Nebraska. Grandpa and Grandmother, Mary Caroline Hill Trauger, maintained correspondence with family members and friends in New Jersey and with many acquaintances across the country.

Grandpa took pride in the family history, but did not share those stories with me as did Grandmother Downey. Instead, he delighted in asking questions. He often asked me how to spell an unfamiliar word that he knew very well and even embarrassed unsuspecting visitors with such questioning as: "How is Chautauqua spelled?" or "how should I spell *charivari*?" Even when I thought I knew the English spelling (*shivaree*), Grandpa would insist on the original, French spelling. As a small child, I sometimes attempted to avoid him, but upon venturing too close, Grandpa would catch me in the crook of his ever-present cane, necessitated by arthritis. Later, in his mid-eighties, he suffered a stroke that confined him to a wheelchair. Grandpa was not very tolerant of toys or noise when they interfered with his mobility or concentration, but he took great interest in my school lessons and childhood accomplishments.

Grandpa would have been interested in the confirmation, after his death in 1934 at age eighty-nine, that his ancestry was traceable to Germany. Although Aunt Minnie Trauger Lewis, my father's older sister, had researched family history to achieve membership in the Daughters of the American Revolution, only recent genealogical research of the Trauger family history has confirmed belief that the family name in Germany was Dracker. The American name may have resulted from the ship captain's reading of poorly written German script, interpreting its *D* as *T* and *k* as *g*, while poorly understanding the German pronunciation. The Evangelical Church records of Bickenbach, Darmstadt Hessen, Germany, translated, include the entry "John Christian Dracker emigrated to America on 9 October 1747 on board the *Restoration* from Rotterdam [to Philadelphia]. Descendants now call themselves 'Trauger.'" When Christian's brother Ludwig came a year later, he had to learn his new name. After serving a term as indentured servants, the brothers purchased farms in Nokamixon Township, Bucks County, Pennsylvania, in 1767. Christian's farm continues to this day in a descendent's ownership. (We enjoyed a family reunion there on July 24, 1993, attended by 398 of the more than 15,000 known descendants of the two brothers.)

Grandpa, having grown up in mountainous northern New Jersey, lacked experience with flatland farming. Consequently, the family often deplored his choice of a poorly drained site for the Nebraska farmstead. It may have looked good to him because it was level and perhaps the tall grass obscured its tendency to be wet. Even in that relatively dry climate, lack of drainage has often made boots essential when rain creates nearly bottomless mud. The farmyard was quickly denuded of the native grasses by the many chickens that my grandmother (and later my mother) raised to provide "egg money" for groceries and other needs. The cattle and hogs produced other problems in the low-lying, often wet barnyard: an unpleasant aroma and plagues of flies were ever present in summer. Now that chickens and pigs are grown commercially by others, grasses again cover most of the fertile farmstead soil.

Grandpa had built an attractive tee-shaped, two-story farm home decorated with interior bric-a-brac typical of the Victorian era. He did much of the work himself and completed construction in 1888. I recall clearly his story of having worked on one cold winter day to finish rooms of the interior. That evening, as Grandpa traveled by horse and wagon to their temporary home, a small cabin on an adjoining tract, he reported having "a strange feeling about the new house." Quickly returning, Grandpa found that the coal-burning stove on the second floor, with its fire banked under ashes, had exploded. Red hot coals on the floor were burning deep patterns into the white pine flooring. The story of that Nebraska farm might have been quite different if he had not returned to extinguish the fire.

The Trauger farmhouse, circa 1895.

The house was patterned after Grandpa's parents' home in Flatbrookville, New Jersey. In February 1946, my wife and I searched for that house on a Sunday auto trip from our first home in New York City. After looking around the area, we noticed a house that met the description. Was it once the home of my great-grandparents? Neither this nor any of the nearby summer homes was occupied, so we were unable to verify the Trauger connection until we saw smoke issuing from the chimney of a log house some distance down the road. Responding to my knock, a very old man came to the door and on learning of my mission exclaimed, "Ed Trauger! We played together as boys." He was pleased to meet a grandson of his friend and welcomed me into his home. The conversation was enchanting in that room of walls darkened by smoke from the open fireplace, with its glowing bank of coals providing comfort on that cold winter day. This gentleman, whose name I no longer recall, was younger than my grandfather, but they had played together in sand-lot baseball games and other activities some seventy-five years earlier. He confirmed our identification of the house.

The Nebraska family, not satisfied with their somewhat modest house, soon initiated expansion. They added an enclosed porch that was later encompassed into the house as an entrance hall, then a dining room, and finally a kitchen unit with a pantry and porch, all in a linear arrangement. This made for a more livable house, but the additions with their flat metal roofs, presented a stark and utilitarian outline behind the Victorian front. However, I recall the beautiful formal lawn bordered by beds of peonies and roses along brick-lined cinder walkways with stately silver maple trees as a background. The house and gardens, viewed from the roadway, were among the most attractive in the neighborhood. Unfortunately, the trees and lawn were to perish in the severe drought of the 1930s. Remodeling of the house has continued to this day with the twelfth major reconstruction or revision now completed by my brother, who has restored much of the original architectural beauty in a different exterior configuration.

The family, consisting of Grandfather, Grandmother, Aunt Minnie, and Father, was often extended by young people who were given temporary shelter. By the late twenties it was the home of Grandfather, my parents, my brother Robert, and me.

I have always thought of the house as an interesting structure that had grown in strange ways. Some have even thought that it might be haunted, several occupants and visitors reporting strange noises and one reporting having seen Grandmother Trauger on the stair long after her death. Reinforcement of the aura came from secret places accidentally created by the amateur architectural designers, who had not fitted all of the space into living quarters. Some of these "imaginary caves" were great childhood hiding places; others provided summer storage for mundane items such as black stovepipe pieces for the wood-burning heaters of winter use.

A water piping system added to the house and farmstead about the time of World War I functions effectively today. The water was pumped from the well by windmill power (now by an electric motor) into a poured concrete tank six feet in diameter and nearly thirty feet tall. The water is protected from freezing in winter and kept relatively cool in summer by an outer enclosure formed by a concentric wall of hollow concrete blocks. From the tank, water is distributed through underground pipes to the household taps and to barnyard watering troughs.

Water piped inside the house was available both hot and cold through faucets at a large kitchen sink. The hot water was heated by natural circulation through a pipe loop connected to a coil in the wood-fired kitchen range, then through a hot water tank of conventional design. The kitchen sink was also equipped with a hand-operated pump connected to a cistern filled with rain water collected from the various roof configurations of the rambling house. This well-equipped sink provided for a multitude of activities, from hands and face washing to hand-washing of clothing to cleaning kerosene lamp chimneys. It also was large enough to accommodate two dishpans for sanitary food preparation and washing of the dinner dishes.

The sink drained to a nearby cesspool constructed in 1923. That is the year my parents moved from their first farm home to live with my grandfather, after the death of Grandmother. The time is fixed not only in mind but also in concrete by my three-year-old hand impressions and the date inscribed on the cesspool cover.

The kitchen addition included a separate "bath room," but its only fixtures were the hot water tank and a "clothes shoot" to the basement laundry. Bathing was in a portable, circular, galvanized iron laundry tub filled with "soft water" obtained from the cistern. Water was lifted by the hand pump and transported in buckets for heating in the reservoir of the stove. (Water from the well had a very high mineral content and hence was "hard water." Effective water softening devices were not then readily available or affordable.)

In my youth the bathing routine was fairly complex: carry the tub from the basement to the bathroom, dip hot water to partially fill it from the stove reservoir, add cool water from the pump to adjust the temperature, and then contort body and legs into the circular tub. This entailed a substantial effort, and the stove reservoir capacity was limited; multiple baths in the same water were often in order. The person in the last position not only encountered water of dubious quality but was required to carry the water outside the house for discharge and then clean the tub.

On cold winter days, bathing took place in the kitchen beside the warm stove reservoir and screened by the dining room door that opened into the kitchen. Washing was hurried in this seclusion, for one was exposed to rapidly changing drafts of cool air. Even this rudimentary system was better than that of most farm homes of the time, although the Downey framed home of 1906 had a similar water system with a permanent bathtub and drain.

A pleasant feature of the Trauger home in the 1920s and 1930s was the east porch. The family sat on the edge of the narrow porch with feet on concrete steps. The dog, or dogs, no matter what other housing had been provided, always inhabited the east porch crawl space. Of course, they joined the party and responded to attention when family members were present. The porch also afforded a sweeping vista of the buildings and fields of the farm, even providing a view to the water tower and grain elevator of Friend, the town eight miles to the east. On occasions mostly limited to summer evenings, the porch was a favorite place to reflect on events of the day, plan for tomorrow, and resolve family problems. However, the horizon silhouetted by the full harvest moon, spectacular sunrises accompanied by mirages, and cloud patterns illuminated by the setting sun were exhilarating to experience at many times of each year. Watching the activities of farm animals, rabbits, pheasants, cardinals, crows, and hawks were year-round pleasures associated with the kitchen window over the east porch.

Despite these many positive features of the house, it had no electricity and would not have an indoor toilet until 1947. The latter had been thought impractical because of the flatness of the terrain and the nature of the soil, but when installed, the septic tank and drain field functioned well. The toilet facility up to that time was the traditional small square building at the end of a brick pathway. It was fitted with board seats of two heights and three holes of graduated sizes to accommodate the posteriors of individuals young and old. Sears Roebuck and Montgomery Ward catalogs of the previous season provided both reading material and toilet paper. The catalogs, fortuitously sized to last until the succeeding issues arrived, were in continuous supply. There was always a roll of soft toilet paper for guests, but household frugality restricted family use to the catalogs. Vivid memories of that facility relate to icy-cold exposure during Nebraska winters.

The farmstead consisted of an open farmyard bounded by a cattle barn and fenced barn lot to the south and a horse barn at the north. The horses were trained to proceed to the barn without roaming over the countryside, or trampling the garden and field crops. Since the horses and cattle grazed over the same pastures and were corralled in a single lot, one also had the problem of releasing the horses without allowing a frisky calf to escape. While sitting on the east porch, we sometimes pondered my grandfather's logic in choosing the farmstead layout. It received more expressive criticism from those attempting to corral particularly venturesome horses or cattle.

The horses were seldom a problem in this regard because they associated a trip to the barn with oats in the feed troughs and deep mangers of prairie hay. We loved the horses as work companions and as pets. Each of the nine horses of mixed quarter horse and Belgian heritage had a distinct and interesting personality. For instance, Old

Harry, as we called him, was an expert at opening gates and doors. His sensitive upper lip could manipulate a spring-loaded latch, a locking bolt, or sometimes even a chain hook. We could admire this talent because he seemed to appreciate being free and stayed within the farmstead causing no damage. However, when his skills also released other animals, his popularity rapidly declined.

The farmstead was a beautiful place in the 1920s. Trees planted by Grandfather had grown to maturity. Wheat and oats waved in the breeze, corn stood tall, and grasses shimmered in summer pastures. The part of the farm that I came to love best as a boy was the twelve acres of native prairie that was never plowed or pastured. In wet years, the grasses grew as high as a grown person's head and waved beautifully in the wind, much as we envisioned ocean waves. In the fall, when the grass was cut and stored as nutritious hay for the horses, even its harvest was a pleasure. The drying grasses were clean, had a wonderful fragrance, and endured handling without shattering leaves, unlike the alfalfa hay grown for cattle. Dry alfalfa is terribly fragile and shatters easily, its leaf and stem fragments entering openings in clothing and mixing with the sweat of the hay handler to cause great discomfort.

I am proud that this bit of native Nebraska prairie has been preserved to this day. The age of tractor farming, with fewer horses to display shining coats of hair when fed this hay, has largely eliminated the native prairies of eastern Nebraska.

My earliest recollection of the Downey grandparents' farmhouse is of Christmas 1923, their last year on the farm. The family had gathered in the living room around a large-base burner stove. Its mica windows revealed hot glowing coals giving warmth and a holiday atmosphere to the room. It was time to exchange gifts. I have long since forgotten the presents I received, but I recall my parents' excitement on receiving their gift from Uncle Harold—a gasoline-fueled, thorium-mantel-type lantern. It was much brighter than kerosene flame units and, in addition to improved illumination for farm chores, provided good light for many outdoor projects and for nighttime trips to the outhouse.

Several successive Christmas gatherings were at the grandparents' home in Exeter after they moved from the farm. On each occasion a large tree was decorated, and on Christmas Eve candles were lit on its branches while we all sang "Silent Night." The candles had been retained for tradition, even though the house had electricity. When freshly cut trees became hard to find, the candle fire hazard was deemed too great even with buckets of water standing by; so electric lights were used. Even though we were more relaxed by the increased safety, the electric bulbs never produced quite the same ambiance as that from the soft, wavering glow of the candles.

The Exeter that I knew had grown from the lone residence of Dr. H. G. Smith to the 1930 census of 941, proudly proclaimed by a highway sign. The gravel road known first as the DLD (Denver-Lincoln-Davenport) had been renamed U.S. Highway No. 6. From the east, it followed section lines to Exeter Avenue, turned north through the town, then west at the next section line and on to Fairmont. It successively crossed the tracks of the Northwestern and Burlington railroads and became the principal business street of the town.

Exeter was a prosperous small town that harbored the world's first index tag factory. Charles Smith, son of the founding doctor, inherited his father's entrepreneurial spirit. He noted that the indexes of dictionaries, then indented into and printed on pages, were awkward to manufacture and use. He devised and patented printed tags to be attached to pages of account books and other business records. Their manufacture in Exeter and sale around the world brought good jobs to many people. As a high school boy my father fired the steam boiler that both heated the factory and provided steam for an engine that drove a direct-current electric power unit. The latter supplied motors to power factory equipment and also fed the initial electric distribution system for village homes and stores. Father rode a mule to Exeter early each morning to tend the boiler and electrical equipment.

The town also accommodated the Spitz Foundry and machine shop that melted and recast iron into parts for machines used on farms and in factories. The equipment of this early plant was powered by an engine outside the building that transmitted its power through a long shaft that reached the far side of the building. Wheels to accommodate belts then conveyed power to each machine. This arrangement was effective engineering before electric motors were available but quite dangerous, as the belts were unguarded and severe accidents were common in such factory power systems. The Svec repair shop, a wagon factory and a machinery dealership and livery stable enhanced the thriving agricultural community.

In Exeter, the relationships between Protestants (Baptist, Christian, Christian Science, and Methodist churches) and the predominant Catholics were more or less respectfully divisive. Most

Catholics had settled north of the Burlington tracks, where their fine church and school were located; Protestants populated the south. Even businesses on Exeter Avenue were partially segregated by religion. One grocery was operated by a Catholic family, the other by Protestants, with similar separations for other stores. There were exceptions; some fields of commerce would not support more than one enterprise, but where the option was available both Catholics and Protestants tended to patronize their "own" businesses.

This separatism was illustrated when a storm damaged the Catholic church's tall bell tower. The congregation had no one who dared climb the tower to make repairs and was too poor to hire a contracting firm. My father volunteered to assist and spent several days restoring the tower. It again stood as part of the village skyline, together with the grain elevators and a water tower's tank. As a consequence of his service, Father was criticized by some Protestant friends for helping the rival religion.

Today, the town and community are more coherent; Catholics and Protestants often work together on ecumenical projects and share in special worship services. My father was pleased to see that relationship develop, for he never fully understood why there should be more than one Christian faith, even though he perceived value in having different ways for people to express the same basic religion. As an active Methodist he followed his father; in succession they served as secretary/treasurer of the Sunday school for fifty years, starting with the church's founding.

In 1904, the Methodists split their rectangular building in two, separated the segments and built a wider nave to join them. The new pews were curved to focus on the pulpit from which rousing sermons were said to have been delivered in early years. The large, figured stained glass windows at either side often provided me, as a boy, with a welcome diversion from the sermon.

My parents always encouraged me and contributed greatly to my early education about farming and local history. I first called my mother Mama, but I think of her as Mom, my name for her through the years. She was particularly beautiful and as a young woman had long, soft auburn hair that fell to her knees when not braided and piled on her head or arranged in a bun at the back. One of my early memories is of her first haircut in response to the bobbed hair trend of the twenties. I liked her long hair and she did too, but fashion and comfort on hot summer days prevailed. She had her picture taken with the hair

Don with Mom while her hair was still long, circa 1923.

streaming down her back before going to the beauty shop. I insisted on being photographed with her, also with my back to the camera.

Mom was a study in energy. She maintained the house, fed the family (relying substantially on the garden and chickens she tended), and actively worked in the community and church, where she often worked in the kitchen to prepare for special events and where, though aware of the limitations imposed by her eighth-grade education, she studied and faithfully taught young children. She was a kind and caring mother who could always find time to console or to help solve a problem. I spent many happy hours with her, playing with toys at her encouragement and listening to her read stories for children. She helped me memorize many nursery rhymes, stories, and poems.

Mom may have saved the farm from financial failure in the drought and depression. Father earlier had thought they should purchase an adjoining property offered for sale. He probably wanted enough farm acreage to support two sons (we will meet my brother later), but Mom would not accept going deeply into debt. Had that purchase occurred, our farm might have become the property of a bank in the depression years, as did those of many neighbors. Mom also was courageous in taking stands on community affairs. For example, she confronted a couple who had adopted two children, seemingly for the farm work they could provide rather than for the good of the children and a balanced family life. Many in the neighborhood were concerned about those children, but it was Mom who counseled with the family on several occasions and made a difference.

Mom also was a teetotaler, a strong supporter of prohibition and was indignant when bootleggers found hiding places on the farm for "drops" of bottles full of illegal liquor. We had hedges, deep ditches, and woods that often were favorites of the trade. In her self-appointed role, Mom became very alert to any suspicious vehicle on our country roads. She would follow up by surveying the potential drop locations, and she often found the bottles. I recall as a young boy with a sturdy stick flailing a cache of bottles to their total destruction. Mom also sold eggs and cream to individual village residents as well as to the local market. She ministered to the sick with food and counsel, supported the schools, and was generally the kind of person who made a beneficial difference in the community.

Pop, as my father was known by us, was different. He was more content to let the community come to him. Because of his success as a farmer, neighbors frequently asked about his choice of seeds, the best time to plant, and the equipment to use or buy and later about the federal farm program. He was elected to the board of directors of the local Exeter Co-op Grain Elevator and took great pride in its successes. When the privately owned grain elevator in Exeter was purchased by a large multi-state company, they attempted to force the Co-op out of business. The new firm's tactic was to pay a little more for grain, even to sell at a loss and thus unfairly attract new farm business. The farmers, seeing their potential plight at the hands of a single, privately owned market for grain, stood by the Co-op until it eventually purchased the other elevator. Of course, the Co-op was then, again, the only market, but its profits were returned to the user-owners.

Pop had grown up with a strong work ethic and early took responsibility for the farm from my grandfather, who was not an enthusiastic farmer. This necessitated Pop's dropping out of school in the eleventh grade, despite the apparent inconsistency with Grandpa's other priorities

Pop, Charley Trauger.

as student and one-time teacher. Pop thus had no experience in sports but on occasion would valiantly join in ball games with me and my cousins when another player was needed. My feeling toward him on such occasions was a mixture of gratitude and sympathy, sometimes almost embarrassment for his awkward but undaunted efforts to learn.

Pop, also called Uncle Charley, was a strong man, both emotionally and physically. He befriended and counseled my Cousin Caryl through a rebellious youthful phase that frustrated his mother (Aunt Pearl) and distressed others. Pop also assisted family members with work or money as resources permitted. His physical strength was demonstrated when we took a team and wagon to obtain some discarded slabs of Exeter's concrete sidewalk for use on the farm. Uncle Harold and Uncle Chester, each weighing about two hundred pounds, had been engaged to help lift the heavy slabs into the wagon. When the heaviest pieces were thought to have been loaded, Pop walked to a store for some purchases, leaving me in charge of the horses and my uncles to heft the remaining pieces. They loaded all but one slab that they judged too heavy for them to lift. Pop was frustrated upon return to find the slab on the ground with my uncles having departed for their homes. He, who never weighed more than about 150 pounds, proceeded to lift and put it in the wagon. Those pieces of concrete became something of a compulsion with him, as he pursued a partial solution to the barnyard mud. The slabs were fitted together and covered with reinforced concrete to form, at minimum cost, a slightly sloping, firm floor for the cattle feeding station.

When Highway No. 6 was paved with concrete as a needed improvement, but also as a federal work project of the Great Depression era, it no longer traversed Exeter Avenue. The new road closely followed the Burlington rails from Friend to Fairmont and beyond, crossing Exeter south of its business center. In response, the town erected a large sign in neon lights of red, white, and blue over the highway to proclaim its existence. Large letters spelling EXETER were encompassed in an arrow that futilely pointed to the business section as autos passed by at or above the city speed limit. The town population steadily declined as the bypass, the shortened road to Lincoln, the drought, and the economic depression took their toll.

Construction of the new highway directly benefited the Trauger farm because traffic was temporarily detoured past the farmstead and a gravel surface was applied to the road. No longer did the family have to contend with slippery mud on rainy days. However, the highway detour was a mixed blessing; traffic was fast and concern for safety rose, particularly for my younger brother. Gravel-surfaced roads are dusty and with the heavy traffic, Mother had problems with laundry becoming soiled on the drying line and with housekeeping in general. One windfall was the nearby wreck of a truckload of shelled pecans. Although the trucking company recovered most of the load, there were enough pecans remaining for neighbors to scavenge a year's supply. One, however, Giles Hanson, was disappointed; he would have preferred either bananas or walnuts.

The community has now changed in many ways, in part because of the traumatic days of the thirties. In my earliest memories, the two-mile stretch of road past our house had four active homes; today, our place (now my brother's) stands alone. The land is tilled but by farmers with much larger acreages and highly mechanized equipment. Diesel and gasoline engines have replaced the horse power of earlier times. Exeter's population also has dropped to about six hundred, and the sign over the highway is gone. Long-distance traffic finds it more convenient to travel on the parallel I-80 highway twelve miles to the north; so there are few passing motorists to observe the small sign identifying Exeter. The flat terrain, large agricultural fields, and rectangular grid of roads still characterize the area: our country road has no turning from the Kansas-Nebraska border north to the Platte River.

Avery tractor and threshing machine with Grandma Trauger and Pop, circa 1920. Photograph by Ethel Trauger.

Two

Farm Life and Energy in the 1930s

L ife on our Nebraska farm during the 1920s was simple. So were the energy systems on which that life depended. As horse power gave way to gasoline and remotely generated electricity in the 1930s, farm life did become more complex, but viewed from the perspective of eight decades, my early life still seems remarkably uncomplicated.

My parents, Ethel Leone Downey and Charley C. Trauger, were married August 30, 1919, and settled in the "House on the Corner" of the farm they had purchased. I was born there prematurely on June 29, 1920, weighing less than five pounds and with a still heart. Old Dr. Stratton, apparently in touch with current medical technology, successfully administered a new, strong stimulant directly to the heart. This weak start may have contributed to a sickly childhood of long bed vigils with smelly chest ointments, foul-tasting medicines, and a thermometer in the mouth, or worse. With fever, the patterns of the ceiling wallpaper of my room frequently seemed to move in madly changing configurations. Even now, a room papered with repetitive

Mom and Pop together at Arbor Lodge, home of J. Sterling Morton.

patterns can transport me back to those disturbing days. The papered ceiling and walls was a solution for problems with plaster that would not retain paint or its sand, but the choice of ceiling patterns remains a puzzlement.

My earliest memories, no doubt reinforced by repeated telling, are framed by two events that took place when I was three years old. The first was the day my Grandmother Trauger died. In the hours before her death she lay on a "fainting couch" while I sat nearby on the floor administering "medicine" from an empty toy cup. Her typical grandmotherly response was, "This is the best medicine, better than that of the doctor." The other event was a short drive in my Downey grandparents' new Nash touring car. They chauffeured me out one driveway of our House on the Corner and returned via the other entrance. I also remember Grandmother's pride in her new, tall, flowered, velour hat.

At age four my fascination with the attachments for Mom's new sewing machine became a problem for her. I loved to remove them from the drawer, spread them on the floor, build intricate toys by attaching one to another, and "make adjustments" wherever knobs would turn. Unfortunately, they were consistently unusable when Mother returned to sewing. In desperation she placed a large, ragged, very black feather atop the attachments. This terrified me and Mom's problem was solved; I would no longer touch the drawer. That feather made such a deep impression that even today I have an aversion to feathers that are not part of a living bird.

Life on the farm was lonely, but I had a feathered companion. An aging Brahma rooster adopted me when we moved to Grandfather's home from the House on the Corner. "Rooster" followed me around the farmstead and clucked approval of my embrace and stroking of his long, feathered neck and back. He was a large and formidable bird with a fearsome, sharp spur on each leg. When a stranger approached me, Rooster protectively flew toward the person's face with spurs displayed; a hasty retreat invariably followed. I was stricken with grief a few years later when the old bird died. Rooster was given a ceremonial family farewell at his burial in the peach orchard behind the horse barn.

Don with grandparents Annie Coates and Robert H. Downey, circa 1923. Photograph by Ethel Trauger.

My brother Robert's arrival in 1928 was a surprise. I had observed something abnormal about Mom and had been worried for some time. Although all of my eight years had been spent on a farm where animal birth was a common occurrence, I was uninformed about human reproduction. The trauma and excitement of the doctor's coming and the ensuing commotion disturbed, even frightened, me. After being introduced to the new arrival, I was insulted by the doctor's answers to my questions. He told me, "The baby came in this black bag." It was a great relief to know that my mother was to be well, but I really preferred to continue life as before. The world had changed; my parents had less time for me and my unwanted role as baby-sitter soon followed. The relationship with Robert grew slowly to reach a fully rewarding realization only in adulthood.

Mom, Ethel Downey Trauger.

As a young boy I met Mrs. Doyle, a friend of my Downey grandparents, who had celebrated her one hundredth birthday. She lived with her family, who then owned the dugout homesite and the then-abandoned early horse farm. This pioneer woman smoked a pipe, used snuff, and perhaps occasionally had a nip of illegal home brew in that time of the Eighteenth Amendment. Her wicker-work wheelchair was pleasantly situated on a porch with an adjacent table that supported her tobacco, glasses, and books. It was fascinating to hear her stories of early days, many about my grandmother and her family; but I was apprehensive of this ancient, wrinkled woman in a dark-gray dress, flowing white hair, and crepitant voice. My distress intensified when she generously insisted on sharing her "candy," sweet morsels liberally mixed with tobacco juice and pipe ashes. She simply could not understand my refusal.

I spent many happy childhood days on the Coates and Downey farms. The Coates farm was occupied by Uncle Harold and Aunt Louise (nee Diekman) with my cousins, Glenn and Doris Downey, while at the Downey farm were only Uncle Chester and Aunt Edith (nee Kail), who had no children. Spending a night with my cousin Glenn was an experience in family closeness, for the frame house that replaced the Coates's "soddy" had only three rooms and one large closet. I slept on the living room sofa, displacing Doris to a pallet in her parents' bedroom. Glenn was tucked into the closet. My chores, as a guest, were to gather eggs from the hen house and bring buckets of water hand-pumped from the well, unless the windmill was turning, to the house for drinking and washing. I remember relaxed evenings around the kitchen supper table. We discussed our different schools, friends, family, and farming and then focused on sports, since Uncle Harold in his youth had been a star baseball player on a local team. These unhurried evenings differed from the more urgent lifestyle of my home, where dinner was often late or farmyard chores required after-supper attention.

Also in contrast, days spent in the house of Uncle Chester and Aunt Edith were shared with their many cats. Savory home-canned pork and beef at Aunt Edith's dinner table remain vivid in memory. Uncle Chester always made me feel that he delighted in my company as he let me play their hand-cranked Victrola. Its turntable was equipped with a device to support several "tin dancers." As the disc turned, the metal hinged dancers were activated to gyrate with flailing arms and legs, seemingly in time with the music. For my entertainment Uncle Chester also had building blocks made of polished stone, and toy wagons, horses, and soldiers made of cast metal. The blocks dated to Granddad Downey's youth and had fascinated my uncles as boys.

Aunt Pearl and her husband, Elzie Steyer, also lived in the "Downey Community" where they had purchased an eighty-acre farm. She was a favorite aunt, perhaps because she treated me so kindly. Their son, Caryl, was five years younger than I and we therefore had little in common, but we since have become very close friends. The Steyer farm was too small in the 1930s to support a family of three, so they planned to raise and feed cattle, investing in a large forage silo and feeding stations. Their plans were cut short when a brain tumor took Uncle Elzie's life in 1931, ending a remarkably brave and terribly painful battle with the cancer. The treatment with radium he received might not be considered acceptable today; however, it prolonged his life, for which he was grateful, even as he suffered.

One of my most poignant memories is of a visit to Uncle Elzie's hospital room as he lay dying. He took my hand as though to say goodbye, lost consciousness, and, still holding my hand, partially awakened to think that I was Caryl, who also was visiting but not then present in the room. Emotionally tense, I could only remain quiet, honoring the moment, as he spoke final comforting and challenging words intended for his beloved son. Tears stream down my cheeks today as I write these words. He was a very favorite uncle and the first person to die whom I had known in a close relationship; I had been too young to understand the significance of Grandmother Trauger's earlier death.

As a child I was never as close to my father's sister, Aunt Minnie, and Uncle Frank Lewis as to the Downey families, perhaps because Aunt Minnie was sixteen years older than Pop and had married quite young. However, I recall stories of Uncle Frank in his youth with his early automobile, a sporty Hupmobile that Pop thought resembled a cricket and referred to as the cricket car. This and the nickname the "Cricket Man" came into common usage. (I thought as a boy that the car was actually of a company or a model named Cricket; it was only recently that I learned its correct name.) The Lewises's daughter Celia was much older than I, and it is with their granddaughter,

Don, circa 1929.

Frances, and her husband Eldon Goble that I have developed a very close family friendship.

There was little opportunity to develop casual childhood friendships on the farm; nearby neighbors did not have boys or girls of my age. My principal relief from isolation occurred at church, but except for the annual picnic the church school provided only a single, somewhat structured hour each week. In compensation and for pleasure, the Downey families frequently spent Sunday afternoons together. A favorite picnic spot was near the site of the dugouts, then as now, identifiable as two depressions in the upper bank of Turkey Creek. I remember those picnic lunches under the trees as repasts to compete with the delicacies of fine restaurants in New York and Paris. Crisp fried chicken, superb salads, and homemade apple or cherry pies were favorites. Each family brought a dish for a prearranged menu that sometimes included hand-cranked freezer ice cream as topping for the pie.

The spring-fed stream still ran mostly clear and always cold. Swimming was a problem on those hot summer days because none of the adults could swim; we learned a dog-paddle technique. My three cousins and I (and later Robert) would wade toward a deep portion of the otherwise shallow stream to splash furiously against the current. I have often pondered the dangers of those maneuvers, particularly when we engaged in a bit of "horseplay." Perhaps because of this beginning, I have never learned to swim efficiently.

In later years another Turkey Creek site displaced the dugout location as a favorite spot. The owner was a family friend who welcomed our frequent visits to this more picturesque setting. Centuries of erosion had carved twenty-five-foot deep "canyons" into the sand-clay soil that had been deposited in the floor of the ancient sea, once part of the Gulf of Mexico. Beyond the canyons, the stream had wandered widely between banks nourishing a large grove of tall trees and grassed areas where blankets could be spread or tables and chairs put in place. We had many good times there and came to call it "Happy Hollow."

Once while our family was camping and fishing on Turkey Creek, cousin Glenn and I strayed beyond the fence boundary of Happy Hollow to an unfriendly neighbor's farm. There we found an illegal fish trap, proceeded to remove its catch, place the fish in our container, and beat a hurried retreat. Alas, the farmer had observed us from the distant farmstead and shouted a stern order to replace the fish. Pursued by this angry man waving a pitchfork, we ran for our lives, burdened by fishing poles and the fish container. I thought he would catch us when Glenn slipped from the log over Turkey Creek, but we escaped when the farmer chose not to follow beyond his farm's boundary fence. In accusing us, he also would have confronted Pop who would have been sternly critical of the fish trap. Of course, we should have released the fish to the stream, but they made a tasty dinner that evening.

Family picnics and Sunday school events were often held in what came to be known as Trauger's Grove at the House on the Corner. Great cottonwood trees stood on either side of Johnson Creek in the grass-covered pasture below the farm buildings. My father had scaled two of the trees, nearly twenty feet above the turf in one and even higher in the other, to attach strong steel cables for tire swings. The longer cable swung the tire and rider high over the usually dry creek bed and into the branches of other trees. That swing provided a thrilling ride for an effective "pumper," one who could combine an energetic push with a good sense of timing. For years many families enjoyed picnic outings in Trauger's Grove, but the swings are now gone and most of the original trees are either dead or infirm.

On the farm my parents expected from each son a reasonable amount of work, gauged to age and experience but always to be performed with excellence. An early task was to pull plantain weeds from the lawn, for which there was pay, five cents for each one hundred plants if the tubers had been fully extracted. The same pay was available for each #3-size tinned can filled with nails that were picked or dug from the ground around buildings. Careless carpenters had left some, but most of the nails had been forced from the boards by action of severe weather changes. Recovery of the nails was important to protect automobile tires and bare feet. Later, more demanding chores were expected, but there was no direct monetary compensation or cash allowance. However, a pig or other animal often was deemed to be mine, and when I had cared for it along with the others, the reward came at market time.

Our Model T Ford touring car was important in family life. It served to transport produce to the local market and to carry us to church, social events, and the doctor. On wintry days it was both uninsulated and breezy even with the side curtains firmly fastened. The only source of heat was a small, portable charcoal brazier stoked with coals from our home's wood-burning stove, which helped to warm one's feet on long trips or while driving to church. Heavy sleigh blankets covered legs and laps of all, except for Pop, who required freedom

Sand load arriving at the sandbox, with Don and Pop. Photograph by Ethel Trauger.

to manipulate the three foot-pedals that controlled its planetary-type transmission and brakes. In summer, with curtains removed, the car was "air-conditioned" by wind blowing from every side. On a trip to the neighboring town of York, I determined the Ford's capability by timing its passing of crossroads one mile apart while tooling along at top speed. I calculated that the car was making the remarkable speed of forty-five miles per hour. That indeed was very fast, considering the condition of the gravel-surface highway on which we bounced and swerved. As was the accepted custom, I learned to drive the Model T with its hand throttle and uncertain steering long before reaching the legal age. That driving was limited, wisely, to farm roads.

The Ford seldom traveled at top speed as it carried Pop and me to the Nebraska State Fair of 1929. We started before daylight and completed the fifty-mile trip in three hours, including a hurried picnic breakfast eaten in a country school yard. It was a hot September day shared with a record crowd of more than a hundred thousand attendees. After viewing the farm-products exhibits, the displays of modern farm machines, some sideshows, and fireworks, we were a tired pair as we started home in the Ford. I have no recollection of the return trip, no doubt for having slept all of the way.

The farm sandbox offered a remarkable place to play. It was about six feet wide, twelve feet long, and often two feet or more in depth. This luxury was not provided solely, or even primarily, for childhood pleasure. The sand was stored for use in making concrete feeding troughs and flooring for farm animals otherwise mired in barnyard mud. Its use for play was restricted only to keeping the sand inside the heavy plank structure and free from dirt. The deep sand bed made possible the construction of hills, roads, caves, castles, and cities, molded or excavated much as I envisioned distant children making similar structures when playing on beaches. My favorite toy was a powerful spring-powered crawler-type tractor that could move sand when a homemade bulldozer blade was attached.

The sand came from a pit at Happy Hollow nine miles away, hauled in a farm wagon drawn by a four-horse team, like the teams hitched to stagecoaches in western movies. A sand procurement expedition with my father always started before dawn with both breakfast and lunch securely packed. We rode side by side on the wooden spring-mounted seat of the wagon, the horses requiring little attention. Those rides—three hours with an empty wagon—five with a load of sand—were among the longest continuous conversations I ever had with my busy father. We studied and fancied cloud formations, learned about wayside trees and plants, and worried lest a sudden summer thunderstorm would strike. A load of sand stuck in the mud of those dirt roads would have presented a serious problem.

Later, when I was old enough to handle the task alone, sand procurement became a lonely trek to Turkey Creek. At noon, the horses were provided buckets of water dipped and carried from the creek, and each had a portion of oats for lunch. My effort in shoveling two tons of sand from the pit to the wagon was rewarded with a sandwich Mom had prepared. While pulling this heavy load on the return journey, the horses required a brief rest at one-mile intervals; even so, they arrived home tired and wet with perspiration.

Each trip for sand required advance preparation by soaking the wagon wheels in water or hot oil to swell and tighten the wood against the iron rims to prevent wear and loosening under the unusually heavy load. The water treatment was easiest. Each wheel was placed for overnight soaking in the animal watering tank, but it was less effective than hot oil applied in a shallow trough. This tedious process required periodic rotation of the wheel and maintaining a wood fire beneath the trough without setting the oil on fire. Even these precautions were not always enough. On one trip, a wheel issued a creaking warning that it could soon collapse. Although only a quarter of a mile from home, I had to shovel all the sand into another wagon to complete the mission.

A highlight of play in the sandbox occurred when Mother's friend and her son of about my age, perhaps eight, came to visit. We spent

Don and Verl Churchill on Molly, 1932.

the entire afternoon in the sand. It was exciting to show this city boy how to mold and pack the sand and to form it into fantastic landscapes. We shaped and reshaped the entire surface until both were exhausted. For me, the excitement of new companionship continued into a wakeful night, as I relived the wonderful day. Worried about my restlessness, Mom came to the room sometime in the night to warn that such visits could not be repeated if they prevented sleep, a message that further prolonged wakefulness. Mom did not understand that having more young guests might have been the cure.

My first real friendship developed in 1931. Verl Churchill, who was exactly one month younger than I, came to live with his Churchill grandparents and two of their sons, Verl's uncles, Gola and Laverne. They had rented the House on the Corner. Verl, his brother, Earl, and sister, Marguerite, were to stay until their father could become reestablished following their mother's death. Verl and I soon became close friends, treasuring many waking hours together in summer and sometimes spending the night in each other's home. We played in the sandbox, creating ever more sophisticated patterns of farms, cities, roads, and bridges but soon graduated to riding bicycles and doubling bareback on the gray mare, Molly. That summer was the best of my early youth. It also was helpful to Verl in recovering from the shock of losing his mother. Verl and I were distressed that the House on the Corner was in School District 20 and ours was in District 22, thus limiting our activities to weekends during the school year. Our friendship lasted only sixteen months; his father then could care for the children and moved them to California.

During one of his brief periods of remission from cancer, Uncle Elzie Steyer encountered a problem with the older Churchill boys. He was displeased that the Churchills's greyhound dogs had been allowed to disturb cattle on his and neighboring farms. Laverne was to ride his bike the three miles to see Uncle Elzie and apologize, so Verl and I rode alongside. The discussion had not gone well, with Uncle Elzie insisting that the greyhounds be more carefully constrained. The Steyers's German shepherd, seemingly sensing tension, slowly circled the bicycle conclave. Shep then either mistook Vern's legging for a tree or wanted to reinforce one of Uncle Elzie's points in dog fashion. Verl and I laughed at Vern's distress over a warm, wet sock. Uncle Elzie nearly collapsed in stress between amusement at Vern's plight and his serious effort to clinch the point he was making.

The depth of my friendship with Verl Churchill was demonstrated sixty years later. We had had no direct contact in all that time, although his grandmother and my mother corresponded regularly while both were still living. After that link was lost, I had no knowledge of Verl's location or activities as he attended college and seminary, became a United Methodist pastor, and served in the Navy. When my brother discovered that Verl had retired near Sacramento, California, I resolved to telephone him as a birthday surprise. When he answered, I asked, "Is this the Verl Churchill who once lived near Exeter, Nebraska?" He immediately responded, "Yes, is this Don?" A year or so later we spent a grand day together in San Francisco and now maintain periodic contact by telephone and letters.

As a child I was too shy to participate in many social events of the church, but for one Halloween party Mom had devised a costume in which I was confident that identification was not possible, and I resolved to go. Upon arrival at the pastor's home and before I could speak in my practiced voice of disguise, the hostess said, "Donald, how nice of you to come." The effect was devastating, but there was no escape; I was ushered in for what became a rather pleasant evening. Upon leaving, I inquired how the pastor's college-age daughter had recognized me so readily. She had seen my farm work shoes; other guests were from the village and had more formal footwear. Soon thereafter, we purchased my first pair of Tom McCann dress shoes.

Most farm families have dogs and ours was no exception. Early dogs were Rex and Fido, but the legendary canine was Possie. When found, he had lost most of the hair from his tail while wandering through the countryside. As an emaciated, abandoned mongrel with the face of a collie and long, sharp teeth, he resembled an opossum; thus the name *Possie* was chosen. He grew to be a fine, small dog that resembled a Shetland sheep dog, easily trained and useful on the farm. Possie was taught to go to the pasture a half-mile away and out of sight to bring home either the horses or the cattle. However, I had to watch for his cresting the hill in the lane because he would occasionally stop, having forgotten which animals were wanted. I would then cup my hands to shout "cows" or "horses" and Possie would resume running to complete the mission.

Possie also loved to ride on farm machines. His favorite was the two-row corn cultivator. There, he seemed most happy with his hind legs on the mechanism, which moved laterally as the machine was controlled to follow irregularities in the planted rows, and front paws on the doubletree, which oscillated as the horses equalized the load. Possie would ride for long periods, apparently enjoying the continuous motion of the machine. He would also jump into a farm or coaster wagon to ride; despite my attempts, he was too intelligent to submit to harness and pull a small wagon in the fashion of European dogs of old.

Copying a circus dog act, I built a structure to support a long board about five feet above the ground, with a ladder at one end and a slide at the other. I then taught Possie to climb the ladder, walk across the board, and slide to the ground, an activity he seemed to enjoy. Possie also loved to chase the farm cats but was careful not to catch them, after an early encounter produced a severely scratched nose. On one occasion, Possie pursued the Churchills's cat at the House on the Corner. The cat sought refuge in a tree that had a twelve-foot ladder propped against its trunk. As the cat climbed the tree, Possie scaled the ladder in pursuit. At the top, he looked down, realized his plight, and howled in distress. He was carefully assisted to the ground; the cat descended unaided.

Possie also chased rabbits even though he was unable to outrun even a cottontail. Sometimes he dismayed the Churchills's greyhounds by joining them in pursuit of jackrabbits. Jackrabbits are fast runners and have great endurance, but they tend to traverse a great circle when pursued over open Nebraska farmland. Although not in a running league with the hounds, Possie learned to trot diagonally across the circle, so that when both hounds and rabbit were well spent he could sometimes take the rabbit himself. A ferocious fighter, Possie was not seriously challenged while enjoying the trophy.

My rural school building had only one classroom and teacher, but it was not the traditional little red schoolhouse. Built of a modern design in the early 1900s, it had a vestibule, storage room for supplies, and a basement that housed the furnace and provided room for play during inclement weather. All classroom windows were in the north wall, and student desks faced east to give proper lighting for right-handed students. The teacher also served as janitor, procurement officer, truant officer, safety officer, and interpreter to the school board. Janitorial tasks included arriving early to stoke the coal-fired furnace and preparing the school for students. She was expected to live and board in a nearby home and walk to school; the salary of thirty dollars per month would not support an automobile.

The coal-fired furnace operation and building temperature were controlled by adjusting chains on a classroom wall to open and close dampers that metered the air supply; it challenged the busy teacher to avoid severe thermal cycling. The furnace also provided realistic fire drills whenever exploding coal gas occasionally belched flames into the room from the central floor register. Presumably the fire box was defective, but this was never corrected. On one occasion flames reached to the ceiling. The teacher and an older boy saved the school by pouring the contents of the drinking-water bucket down the register and then rushing to the school's well, hand-pumping vigorously, and bringing more water. The room cooled slowly, but school was not dismissed.

The building stood on an acre of land purchased through eminent domain from a 160-acre farm. Each year at plowing time, we watched the farmer turn one or more furrows of the school's yard into his field. When the teacher was out of sight, we even threw clods of dirt in an attempt to stop him. He was angered but understandably chose not to report us since he was stealing land from the county. This series of incursions came to an end during my school tenure when the farmer's plowing reached the back of the outhouses placed on opposite sides of the plot for maximum separation of facilities for boys and girls.

The privies polluted the water of the school's well, so it could not be used for drinking or washing. Thus an exercise at first recess, for a pair of older boys and sometimes girls, was to walk to a neighboring farm well to obtain a bucket of water. Another task was to attach Old Glory to the rope and hoist it up a tall flag pole. The only installed play equipment at school was a pair of "teeter-totters." Through their improper use, I managed to break my left forearm and won the distinction of a plaster cast. The bones healed properly, but in adjusting the cast on one occasion the doctor allowed his shears to slip and cut my hand. After that, I would only allow his nurse to attend me; the scar remains to this day.

Rural electrification was not then prevalent in Nebraska, so the school had no modern lighting or electrical equipment. Kerosene oil lamps and reflectors mounted on the wall served poorly on cloudy winter days. There was no piano; music came from a hand-cranked, spring-driven Victrola and a set of records donated by the community. As I recall, the records were mostly of band music and folk tunes. Large maps hung from spring rollers on the east-wall front of the room. I had almost completed my early schooling before envisioning the North Pole as not positioned in the east. Textbooks also presented problems. One reader included a story about the east wind bringing rain. In that Midwest locale, everyone knew that storms always came from the west and that the rare east winds never brought rain. Such were the anomalies in the texts written by city folks of the East.

I had only two teachers in elementary school; neither had a college education, but they were caring and understanding young women who demanded performance. Edith Dumpert Geis and Amy Walker Dumpert, in succession, deserve recognition beyond this testimony. Each maintained discipline even with rough farm boys, some of whom were nearly as old as they were. Fortunately, their positions were strengthened by the support and cooperation of parents. Neither teacher resorted to physical punishment, but the parents, including my own, knew how and when spanking was to be applied. Through a combination of ability and dedication these high-school-trained young women, with only short periods of study at teachers colleges, proved excellent teachers.

As a typical grandfather, I now enjoy telling stories of walking to school. My favorite path was down the lane to the pasture and prairie, through a boundary fence, and across the neighbor's hay field and pasture. From there a fenced lane led to the farmyard where we obtained drinking water for the school. Later, when bike riding had been mastered and our dirt road was passable, the route was by road and included a mile of U.S. Highway No. 6, the graveled roadway beside the school. In retrospect, this road seems to have been too dangerous for a nine-year-old to navigate. The loose gravel made vehicle control difficult, especially for the bicycle. Trucks and buses were wide, and their tires threw small rocks like missiles. However, I strongly resisted Mom's urging to travel the other way around the section; it was half-a-mile farther on very rough dirt roads. Bicycle paths were unknown. When the dirt road that passed our house was muddy or snow too deep for the bike, it was stored in a neighbor's garage located on the highway, which was always passable, and the beginning and final distance from and to home traversed by slipping and sliding on foot over the black-dirt road.

My bicycle provided recreation for the older students at recesses. Using the furnace coal shovel, we built a mound of earth for a jump at high speed and later made a daring path for riding up and down the sides of the highway ditch. Motorists must often have experienced sudden panic when bike riders would suddenly top the ditch, perhaps going out of control into the roadway to be struck by their vehicle; fortunately, that never occurred. This contrasted with other recreation supervised by the teacher. Running and contact games such as Red Rover–Red Rover and Fox and Geese played in snow were popular. On one bright winter day, we were granted extra lunch time to make a monstrous snowman. Miss Walker helped construct this nine-foot-tall icon on the shaded side of the school, where we envisioned its endurance until spring. We were shattered the next morning to find our masterpiece destroyed, someone had maliciously driven a truck into the snowman's base to topple the poor fellow.

The principal limitation of the rural school was that the sole teacher could impart only the knowledge that she herself had acquired. Music instruction was largely limited to simple singing sessions. Art was from exercises learned in teacher training, and, as I recall, three prints provided the only exposure to the masters. Stuart's portrait of George Washington and "The Gleaners" and "The Angelus" by Millet were displayed above the bulletin board. Arithmetic and spelling taught and learned by rote repetition have served me well, although I, too, rely on the computer's spell checker. Geography was a favorite subject. I learned the principal cities of Africa, Asia, and South America and to draw a reasonable representation of rivers and the outlines of countries. Later, I had to learn new country names and associations because our geography books had been printed in 1895; so Cuba, for example, was shown as a colony of Spain.

The student body varied widely from year to year in size, age, and interest. District 22 included a predominantly Roman Catholic

Don (on front row beside teacher Edith Dumpert Geis) and schoolmates at District 22 school assembly, circa 1926.

community to the north with more Protestants in the south. The Catholic church maintained a parish elementary school and encouraged attendance there. Apparently, enrollment in alternate years satisfied a church criterion, and tuition charges caused families to place their children in public school to the extent acceptable. Thus the enrollment in District 22 during my tenure varied between as few as six to more than twenty. Prayer in school was not an issue. Even though most families were faithful parishioners in local churches, there was no audible prayer. One winter, with Catholics predominating, students engaged in a mock Catholic altar service in the school basement during recess. The teacher forbade the practice, realizing that a Protestant could not participate comfortably. Incidentally, both teachers were of the Catholic faith.

Recitation and lectures in the one-classroom school were necessarily in the view and hearing of everyone. Serious students learned to concentrate. I never had a classmate, in part, because of the annual restructuring of the student body. Thus, individual attention by the teachers, particularly in the lean population years, enabled me to proceed at my chosen pace. By the middle of my sixth year, I had met the requirements of the eight grades. In addition, I had explored the limited library, and not much else was available to keep me properly occupied. I was to enter high school at age twelve in the fall of 1932.

Mom required a major operation that spring and spent nearly two months in a Lincoln hospital. My brother and I were never told the nature of the surgery, but presumably it was a hysterectomy. Robert and I were sent to live temporarily with our recently widowed Aunt Pearl. She needed help with the farm chores, and I tried to do as much as possible. Perhaps our presence was helpful to her in her bereavement. Nevertheless, feeding and looking after her nephews was surely a burden for her.

We visited Mom in the hospital on two occasions. It was reassuring to see her but distressing to see her confined to bed for such a long time. By current practices, she would have been on her feet by the day following surgery. The hospital and doctor were chosen for their low cost, but I now wonder if it was, nevertheless, an expensive decision. A visit to that hospital during my college years left me with an impression of poor sanitation, inadequate equipment, and incompetence. However, the choice was made during the depression and in an early year of the drought. Perhaps there was no other option in that time of little regulation and no public assistance for medical treatment.

Family vacation trips were momentous occasions. Only three lasted more than one night. The first, a trip to the Black Hills of South Dakota in 1927, was in some ways like a safari. Three automobiles traveled together on primitive roads, with our Model T Ford often in the lead followed by Uncle Harold's "T" and the Diekmans's Graham Paige. Crossing the Sand Hills area of Cherry County, Nebraska, required patience and perseverance. The heavy Graham Paige often mired down in the deep sand of unsurfaced roads and had to be towed out by one of the Fords.

We traveled on what is now U.S. Highway 20, then hardly more than a trail but marked as a national roadway. From time to time the road was blocked by a rancher's fence. We stopped, opened a gate, drove the cars through, and, as disciplined farmers, closed it securely. Each night, we unpacked tents and camping gear from a large wooden box attached to the rear of our Ford and started a fire for cooking. Our family's tent was a lean-to attached to the side of the Ford and used primarily for dressing; we slept in the car. Pop had laboriously sawed through the steel back-support of the front seat and fitted it with hinges and locking pins to hold it in place while driving. For sleeping, the seat was laid back to form a bed contiguous with the back seat.

As "flatlanders" we marveled at the rugged beauty of the Badlands and reveled in the height of the Black Hills that we saw as great mountains. The majestic face of Mount Rushmore and the plans and model for the presidential profiles yet to be carved there held our interest for several hours. (A few years ago I accompanied my wife, Elaine, then a member of the Energy, Environment and Natural Resources Committee of the National League of Cities, to a meeting in Gillette, Wyoming. When a side trip group of fifty visited the Black Hills, I alone had seen the mountain before the marvelous figures were created.)

We drove through the needles area with its spectacular erosion patterns and to the top of Harney Peak. The Fords made it to the top even though ours had lost the fan belt (which was never replaced), but the Graham overheated so badly that its occupants crowded into the Fords. I recall one incident vividly: a picnic on the side of the mountain where Pop dropped a slice of bread freshly spread with butter and jam. It landed face down. After angrily expressing dismay, Pop wondered aloud why garnished bread always seemed to land face down. Recently a technical journal published a study showing that the height of a table is optimal for a flat object the size of a bread slice to make one half turn before landing.

We replaced the Ford with a Chevrolet sedan in 1929. Pop and Mom had selected a Model A Ford, but I observed that vision was obstructed from the back seat for one of my size; the Chevrolet sedan was chosen for me. If the decision to purchase had been planned even a few months later, the depression likely would have restricted us to the old Ford for several more years. Although the depression affected everyone, the drought of the 1930 decade was even more significant to Nebraska farmers.

With minimal funds, a farm family could survive a depression well on fruits, garden vegetables, chickens and a few pigs for meat, and cow's milk. As an adventure in a good-weather year, we set about to demonstrate this capability. Our family lived for two weeks eating only the products of the farm except for salt and sugar. The bread made from wheat coarsely ground in the animal feed grinder was heavy but flavorful and nourishing. Overall, we ate quite well. Also, we continued to use the energy of horses, so the purchase of fuel for tractors was not required. The depression was difficult for us but disastrous for many neighbors who, because of low prices brought for farm products, lost their farms to creditors, principally banks, when mortgage payments could not be made. Fortunately, our indebtedness on the acreage purchased in 1919 had been repaid during the profitable 1920s.

For us, the stock market crash of 1929 was only newspaper headlines and stories. My concept of that market was of large buildings where people bought and sold intangibles, envisioning the process to be something like the sale of our cattle at the Omaha Stock Market. The reports of desperate people jumping from windows made the crash more real, but otherwise the depression had little immediate relevance to me or my family. We had no stocks, bonds, or other securities, except Pop's shares in the Exeter Co-op. Later, as the price of farm products dropped, it became more real, particularly as the crisis was heightened by the drought. Nevertheless, even though prices were low, we planted crops each spring with hopes for produce to sell. A drought is difficult, even disastrous, for farmers.

As the hot south winds of summer continued in the 1930s, the leaves of the corn plants rolled themselves more and more tightly to conserve moisture. When temperatures soared to 120 degrees Fahrenheit and higher, leaves and then the stalks faded to yellow and then brown as they died without producing ears to be harvested. Our spirits faded too as we perspired in the persistent heat of both day and night. Summer nighttime temperatures were often about 100 degrees, so I slept on a quilt under the stars every night through several summer months. This severe drought and depression affected my lifetime outlook in searching for and providing financial stability. It made me acutely aware of many peoples' plight in Africa and Asia who are dependent on agriculture and subject to frequent drought. Other weather-related problems are equally disastrous, but for me drought occurs as a fearsome, creeping reality.

I can never forget one day of August 1934. It started traumatically as we assembled half of our cattle for sale. We had held on to them as long as possible in the hope that rain would again revive the fields, but the land was becoming barren and there was no new alfalfa hay for the winter. After lunch, Pop accompanied the animals in a large transport truck to the Omaha market. Mom and Robert, as my brother Bob was then known, were to visit friends in Exeter. My role was to ride the quarter horse, Molly, as a cowboy in a western saddle, to herd the remaining cattle over the fields and roadside ditches to eat whatever plants could be found.

Molly and I moved the herd to some remaining forage in a low region near Johnson Creek nearly half a mile from the house. Upon reaching the site, I turned in the saddle to survey the area and welcomed a dark cloud on the horizon; it might bring an evening rain.

Later, the cloud seemed to be approaching rapidly and I estimated that it would reach the farm in late afternoon. Soon, there was a stillness before the storm, and I could hear distant voices and laughter of girls as they enjoyed an outing in a neighboring farm grove. The clouds continued to build and approach, and I tried in vain to move the cattle back to shelter, but they refused to face the storm.

As the great cloud reached a neighbor's farmstead, I realized with horror that the storm was not rain but dust. Massive, black, turbulent clouds became laced with lightning and the increasing roll of thunder intensified my fear. I pressed my heels into Molly's side, and she galloped to the farmstead, leaving the cattle to fend for themselves. We reached the barnyard gate just as the swirling dust encompassed us, and I abandoned poor Molly at the gatepost.

It mattered not that it was suddenly as dark as night, for I could not open my eyes to the strong wind that carried biting dust that pelted face and hands. Like the early pioneers during blizzards, I groped my way along the fence for orientation, and while passing a straw pile from an earlier year's thrashing of grain, the stem ends of straw pricked my face and were even painful through a denim shirt. The fence led to the stock watering trough from which an overhead pipe guided me to the water storage tank. From there, after taking careful bearings, I reached the house. Although there was no choice in abandoning Molly, I was concerned over possible damage to her left eye. It had been destroyed before we bought her and could not be closed. The dust must have caused severe pain.

Fortunately, Mother had seen the impending storm and had closed the house before leaving, otherwise it would have offered little refuge. I had never before experienced such total darkness; the only light came from an increasing glow on the metal screens of doors and windows. Static-electric charges carried by the dust had built very high voltages on the screens to cause an electrical discharge of an eerie blue color. (Decades later, I recalled it as resembling the Cherenkov radiation from nuclear fuel in operation under the shielding of water.) I lit a kerosene lamp and continued in that dim light to watch the screens with fascination.

As the storm abated, I could see the glow of auto headlights in the driveway and thought Mom and Robert, having reached that point, were afraid to come to the house. When it seemed safe to venture out, I discovered the occupants to be the teenage girls whom I had heard earlier at their picnic site. Interrupted by the storm, they had not eaten. I found some of Mom's cookies, and we arranged a party around the dining room table. Shyness would have precluded my inviting these older girls into our home under less dramatic circumstances; the role of impromptu host was an exhilarating experience.

The day ended in relief when family members and animals were accounted for, free of serious injury. Molly's eye socket showed severe redness, but it slowly recovered a normal appearance. Reality set in on the morrow as we surveyed the aftermath. Dirt had been forced through the ill-fitting windows and doors of the house, necessitating a thorough cleaning. The storm also had contaminated animal feed and water. Pop returned with news that the cattle had not sold well; the market had become flooded with livestock from the many farms that were short of feed. The sale had netted approximately $350, which would represent about one-half of the family funds for the year. Only a small part was income, because most of the animals were properly classified as inventory; mature cows would have provided calves for several years to come. Times were hard on the farm, perhaps no worse than elsewhere during the depression, but that dust storm had cast a very dark shadow.

We were impressed to learn later that fallout from that storm descended on ships in the Atlantic Ocean. Other dust storms were to follow as the over-tilled and heavily grazed fields of the West were eroded. As if the drought was not enough, hordes of grasshoppers ate much of the vegetation that had survived dry weather. The destructive "hoppers" even ate holes in leather gloves and well-worn hickory tool handles, damaging them beyond use; the insects were apparently searching for salt deposited from workers' perspiration. Many neighbors and even government agencies spread poisoned bran to kill the orthopteran insects. Unfortunately, birds that were helping to reduce the infestation were killed both by ingestion of the bran and by eating poisoned grasshoppers. The great flocks of blackbirds that earlier had blackened the skies during their spring and fall migrations soon disappeared.

We chose not to use poison and fashioned a baffle and trough of kerosene oil designed to catch and drown the hoppers when carried rapidly across the fields on the Chevrolet's front bumper. We caught and drowned many deep trays of the offending insects and burned them with the kerosene. This provided some satisfaction but, unfortunately, did not noticeably reduce their numbers. Grasshoppers were even a hazard on paved highways, where, attracted by nighttime warmth, at times they congregated in numbers sufficient to render the surface slippery.

The atmospheric dust from the drought also produced remarkably pleasant effects. Sunsets were spectacular, with colors of every hue spread magnificently across the sky from west to east. We sometimes saw brilliant displays of the Aurora Borealis, not as a result of the drought but perhaps of the same origin, both having been caused in part by sunspots and solar emissions.

By destroying the pasture grasses, the drought also revealed the ruts of a historic road used by early settlers. The trail crossed Johnson Creek where it could be forded at times when the creek was up. Erosion had deepened the tracks where wagons had been pulled up the creek bank slopes, but they also were detectable on level areas. The road led clearly toward Exeter and appeared to come from the farm community where my maternal grandparents had lived as children. By studying topographical maps, we concluded it likely that my grandmother had passed that way before either the rectangular road pattern was in use or the Trauger farm established.

Unfortunately, the drought did not reveal the road until after my grandmother died in 1934. I would love to have asked if she recalled the old road and perhaps to have evoked anecdotes of trips to the village of Exeter. Grandfather Trauger is reported to have constructed a barbed wire fence to divert traffic from this shortcut alternative to the rectangular road pattern established before his arrival in Nebraska. Further wind erosion and restoration of pasture grasses have obliterated the trail.

The love and dedication of my parents was severely tested in the fall of 1934 when I became obsessed with the reported wonders of the Chicago World's Fair. Exeter High School's *National Geographic Magazine* issue describing the fair was worn from my handling, as I enjoyed the descriptions and photographs and sought strategies to see the fair. That seemed impossible in that year of depression, drought, and dust storms, but my parents had never positively said no, so family discussions intensified when the final week of the fair coincided with a school holiday for a teachers convention. Pop was noted for not making decisions until the last moment; I left for school on Thursday morning not knowing if we would go to the fair. I can still feel the excitement at the close of school as I ran across the athletic field to the Chevy and my waiting parents. This second family trip was to be an adventure, a test of endurance, and an example of depression-era frugality.

We drove until dark reaching some point in Iowa, perhaps halfway to Chicago. After our picnic dinner previously prepared at home, Pop and Mom rented a room for two dollars. This was before the term *motel* came into use, but the small cabin hardly deserved a title. It included only the bedsprings and mattress, a table, and a tiny bathroom. Mom made up the bed with sheets and blankets brought for the purpose, and Robert and I slept in the car. After an early start, we arrived in the "Windy City" in late morning and promptly lost our way. A friendly Irish policeman was helpful after surveying our shamrock-green Chevrolet and remarking that we were certainly "green in Chicago." We found the lakeside fairgrounds, chose a parking lot near public rest rooms and selected a space that was to be our home for two nights. We bought passes valid through the closing day and were ready for the marvelous world's fair.

We first ascended one of the twin steel towers to gain a view of the fairgrounds. I was frightened because the tower swayed wildly in the lakeside gales, but the views of the fair, the city, and Lake Michigan were thrilling on that clear October day. I can no longer recall the sequence of activities, but we visited many of the free exhibits, ate lunch sparingly on the grounds, and returned to the car for dinner prepared on a small gasoline-fueled camp stove. Back at the fair each evening, we were thrilled by the great fountains bathed in colored lights and by the spectacular fireworks display.

I must have frustrated my parents by tarrying so long in the science exhibits and those of new engineering devices. A display of "working perpetual motion" machines particularly held my attention. Each had been fashioned from a design submitted to the U.S. Patent Office and, of course, rejected. Neither Pop nor Mom could explain why they really did not work perpetually, as they seemed to do. I peered at each device from every angle possible, but they were so cleverly arranged that the motive force was not detectable. Particularly fascinated with one, I memorized its mechanical design and later constructed a model at home. It would run for quite a long time after starting, but always stopped. It was in the high school physics class of the following year that I learned the principle involved and finally understood why that machine did not work perpetually.

We all slept in the car during those two nights in Chicago. Our parents tightly shared the back seat, extended by boxes padded with quilts, but they had only as much space as seat adjustment permitted. Robert was curled among the brake and accelerator levers while I slept well on the front seat. We were warm under blankets despite the October chill from damp lake breezes. Today, persons sleeping as we did might be robbed, arrested, or worse, but those were depression days. Fortunately, the notorious Chicago gangsters of the 1930s were seeking more affluent targets. The total cost of the trip for gasoline, parking, lodging, purchased meals, and entrance tickets was thirty-six dollars.

The return trip was made hurried by my tarrying on the closing day, on the one hand, and, on the other, by my own foolish insistence that I not be absent from school. Robert and I were exhausted and slept essentially all the way. Our tired parents were burdened with driving, each relieved only by naps as the other drove through the night. In the wee hours of the morning, somewhere in the Iowa hill country, they stopped at a coffee shop leaving us boys asleep in the car. While sipping her coffee, Mom was terrified to see the Chevy moving on its own. She bolted from the shop and barely stopped it from rolling to disaster. Accustomed to the flat terrain of home, my parents had neglected to set the parking brake.

We arrived at the Exeter High School fifteen minutes ahead of the bell, so I could wash and meet the first class. I often have pondered the folly of maintaining perfect attendance on that occasion. We could have learned much during additional hours and enjoyed the closing exercises of the fair. Even a leisurely discussion of the exhibits during the return would likely have had greater value than any typical school day.

The third family trip was in the summer of 1935, again to the Black Hills. Aunt Pearl and Caryl accompanied the four of us, usually camping, this time with our gear stowed in a commercially made box attached to the rear of the Chevrolet. We retraced much of the previous route on graded and graveled roads unobstructed by fences. This venture extended to the Devils Tower of Wyoming and to Chimney Rock near Scottsbluff, Nebraska.

At the Chimney, I insisted on climbing to the base of the great rock. As the family waited, I trudged up the slope and with my back to the vertical monolith, surveyed the countryside. I noticed an automobile parked far from any road that seemed in an odd location for a lone car. I looked about for a possible camper or fellow explorer but could see no one. Choosing an alternate path of descent, with every step primarily considered for avoiding the region's many rattlesnakes, I suddenly found myself directly beside the lone vehicle. To my horror, a man slumped in the back seat was dead. A hose attached to the exhaust and into a window indicated suicide; the generally revolting condition of the corpse suggested that death had occurred some days earlier. I ran madly to our Chevrolet with less concern for the snakes, and we then drove to the sheriff's office to report the find. They detained us only an hour to determine the location and description of the automobile, but the delay was a costly one for us.

Upon reaching a tributary of the Little Blue River in southern Nebraska, we found it flooded from a torrential rain somewhere upstream. The bridge we were to cross was under water and declared unsafe by officials just before we arrived. We camped nearby and slept in the tents; that is, some slept even though a second storm buffeted the top, sides, and flaps of the tent. I could only toss and turn while reliving the shock of discovering the decaying suicide. Even consoling words by Aunt Pearl, who shared the tent with Caryl and me, were to no avail as the night became entirely sleepless. We packed up early, found an alternate bridge, and arrived home at mid-afternoon. I dropped to the floor to rest, for it was too hot for bed. I awakened in a dim light, ready for a late supper, to find the next morning's breakfast nearly ready.

As a boy, I was fascinated with transportation systems. Pop, thinking as early as 1930 that rail travel soon would be discontinued in favor of automobiles and buses, took me for a train ride as a kind of historical lesson in the use of a passing form of energy application. We boarded the local train in Exeter for the nine-mile trip to Friend as Mom drove the Chevrolet to meet us. An excursion ride on the first Diesel streamlined train, the Pioneer Zephyr, followed a few years later. My interest in air travel was heightened at an air show in Omaha where we saw a Ford Trimotor plane and took a sightseeing ride over the city in a small aircraft. I often daydreamed that an emergency landing might occur in our pasture and speculated on what kind of people traveled by air.

The landing never happened, but on a shopping trip to nearby York we saw a most unusual aircraft in the distance. It first appeared to have no wings or other support, but as it approached from the right, we could see a rotor on top and then a propeller. The Autogiro crossed over the road at a low altitude in front of our car as it descended to land in a nearby pasture. We rushed to see this strange machine. The pilot who emerged was my heroine, Amelia Earhart; she was even more charming than my imagination had afforded. She was in the midst of a cross-country flight to promote air transportation and to advertise Beechnut Coffee, with its merits proclaimed in painted figures and print on the side of the fuselage. It was a great privilege to be among the dozen or so who had gathered to greet this heroic aviator; I was thrilled beyond description. The mystery of her disappearance several years later over the Pacific Ocean on July 2, 1937, was grievous for me, as for the nation.

The acceleration of elementary school years placed me in high school only three months after my twelfth birthday. I liked to study and looked forward to the experience. Unfortunately, introduction to the larger school was a shock. I was very shy, and classmates were between two and three years older, an important span at that stage of life. Although it was a relatively small school, lecture and recitation classes seemed large and intimidating. The faculty of college graduates appeared formidable; I had not previously been closely associated with any person having a college degree. Compounding my problems was a debilitated physical condition not easily diagnosed. My parents were concerned about my lethargy, but they attributed it to my stage of life and not to any abnormal condition or pathology. Proper recognition occurred only after the last day of school, when I made plain my physical inability to accomplish a strenuous chore. They were shocked when I told them that fatigue had caused me to lean across my bicycle and walk it home from school that afternoon. I was rushed to a doctor in Lincoln, who pronounced me severely anemic. The prescribed medicine was most distasteful but not the raisins recommended as a source of iron, which I still enjoy today.

In whatever ways that freshman year was remarkable, my grades were not. Even Manual Training (woodworking) that should have been a snap for a farm boy familiar with hand tools, was a problem; we were to learn cabinet-making of a precision not required for farm machines. I measured the boards so carefully before sawing, that professor Knapp paid me a compliment before the class—probably for lack of anything else to say about my accomplishments. That compliment generated the nickname "Trysquare," which persisted through most of my high school years.

Mr. Knapp was kind by nature and tolerated many tricks. He was routinely forced to carry keys because students frequently locked him out of the room. On such occasions further mischief occurred, including rearrangement of the stock room so that boards normally stacked for use in shop projects would fall unexpectedly upon opening the door. Mr. Knapp was a good teacher, far ahead of his time in expounding on farm safety, soil conservation, and concern for the environment. I was reminded of his advice to plant black walnuts along fence rows when walnut wood became so expensive a few years ago that even trees in the city became subject to theft.

One male teacher was a woodworking hobbyist who sometimes used the power tools when the class was otherwise occupied. I once watched him attempt to cut a slot in a small block of wood by hand-holding it against the bench circle saw. The saw jammed, as should have been expected, and a corner of the block cut a deep gash in his hand. He swore at the saw, applied first aid, and tried again; once more the saw responded to misuse and inflicted a second gash. The event strongly reinforced Mr. Knapp's teaching of safety and proper use of tools. For me, it instilled respect for power equipment and even a preference for hand tools, which was just as well because electricity had not been extended beyond the town limits to the farm. Although sharp hand tools could be hazardous, they did not have the destructive potential of that teacher's lack of respect for the energy driving the saw on which he was injured. As energy sources have become more available, and ever more powerful, the potential for serious accidents has also increased. Thus, measures designed to insure proper respect by those in the vicinity of a hazardous activity have become more important, requiring development of procedures, regulations, and laws assuring such protection. This need was already evident in the 1930s, from experience with automobiles and buses and in newer factories. Accidents with horses were frequent, but seldom fatal.

I was initially excluded from athletics by size and immaturity until my senior year, and then by afternoon farm chores. However, my peers did recognize a few of my physical talents. One of my heightened skills involved precisely pitching erasers so as to lodge on high moldings in the math classroom, out of the teacher's reach. As a quiet, model student, faculty suspicion was never directed my way, and students apparently admired the talent enough not to tattle, even under threats of extra class work. Other high school disturbances were limited to similarly devious but minor ways of keeping the teaching staff alert.

Serious problems, such as those reported in schools today, were unheard of, even though some kids were tough and a few later served time in prison. However, the seriousness of high school antics had intensified somewhat by the time my brother attended. The school was closed for three days when a skunk was found one morning in the building's hot-air heating system. Robert was a prime suspect because he raised skunks on the farm, both for the novelty and to sell the pelts, but as far as I know the allegation was never proven. Incidentally, he was remarkably successful in keeping the farmstead mostly free of their malodorous secretions.

Two high school teachers were outstanding. Miss Marie Minnick, the school principal, was a strict disciplinarian. Small of stature, she occasionally reached up to grasp the ear of a tall errant boy and lead him in a bent position to a seat before her desk. Such embarrassment was an effective deterrent for misbehavior. She also taught history vividly enough to initiate my lifelong interest in the past. Mr. Owen Zook taught General Physics using a combination of lecture, demonstration, and hands-on experimentation. For me, the course explained many mysteries of the world: why ice forms on the surface of water, why a lantern was lowered into a pit before entering in daylight, why perpetual motion is not possible. Mr. Zook patiently explained the fallibility of my attempt.

That class also taught us the concept and practice of experimentation. I previously had built simple electrical circuits using discarded telephone batteries, had smelted lead from car batteries to cast objects in crude sand molds, and had worked with levers and pulleys on the

Don's Exeter High School graduation portrait, 1936.

farm. This class clarified for me many aspects of familiar but confusing devices. It also reduced the frustrations I encountered because of problems with my self-designed toys and the more complex farm equipment. More significantly, it helped me see that advanced schooling in science or engineering might be interesting, challenging, and achievable if money could be found for college.

I set a secret, seemingly unattainable goal to become an electrical engineer. That inspiration also instilled a new and lasting zest for studies and learning. Mr. Zook is one of the people to whom I am forever indebted, for his excellence as a teacher and for his opening of new horizons to me.

Exeter High School offered extra-curricular programs in instrumental music and chorus. Regrettably, I did not pursue those opportunities, probably from shyness and from a nearly total lack of musical experience. Although the need to return home promptly to feed horses, chickens, pigs, and cattle was a factor, Mom and Pop would have endorsed some music education. Regardless, high school had been a great experience that ended much better than it began. Three years had made a physical difference; by my senior year I was nearly the tallest, instead of the smallest, person in school. Roles in both the junior and senior plays built some confidence and stage presence, but both shyness and reticence were continuing problems.

I graduated at age fifteen, fourth in a class of thirty-six. Ironically, I was fortunate not to have graduated higher in the class because I might have earned a scholarship and enrolled in college. Another disastrous first year could have ensued because I still lacked both the confidence and maturity to succeed in a college program.

On graduation night, other class members had planned parties in celebration, but I did not join them. I was depressed and recall intense sadness in contemplating a life without further education. During the intervening years I have, of course, experienced many other disappointments, but my reactions to them have faded from memory. In contrast, the feelings on that graduation night remain vivid. After a long, traumatic evening I fell asleep, exhausted. Recovery came slowly during the ensuing days, and eventually the feeling of depression disappeared. Although my despair may not have been clinically significant, the intensity of those feelings have made me appreciate the problems of those who suffer from prolonged and severe depression.

Farm work gradually became boring, even drudgery. With a six-horse team for energy and a two-bottom plow, it took about thirty

Grain binder with wheat shocks in the field. Photograph by Robert Trauger.

days to prepare the ground for sixty acres of wheat. During the first hours, it was fascinating to sit on an old automobile seat mounted atop the plows and watch the black loam turn over. The seat had been salvaged from the 1911 Jackson automobile that my father had driven when courting Mother and was more comfortable than the cast iron seats with which most farm implements of the time were equipped.

At first, watching migratory sea gulls follow the plow and fight needlessly over the multitude of exposed worms and insects also spurred my interest. However, during the second and following days, interest devolved into the routine made of caring for the horses, adjusting the plow, checking for vegetation entanglement, and daydreaming while totally bored. Frequent stops were required for the horses to rest. At the lunch break we would return to the farmstead and barn to feed them. The first stop was to the stock water tank, where the horses were released to drink. Queen would cool herself by first immersing her long proboscis into the water up to her eyes. The horses then proceeded to the barn to feed on a measure of oats and the ever-present prairie hay in the mangers. Mom always had prepared a delicious and filling lunch, which we called dinner. Then we returned to the field until supper.

Harvest time was more interesting and challenging, owing to the more sophisticated technology of the equipment used. Whereas the process of plowing had developed only marginally since its first employment at the dawn of civilization, our harvesting in the mid-1930s used much technology and energy systems that had become prevalent only in the late nineteenth century. Although combines that cut and thrashed the grain from ripened standing stalks had been available for several years, they were not perfected or in general use in our area. When the grain was determined to be properly ripened, the stems were cut, bunched into bundles, and tied with twine. The machine for this step was called a grain binder—operated by means of chains, eccentrics, lever arms, and gears driven from one master gear attached to a large lugged wheel—and was powered by three or four pulling horses. The bundles disgorged by the binder were dropped into a carrier and released by the operator at spacings convenient for standing them in rows of shocks for drying.

Occasionally the binding mechanism failed to tie a bundle. Other farmers carried extra twine to secure the loose sheaves. Pop had learned to bind such grain bundles with a few stems, thus securing the bundles without the need for extra twine for subsequent transport to the threshing machine. His father, who had harvested grain with a scythe before the age of grain binders reached his mountainous area of New Jersey, had taught him the technique. I learned it as a more convenient and quicker method than cutting twine, wrapping the bundle, and tying a knot.

The thresher required servicing each year. This large and complex machine needed lubrication, re-nailing of the wooden straw racks, and installation and adjustment of its many belts. Earlier, at about age thirteen, I had taken a team of horses to the shed near the house on the corner where the thresher was stored during winter and was to bring it to our shop for service. Having properly hitched the horses, I stood inside the building and ordered the team to pull it out. As the machine moved forward, it shifted on the uneven earthen floor and pinned me at the chest between a post and the main drive belt pulley. With a gasp, I called for the horses to stop and they did so immediately, but I was unable to speak a word more.

Fortunately, I still held the reins and could signal them to back the heavy machine. I am forever grateful to Bill and Old Harry for working with great difficulty to free me, for I might have perished had they not been so responsive. My chest was very sore for weeks, and I later concluded that a rib had been broken; but I never told my parents of the incident, both to avoid embarrassment and for fear that they

Early Avery gasoline tractor of Robert Trauger's collection with Tom and Byron Trauger standing in front in 1955.

might restrict such activities, which were the only aspect of farming that still held out challenge and interest for me.

The monstrous early tractors that powered the threshing machines by means of a long belt were used primarily for stationary equipment. They could move themselves and pull the thresher to a new site at a snail's pace but were much too cumbersome and heavy for plowing soil or for other farm work. Our heaviest tractor weighed 12,000 pounds. Those early engines were reliable, though often balky in starting, requiring much effort applied to a long and heavy crank. They also were difficult to steer, usually requiring the driver to stand to obtain leverage in turning the cast iron steering wheel. Those early engines were also inefficient users of fuel, although many would burn kerosene, then less expensive than gasoline. As late as the 1930s, several neighbors still powered their threshers with steam engines, inefficient users of coal as fuel; the boilers, for example, were uninsulated. Nevertheless, the steam whistles, similar to those of railroad locomotives, when sounding over the countryside seemed more romantic than the noisy gasoline engines. It was necessary for the steam engineer to rise early in the morning to stoke the boiler so that it would be fully heated when threshing was to start. Each engineer vied to engage his steam whistle earlier than his neighbors. Those of us who arose early for harvest activities had no problem with the pre-dawn mayhem, but

Reeves Steam Threshing Engine of Robert Trauger's collection.

our friends in the village objected to that unwanted and recurrent awakening at five o'clock or even earlier.

Before grain combines took over, my role was to bring the bundles to the threshing machine by use of a large wagon called a rack, drawn by a team of horses. Although the process of loading and hauling bundles was among the more strenuous tasks of the harvest, it was something that I enjoyed. I liked the challenge of placing the bundles into a stable arrangement that would not spill as the load grew high above the sides of the rack and the wagon tilted in crossing a ditch. The activity fell somewhere between a child's building structures with wooden blocks and an engineer's ordering of items and activities.

To haul grain bundles, I usually chose the horse team of Bob and Bill. They learned to follow the row of shocks without guidance from the reins. Bob was as adept at removing his bridle as Old Harry was at opening gates. As a convenience to Bob, I would remove his bridle and hang it on a hame of his harness. He then would turn his head without restraint of a bit to watch my progress and advance the wagon each time a shock had been loaded. My colleagues first warned of danger in driving a horse without a bridle, but they came to envy my freedom from the task of guiding the team. The bridle was used after loading and in travel to the threshing machine. To build the largest load of bundles required more effort than other members of the crew were willing to make, but my extra effort paid off when I could charge $3.50 instead of the usual $3.00 for a day of "pitching bundles" as an extra hand in other harvesting crews. (The worker provided both wagon and horses.)

Threshing crews like ours were noted for playing jokes. I recall one member who could drink coffee hotter than anyone else I have ever known. He enjoyed pouring two very hot cups, one for himself and the other for an unsuspecting newcomer. He then would proceed to drink his cup immediately, and his buddy, assuming the coffee to be drinkable, would soon be spewing hot coffee onto the ground. Once as I was unloading the grain bundles into the ravenous thresher, a rattlesnake appeared under a bundle near the top of the load. I quickly observed that the rattler was dead and saw that those waiting to unload their wagons had an unusual interest in my activities. Someone obviously had placed the snake there to startle me. Remaining nonchalant as if not noticing the snake, I managed to maneuver the dead critter subtly until I had solid footing on the floor of the wagon. By balancing the snake on the tines of the fork, I tossed it high into the air in the direction of the interested onlookers who were resting in the shade of a loaded rack. Remarkably, the snake landed in front of the practical joker. Startled, he jumped up and struck his head on the wagon structure, but his straw hat cushioned the impact and prevented serious injury; no harm was done on either side.

Another incident involved an elderly neighbor, Mr. Churchill, who could not do heavy work but often assisted with tasks such as monitoring and leveling the threshed grain as it flowed into wagons. This dignified Southern gentleman with snow white hair and whiskers chewed tobacco with his front teeth to protect defective molars. The resulting movement of his gray mustache and beard, as viewed by the threshers, resembled a rabbit's nibbling on lettuce. He thus was known as "Old Rabbit," but the term was never used in speaking directly to the proud old man. One noon, Mr. Churchill arrived late for dinner. As he was washing at the outdoor cistern pump near the open window of our dining room, the conversation between Mr. Churchill and my four-year-old brother was clearly audible. Robert opened with, "Hello, you old rabbit." The response was, "What did you say?" When the greeting was repeated, the old man stopped washing and disappeared, not to be seen again until the next day. I have seldom been among a more embarrassed group than those farm hands who finished their meal in silence.

I vividly recall the comradeship among farmers who banded together for the small grain harvests of oats and wheat and for shelling kernels from ears of corn as part of that harvest. Such community activity had begun with the earliest settlers, who needed help in building barns and other farm structures. The men cooperated for the heavy outdoor work, and the women often worked together to provide outstanding dinners and suppers to support their men. These group efforts brought neighbors together to develop familiarity and understanding among families. Today, with large farms and correspondingly large internally powered equipment, each farm can be self-sufficient. This isolation in the activities and energy systems for farming also has changed the social structure to more closely resemble that of the village or city.

In many ways, I did not fit the farmer stereotype. For example, following my Grandfather Trauger's death in 1934, I fell heir to white shirts with detachable stiff collars that he had worn to church and on other formal occasions. The shirts were oversized and loose fitting on me; the smooth, white fabric shed chaff and reflected the sun so as to be cooler and more comfortable than the hot and sticky denim. Even with no collar, I was teased as a "white collar worker," but comfort prevailed. My hat also was different from the traditional wide-brim straw hats then favored by Midwest farmers. I wore a cardboard "helmet," an inexpensive model of the headgear traditionally associated with African explorers or former British rulers of India. The hat choice resulted from a frightening experience when I was building shocks of bundled wheat stems on a very hot and sultry harvest afternoon. An hour or so after lunch, I experienced nausea, apparently related to overeating, then dizziness. To me this signaled a more serious problem; on growing faint, I decided that it was heat prostration. What was I to do?

The distance to a shade tree or help was too great to attempt leaving the field under the blazing sun. Fortunately, there were long shocks nearby that had been set up during the previous evening. The stalks had cooled by evaporation, so I carefully crawled inside a shock to rest quietly. The symptoms disappeared with time, confirming my amateur diagnosis. After I felt able to walk to shade, I spent the remaining hours of the afternoon resting under a tree. This led to my purchase of the pith-helmet-type hat and reinforced the choice of shirts, despite the comments they engendered. My colleagues were further confounded by my carrying a jug of unsweetened iced tea instead of water as a thirst quencher.

The drought continued relentlessly for several years, and farm product prices remained low. Small wheat crops brought little, if any, income above the cost of harvest; often, the corn crops were even more limited. In 1936, when Pop reported the corn harvest yield to the U.S. Department of Agriculture's farm assistance program, he entered "15 pints." With its implication of corn whisky, this report tantalized his friends in the office while prohibition was the local law. The yield referred to home-canned corn harvested as roasting ears from a location near Johnson Creek before even those stalks perished.

We had some fun, even at what would now be a poverty level, and hoped for better days.

Queen and Mae at the House on the Corner, circa 1938.

THREE

The Tractor
Displaces Horses

O ur 1924 radio served us well, despite its persistent howls and static, as we listened through headsets and watched its exposed condensers move visibly as the station was tuned. That old radio brought us timely news, including reports of Charles Lindbergh's solo flight from New York to Paris. I remember listening to the headset, ear-to-ear with Mom, to learn that a ship in the mid-Atlantic reported an aircraft had passed overhead in the darkness. Later we were thrilled to learn that Lindbergh had landed in Paris at 10:00 P.M., May 21, 1927. In its later years, that radio could consistently garner signals from only two stations, one nearby and the other owned by a seed company in Iowa. They provided news and farm market information, but music was limited to country-style tunes. A favorite girlfriend once asked, "Which radio orchestra do you like best, NBC or CBS?" Embarrassed not to know of either, I chose NBC with good fortune; it was her preference.

In 1936, Pop surprised us with a new radio that came in a large cardboard box with vacuum tubes and a multitude of wires packed separately. The circular tuning dial encompassed many wave bands including stations overseas. I rushed to assemble the set, making proper connections for "A," "B," and "C" batteries and installed a long antenna wire to the windmill tower, including the lightning protection important in the storm-prone Midwest. The wiring was completed just before six o'clock one evening, in time to set the dial to the BBC station in London. The sound of Big Ben tolling the midnight hour from the Parliament tower still rings in my memory. Pop had given us a great gift. I quickly learned about the big band orchestras, Mom found that "soap operas" seemed to lighten her housework, and Robert regularly enjoyed children's programs. Pop's reward: to hear some of the evening news from a more comfortable chair but to soon fall asleep.

For me, this radio provided a new understanding of world events. The long antenna, Midwest location, and steady battery power, enabled the set to bring distant news directly to the farm. Language was a limitation, but the news in English from many sources provided exciting information including my first exposure to government propaganda. Broadcasts to sway British, Canadian, and American listeners toward prejudiced points of view seemed endemic for Berlin, Moscow, and even Tokyo, on the rare occasions when we could receive a clear signal from the Far East. The Spanish Civil War and the war in China, both vicious preludes to World War II, were constantly in the foreign news. Spanish partisans solicited help, while Russian and German government broadcasts attempted to justify their involvements. My understanding of the Spanish "Civil" War's international significance increased as I listened night after night.

The radio gave me a new perspective on war. My school textbooks had made the American Revolution seem almost romantic, had glossed over the tragic carnage of the U.S. Civil War, and were too old to include World War I. Radio reports of a million Spanish deaths provided my first significant exposure to the horrors of war. Broadcast after broadcast also brought an understanding that the USSR and Germany represented serious future threats to international stability. The United States and Europe, shaken by ravages of the Great Depression, seemed unwilling to face that major challenge. Hitler's Germany, formed from the ruins of their post–World War I depression, and the USSR, created by the Communist Revolution, seemed formidable even though their economic bases were weak.

Short-wave radio broadcasts, with their national perspectives and widely differing angles of vision, often made it difficult for me to tell the "good guys" from the "bad guys," but I recall deciding that the USSR of

Joseph Stalin was a potentially serious long-term threat to Europe and world stability. My opinions were heavily influenced by the BBC's news reports, which seemed the most balanced and reliable of all. This array of contrasting opinions from around the world stood in sharp contrast to the relatively benign and dissociated positions of local news from the publications and commercial radio of the U.S. Midwest. The isolationist attitudes of our neighbors seemed more extreme than those of the country as a whole, but none of us anticipated the impending world war only three years away.

The new radio made me hungry for greater knowledge and experience. My father hoped that this would make me more comfortable with farm life. It did make life more interesting and helped prevent further feelings of depression, but in the end the radio brought siren songs that lured me away from the farm.

The radio also introduced an appreciation for classical music, although my taste was only slightly beyond the big band swing tunes of the time. On Saturday nights we often listened to the *Grand Ole Opry* program of country music and corny humor. During those years, I could not have imagined that the music of Haydn, Beethoven, and Mozart would become my favorites. Overall, the radio introduced new thoughts and experiences to a boy of limited perspective.

A visit from the local manager of the Iowa-Nebraska Light & Power Company was the next exciting event of 1936. Pop and I stood in the farmyard beside his automobile as plans were presented to bring an electric power line to the farm. The benefits were too valuable to forego, even though money was short for the necessary wiring and appliances. The year 1935 had brought only a partial crop and though not yet apparent, the 1936 crop failure was destined. We were the first farm family in the Exeter area to sign as rural-electric customers, but the company assured us that our neighbors would soon have the same service.

Our enthusiasm for this forward step dampened when we discovered the true motive behind that rural electric line. The company had selected the most favorable routing to gain enough customers at minimum cost to discourage a national Rural Electrification Administration project from coming to our area. The gambit succeeded; there was to be no extension of the power line, and our neighbors were to be deprived of cooperative electric power for more than a decade. We perhaps should have recognized the ploy, since private power-company fights with government-sponsored electrification were well known.

One of the most important of these fights resulted in creation of the Tennessee Valley Authority (TVA), established to convert the Tennessee River from a flood-prone stream to a navigable waterway and source of electric power. Hydroelectric public power programs of western states had already shown that their energy production was inexpensive, and this project in the Southeast seemed threatening to private utility companies. Nebraska's own forward thinking Senator George W. Norris had become interested in the Muscle Shoals project of the First World War and the plight of residents in Appalachia. Because he saw the value of more plentiful and less expensive energy, he sponsored and shepherded the controversial TVA legislation. I had become interested in the construction of Norris Dam, TVA's first new power unit, and followed the project's news reports. That dam was a major pioneering step by government to create a regional system of electric power generation that included flood control, preservation of natural resources, and the development of improved agricultural fertilizers and practices. It also was to be a future neighbor that would supply electric power to me for many years.

Pop hired an electrician to wire the house, provide a floodlight over the farmstead, and install lights and receptacles for the cattle barn. I served as an apprentice to minimize the cost and learned to install wiring in other buildings. In the house, wires were pulled through the attic, between ceiling joists, through the fruit and vegetable cellar, and to the basement laundry; all rooms soon had fixed ceiling lights and electric outlets. The outside light, anchored to the top of the water supply tank, illuminated most of the working area between buildings.

I later wired the corn crib, the pig feeding pens and the horse barn. All was done according to code and served well, except for the fixture at the entrance to the barn. When a suitable electrical conduit to protect the wires from weather was not available, I substituted plumbing parts. In due course, this shortcut was to cause trouble, embarrassment, and a costly lesson on the importance of codes and quality control.

My fascination with experimentation proved painful on the very first night we had electric lights, as I repeatedly switched on various sizes of electric bulbs to compare their illumination with that from the kerosene lamps. In rapidly changing from one source to the other, I grasped a lamp chimney before it had cooled and painfully burned three fingers. Although injuries and fires were common with kerosene lamps, I had never before suffered injury or accident in all the years of using oil lamps. How frustrating to have erred so painfully at the very moment of casting off dependence on the malodorous kerosene units. Incidentally, the observations showed that the kerosene flame lamps, with wicks properly trimmed, were not quite equivalent to a 25 watt bulb; however, the Aladdin lamp with its thorium oxide mantel (that also used kerosene fuel) compared with a 100 watt electric bulb.

Our new refrigerator replaced a primitive cooling system contrived for food preservation in summer. We had made a cylindrical, shelved container that was lowered by a rope to the bottom of a board-encased hole dug in the earth near the kitchen door. At about fifteen feet depth the temperature was 58 degrees Fahrenheit; milk stored there was passably cool for drinking. The unpasteurized milk soured in about one day, but the supply was replenished by milking one or two cows each morning and night. Really cold milk from the refrigerator was a greatly appreciated pleasure. With the advent of the refrigerator, ice cream could come from the store in cartons instead of from a major project with ice, salt, and a hand-cranked freezer.

In due course, Mom had an electric washing machine. It was much more convenient than the previous washer powered by its built-in small, but noisy, gasoline engine. That old machine, made by a company called Dexter, had been an improvement over the first one used by Grandmother Trauger and later by Mom. That earliest washing machine had a wooden tub and wood oscillating arms to agitate the clothes. It, too, was powered by a gasoline engine, but this was a separate, hand-cranked, one-cylinder unit that drove the washer mechanism by means of a belt. That old engine with its large flywheels

Farmall tractor with Robert Trauger and Caryl Steyer (seated), circa 1945. Photograph by Ethel Trauger.

made a set of sounds something like "chug, chug, chug, bang-chug, chug, chug, bang," repeated ad infinitum. For both, the exhaust was piped outside the house to prevent asphyxiation by carbon monoxide.

Electricity had not solved the water hardness problem, and we still had no electric water heater. For all three washers in succession, rain water was pumped from the cistern, transferred to a stove that burned corncobs or wood, and was heated in large copper containers called boilers. Steaming hot water was then carried in buckets to the washer at some risk to my mother or a helper. The electrically powered washer included a pump to discharge the wash water through a hose extended to the backyard, a welcome improvement over lugging buckets of water upstairs for disposal. Even the stifling heat in the basement—created by the wood-fired stove, hot water, and steam—was made more tolerable by an electric fan.

This new source of energy was changing the household and farm-stead work. Electrically driven tools slowly replaced devices such as a hand-cranked drill press and the foot-powered grindstones for sharp-ening tools. This also made many tasks more efficient.

Our farm was late in converting to tractors for field work. In 1936 Pop purchased a second-hand Farmall tractor with cleated steel wheels, which we used primarily for plowing. Most other operations continued with live horse power. This tractor could pull three plow "bottoms" at nearly twice the speed of the horses. It also did not require rest periods, and work in daylight was limited only by the operator's endurance and permissible weather. Thus, the nearly month-long spring and fall plowing periods were shortened to about one week each. The Farmall was a noisy machine as operated without a muffler, which would have reduced power. I always packed my ears with cotton for noise protection; Pop, who did not, suffered severe hearing loss in later years that he attributed to that tractor.

I never liked to use that impersonal machine, but it was a labor- and time-saving energy provider that could not be ignored. Compare the preparatory efforts involved in filling its fuel tank and lubricating a few bearings to that of feeding, currying, brushing, harnessing, and hitching six horses to an implement for plowing.

After World War II the massive switch from horses to rubber-tired tractor power had two significant effects: it made our and other farms dependent on purchased fuel, and it freed land for cash crops. About 18 percent of our farm acreage had been required to produce feed for the horses when they were the primary energy source. The transition shifted the farm energy supply from the renewable crops of the land to the nation's limited reserves of oil. This recognition initiated my serious interest and concern about energy resources.

Immediately upon reaching the age of sixteen, I went to the courthouse and obtained a driver's license. This new independence made possible my first significant and satisfying entry to group activities. Roller skating was a popular pastime in which I had skill enough to negotiate my way around a commercial rink. Most of my friends had learned to skate on village sidewalks, but walkways on the farm were neither smooth nor long. However, we had a concrete floor poured for a farm workshop as part of a shed planned for farm machinery storage. Hard times had stopped the project, but the floor was smooth and fine for skates. It was too small for distance runs, but I could practice turns and stops. Good times developed with teenage peers at commercial rinks in two nearby towns. With the car I could provide my share of transportation for a small group that made weekly forays to the rinks. Another credit is due my parents for scheduling use of the vehicle to include my recreation.

On a night in February of 1937, I drove the Chevrolet to a high school basketball game in Exeter and started home in a severe ice storm. With no windshield defroster and ineffective wiper blades, visibility soon was near zero. I drove slowly, alternately with my head out the window (until the cold sleet was intolerable), then held a steady course from inside while searching for the glow of a taillight. During one inside portion of the cycle, I suddenly crashed into something and was thrust sharply into the steering wheel. The Chevrolet had plowed into the automobile owned by the opposing team's coach. The coach apologized at the time, explaining that he had stopped on the road to clear the windshield and that his taillight was inoperative, which explained why I had not seen it. In later negotiations, he seemed to have forgotten the admission and the defect. Damage to the coach's car was limited to the rear bumper so that cost was not great; however, we donated a new taillight in the settlement even though there was no evidence of impact there.

Our car had suffered radiator damage and could not be driven without repair. The family decision was to make this an occasion to buy a new Chevrolet. I was delighted, except that to pay cash, as was our inviolable family custom, I had to contribute substantially from my savings. This set back hopes for college, but I had pride in partial ownership of a new auto. Perhaps that minor wreck served as an important lesson, for I have not since been responsible for even a scratched fender.

The new Chevrolet had two taillights, even though only one was then required. The right-side light found in the car trunk was an expensive extra for which we were not charged. I had the satisfaction of installing it, thus further increasing my feeling of ownership. The new car also brought a greater sense of equality with peers. The people of the farming community and those whose businesses were dependent on weather and produce markets generally drove old automobiles. However, many of my friends' families were more affluent; their fathers worked in the Smith Index Tag Factory. Those employees were well paid even in the 1930s. The new Chevrolet was not on a prestige par with their Nash, Hudson, and Buick automobiles, but it was attractive and opened new horizons.

The pioneering Agricultural Adjustment Act of 1933 had offered a new era for farmers, but my parents, like many in the Midwest, were opposed to government intrusion into business or farming activities. Their initial reaction was to ignore this part of President Roosevelt's "New Deal," hoping that it would be discontinued. However, Pop was curious about anything new and soon studied the program in detail. A principal purpose of the act was to reduce commodity surpluses, thus to raise prices and stabilize farm income. Nebraska farmers who participated were to be compensated for reducing their planted acreages of wheat and corn, products previously produced beyond demand. These objectives seemed reasonable to Pop, who had long deplored the lack of an effective economic mechanism by which farmers could control their own destiny. Labor could strike for benefits and companies seemingly could raise prices at will; farmers, who valued their independence, remained at the mercy of markets. By 1934 our practical response to this government program had become consistent with the adage "If you can't beat them, join them."

That proved fortuitous. The reduced acreages of corn and wheat imposed by the program represented little initial loss to us or our neighbors. Owing to the increasing severity of the drought, farmers grew fewer and fewer crops; so the government payment helped to sustain life on the farm. When in 1936 the Supreme Court ruled the 1933 Adjustment Act unconstitutional on the basis that its system of payments constituted coercion to farmers, new legislation was passed. This was the Soil Conservation and Domestic Allotment Act. It authorized payment for practicing soil conservation measures instead of growing staples. Soil conservation was a tradition of our family, so little change in practice was necessary to comply. Overall, the loss of topsoil had become rampant after the Midwest soil lost the root-laced texture of the prairie grasses. Agricultural practices were causing severe water and wind erosion, the dust storms proof positive of the latter. Pop became an enthusiastic supporter of the government farm program.

In that role he started as a volunteer worker helping other farmers design their participation. He then became a paid, part-time employee. I also found this an interesting program and was soon helping with some of the paperwork accompanying land measurements. That experience later was to enhance my college prospects. But in 1937 the farm program had another surprising impact. On a pleasant Sunday afternoon in late June just before the summer harvest was to start, Pop suggested that we take a walk about the farm to estimate the moisture content of the standing grain and to plan and schedule the harvest work. The wheat and oats fields would yield somewhat better crops that year, and it was therefore imperative that they be properly garnered. Pop began our walk with a startling announcement: He was to work full-time in the Farm Program Office throughout July, and I was to manage the harvest. He had hired two itinerant workers to help, but this time I would both work and supervise. How could I properly manage the two experienced men? Protest was to no avail; the grain was ready to cut and these harvest hands, who had worked for us the previous summer, would arrive the next morning.

I also was to take Pop's place in organizing and managing the threshing operation that would serve both our farm and those of neighboring families whose incomes also were dependent on the summer harvest. The threshing machine was owned in partnership with a neighbor, but the engine that would power it was ours. The Farmall F-30 general purpose tractor had replaced the monstrous Avery and Aultman Taylor tractors of previous harvests. This was a considerable responsibility for one who had passed his seventeenth birthday only two days earlier. During the first days of fear and feigned confidence, I found the men responsive, and I quickly learned the value of their experienced advice. We worked together as a team, an important lesson that would later serve me well. The harvest and threshing progressed smoothly and to the apparent satisfaction of all. This accomplishment made farm life seem more attractive; Pop had won on a high-stakes gamble.

The tide toward college turned favorably again as I found two new income sources. One was the government farm program; the other was to work directly for farmers in servicing their windmills. Even after

Farm windmill. Photograph by Robert Trauger.

electric power became available, many farms continued to use the silent, inexpensive power of windmills to pump water for household use and livestock consumption. Many farmers were afraid to climb the steel or wooden towers to lubricate bearings and repair the wind-driven wheel some thirty to forty feet above the ground. I enjoyed the challenge, relishing the relatively good pay, two to five dollars for an hour or two of work. Most windmills needed service at least yearly, but farmers having very old equipment called for help frequently when the mechanism began to squeak or showed other need for repair.

To lubricate, inspect, and repair the mechanism, one stood on a platform located just below the wheel blades. There were many designs to be learned, for manufacturers had been innovative in striving for market share. Each wheel blade that translated the wind force into rotary motion had to be checked for loosening by weather forces: strong winds, freezing rain, hot sun, and cool nights. Icing sometimes upset the wheel balance, further adding to stresses and causing damage. Renailing blades of wooden wheels was a tricky two-handed job, as was rebolting the steel wheels; a third hand was needed to grasp the tower. Extension of electric power lines to all farms finally doomed the windmills to antiquity. Manually or automatically switching on an electric motor to drive a pump to maintain the water level in a tank was simply far easier and, ultimately, more cost effective.

The farm program provided more money than windmill servicing. The work was pleasant and challenging, and it paid $4.50 per day. In the years 1936 and 1937 I did the field work, but my father was responsible for it and paid me, since the federal government could not hire a person under eighteen years of age. My task was to walk and measure the fields as planted each year to ensure compliance with the farmers' contracts with the Department of Agriculture.

I learned much about people in the course of working throughout the Exeter Township. Farmers were surprisingly unobservant about some details of their farms. Once, in planning a survey, I mentioned the location of a tree for reference. The photograph showed it to be in a field labeled "A." The man who had worked these fields all of his life insisted that the tree was on the other side of a fence and hence in field "B." He would not believe that the photograph was correct until I drove him to the field. Aerial photography was new, and in many similar instances farmers unfamiliar with maps and photographs did not recognize features of their farms on those large and remarkably clear prints.

Field measurements were made with a device called a "Walking A." This sturdy, lightweight, wooden frame was shaped like the letter of its name but with a straight, smooth handle projecting about two feet above the apex. Each leg ended in a pointed steel peg so shaped that it would retain an exact position when poked into the ground, providing a pivot on which to rotate the frame 180 degrees by maneuvering the handle. One measured distances by "walking" the frame, just as a navigator "walks" a measuring compass on a map, and carefully counting the number of 180-degree walking turns. The distance between pegs was such that when the number of rotations of the "A" on two adjacent sides of a rectangle were multiplied and the result divided by one thousand, the answer was acres of land. It was interesting to learn more about the topography and farming practices of the neighborhood, but most importantly the job provided money.

One measure of a successful farmer is the absence of weeds in his fields. Most, but not all, of the farms visited would have been judged favorably on that criterion. The worst were those that harbored sand burrs and other prickly or sticky plants that were a hazard for one required to walk in a straight line. The solution came in the form of sixteen-inch lace-up boots. Boots also gave one confidence with the farm dogs attracted to the unusual motion of the "A," giving chase with much barking. I have never feared dogs, and they seemed to sense it unwise to attack one who carried an unusual, sharp-pointed device.

Don with "Walking A," 1999.
Photograph by Robert Trauger.

However, some colleagues of other townships received injurious bites.

Occasionally, I was frustrated to approach the end of a half-mile measurement uncertain that my count of the laps was correct, sometimes distracted by a neighbor's conversation, sometimes by daydreams. Whatever the cause, such confusion always required repetition of the measurement, which invariably showed the first number to be correct. The mind can function in remarkable ways.

One farmstead lay in a low terrain still heavily shrouded with trees. The farmer and his wife were reputed to be fortune tellers able to communicate with the dead. My first encounter there occurred while measuring a boundary line for the fortune tellers' neighbor. I had seen Mr. Specter (as we will call him) approaching in the distance with a four-horse cultivator to clean the corn rows of weeds and to loosen the soil. As he approached and I prepared to greet him so that I could set a date for measuring his fields, I positioned the "A" at lap 151. While he turned the cultivator around, I flipped the "A" frame over to 152 to find a more secure position in the soil to hold it in place. Mr. Specter's greeting alarmed me: "That was lap number 151."

I never counted aloud or knowingly moved my lips. He could not have counted from a distance, for my path was not fully visible to him, and the location had no relationship to landmarks on his farm. Although not believing in fortune telling, this experience left me apprehensive about an appointment with this mystic. When the time came to knock on the door of his tree-shrouded house with walls and window shutters painted a dark green, I was greeted by a beautiful, blonde young woman about my age. Was it an apparition? No, she was real and cordial. Mr. Specter later explained that she was a niece about to return home after visiting for several weeks. It then seemed unfortunate not to have scheduled his farm earlier in the summer. However, I recall that Mr. Specter and his wife were arrested somewhat later for having obtained money under false pretenses in faked "other-world" communications.

There was only one farm in the township that I did not visit. The owner had threatened "to shoot to kill" anyone who came to his place representing the farm program. Enforcement of the government order requiring an offer of program benefits to every farmer also was ignored by the county committee. A few years later the farmer left the area, the place was sold, the farm buildings were removed, and no trace remains of his existence.

In contrast, most of the farmers and their wives were pleased to greet the "government man" who represented new income through the program. Sometimes they would serve lemonade and cookies in addition to the usual cool drink of water from the well. On the other hand, some younger wives were quite shy. I recall one, who on seeing the car enter the drive, dashed into the outhouse toilet. It was a very hot day and after waiting some time for her to reappear, I departed choosing not to torture her further. One of my colleagues reported having come upon two young women "skinny dipping" in a farm pond; my experiences were interesting but never so revealing.

Introduction to several high school seniors in Friend provided new interests. A foursome developed with a farm boy, Willard Engel, who later became my first college roommate, and two village girls. Willard dated Sarah Brown, and I escorted Ruth Sallenbach. Ruth was the daughter of long-standing family acquaintances, distant relatives of Uncle Elzie. Each of these new friends was planning to attend Nebraska Wesleyan University, and we spent much time studying the college catalog. My goal, of course, had been to attend the school of engineering at the University of Nebraska, where I had already studied by enrolling in correspondence courses. After some exploration of the options, NWU seemed attractive, financially possible, and an acceptable choice. Early that summer I met Professor John Christian Jensen, head of the physics and preengineering programs, and was excited about the possibilities he offered.

The year 1938 was truly one of new beginnings. Finally I could plan seriously for a college education.

PART TWO

1938~1957

Pioneering in Nuclear Energy

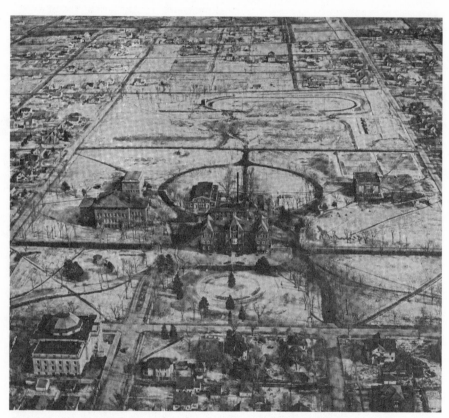

Nebraska Wesleyan University, 1941. Courtesy of NWU Bulletin.

FOUR

Fission Energy Discovered

A bright autumn day in September 1938 marked my departure to college. I had become increasingly aware that this would be a major change in life, but farm activities had been demanding, leaving little time for speculation. The early corn harvest was in full swing and family and neighbors had joined in to fill the tall farm silo with silage that would ensure feed for the cattle as effects of the drought subsided. It would be difficult to break away from that project and from the wheat harvest, both farm neighborhood events that I enjoyed. I also thought that the enrollment at Nebraska Wesleyan would be but an initial step toward an engineering degree at the University of Nebraska, in electrical engineering or possibly agricultural engineering. I loved that farm and still do, but a satisfying life there seemed unlikely for me; thus, I had no regrets in leaving beyond the stretching of family ties.

After a last-minute rush from the dusty, sweaty work in the field, a quick wash in my rigged-up, cold-water, basement shower, and a bit of packing, I barely had time for hurried good-byes before the transportation arrived. Ruth Sallenbach's mother had volunteered to take the four new students to the Nebraska Wesleyan University (NWU) campus. We had anticipated a problem finding room for all the luggage, but mine added little, one small suitcase and my Underwood No. 5 typewriter. The hour-long trip, in contrast to the three hours it would have taken less than a decade earlier, seemed short indeed; we were filled with anticipation of the next years.

Willard Engel and I moved into the second-floor room at the Barrett house one block from the campus. The university had no dormitories, and the Barrett's location seemed to fit our needs and budgets. We soon set forth to explore the community of University Place. The drought had not spared Lincoln, but the city's low-lying land and a fine water system piped from the Platte River had provided moisture for grass and trees. The campus of seven buildings seemed inviting, with its green lawns and trees that contrasted with the barren farmsteads near Exeter.

The designation *university* seems to have been chosen somewhat optimistically at the school's founding in 1888, but a few decades later the institution expanded for a time to five colleges. When our class entered, only the College of Liberal Arts remained, but the term *university* had been retained with the approval of the state of Nebraska and accrediting institutions.

Willard and I chose the Hays Boarding House for our meals, for reported quality cooking and modest price. We were correct, however, only about the price. Meat and potatoes were always good and plentiful, but vegetable dishes all tasted alike—texture, flavor, and no doubt vitamins having suffered from excessive cooking. Quality problems were best illustrated when the dessert tray passed: the smallest serving was taken first. The poor soul in the last position was obliged to eat the largest portion of dry, tasteless cake or another offering. Mrs. Hays carefully supervised the meal and became upset if we failed to finish her desserts. It was a culinary shock to leave Mom's cooking.

Living quarters at the Barrett house were modest but comfortable. Willard and I shared a small room tightly fitted with double bed, dresser, tiny closet, and two chairs at study tables. A nearby bathroom was shared with another student, Luther Powell, a pre-ministerial graduate of NWU who was enrolled in the University of Nebraska Graduate School and who worked part-time. Luther, versed in the arts and languages, served as a mentor of great value to me. As I gained knowledge in the sciences, my sharing may have added a little to his interest in technology. We became close friends and corresponded until his untimely death some years ago.

Although quiet hours were required after ten by both the Barretts and the college, evenings in the house were anything but silent. I particularly recall Mrs. Barrett playing "Alexander's Ragtime Band" on the piano and her middle-aged son singing songs such as "Roll Out the Barrel," with or without benefit of the instrument. Their music was often varied and enjoyable, although at times it distracted from studies. Mrs. Barrett was excitable and prone to physical gesturing. Demonstrating for Willard my imitation of Mrs. Barrett's manner, I threw an arm up in mockery while holding a fountain pen, only to spend the next hour washing ink spots from the ceiling. Willard soon joined a fraternity as a nonresident member and found both recreation and discipline for study there. That helped because it was difficult for two to concentrate in such close quarters at the same time.

My first new friend on the campus was Vernon Dix from Colorado, a farm boy also unaccustomed to city life. We both joined a social organization called Bleu Thonge, and, in addition, I joined the YMCA; a fraternity was far beyond my budget. Bleu Thonge had been in existence since 1934 as a social organization for men and women who could not afford the Greek fraternities. The name was chosen by students who formed the club, but I belonged for three years and never learned the origin of its name. Dues were 25 cents per semester, and a typical party cost each member about that much.

The finances for most of the 450 students impacted by the drought and economic depression, as well as the financial strength of the college itself, were about as meager as my own. It was not even clear that the institution would survive the four years to our graduation. Such thoughts were not greatly disturbing—perhaps we were conditioned to hard times. College buildings were in poor repair and the professors underpaid, with salaries in the range of one thousand dollars to two thousand dollars per year. The faculty consisted of remarkably talented individuals who were dedicated to teaching and loyal to the college. Although teaching jobs were scarce, most NWU professors no doubt could have applied to the University of Nebraska and been accepted at better pay. Each seemed to prefer the small college atmosphere and its more intimate contacts with students. To the extent that we took advantage of that closeness, the benefits to students were enormous. The faculty willingly shared with us both abundant academic knowledge and knowledge of life in those times. The potential cost-benefit ratio for the student was remarkable, with tuition costing seventy-five dollars per semester for sixteen hours of credit.

Chancellor Benjamin F. Schwartz came to the college the year our class enrolled and called himself "Freshman Chancellor," reflecting his friendly nature. "Chan" Schwartz, as we called him, faced financial concerns similar to our own but proved to be a skillful manager with excellent fund-raising abilities. The situation was helped greatly by the national recovery from the depression and later by tuition support for veterans through the farsighted G.I. Bill of Rights. That era was preceded by a severe lack of men on campus during World War II. Today, Nebraska Wesleyan is a thriving college of fifteen hundred or more students, with a student-to-faculty ratio of 13:1. Strong science programs continue within a broad and well-balanced curriculum. Only three buildings of my era remain among the present nineteen, and all are free of debt. Thus, the campus is relatively new, beautifully landscaped thanks to a generous benefactor, and is still growing.

No student body could have had a more supportive faculty than we enjoyed at Nebraska Wesleyan. Other classmates may have a list of outstanding professors different from my own, but I viewed several as valued role models. Professor Claude Shirk, who taught biology and premedicine, had no attendance requirement, but I never missed a lecture of his fascinating class on the History of the Earth. I also recall that no student who completed his premed program was ever denied admission to medical school or failed to graduate therefrom. Enid Miller, who taught speech and dramatics, had winning teams in national debating competitions throughout her career. John Rosentrator, head of the Department of Philosophy and Religion, was a particularly notable lecturer. He had fought and been severely wounded in World War I. His face and body had been distorted by bullets and his lungs severely damaged by mustard gas. Rosentrator's speech was labored and sometimes difficult to understand, but the brilliance of his mind easily compensated for these limitations.

Professor Rosentrator often deviated from the subjects of the course title but never from philosophy. My recollections of his observations on the causes of World Wars I and II are vivid. Perhaps through his war experiences or through the long recovery that had ensued from his injuries, Rosentrator had become a serious student of international relations, economics, and intrigue. He often discussed those subjects in a 1939 course on Great Philosophies of Religion. Rosentrator projected the coming and nature of World War II and the years to follow with uncanny accuracy. He foretold the rise of Japan and other Far Eastern nations to economic prominence and projected China as a major power by the turn of the century. He predicted that the West would become increasingly decadent following periods of devastation and reconstruction resulting from a great war. In retrospect, Rosentrator's informal lectures were prescient; overall, he was somewhat pessimistic about the long-term prospects for Western nations.

Unfortunately, I took notes only on the required material. Those notes, although important for "finals" preparation, are of little value to me today when compared to those I should have taken. Some years ago I contacted several members of Rosentrator's classes, only to learn that no one had taken notes on the important matters. My only consolation is that I periodically visited and corresponded with Rosentrator for as long as he lived. On learning of his death, I inquired about his papers, hoping that they might be made available to NWU. To my great disappointment, his family of rural people—who apparently never understood or appreciated him— had piled his papers and books in the front yard of his home and burned them to ashes.

Similar stories could be told of Professors Rose Clarke in geography and geology, Glenn Callen in economics, B. E. McProud in education, Oscar Bennett in music, Ethel Booth in English, and Gale McGee in speech. (McGee was to become a U.S. senator from Wyoming and later ambassador to the Organization of American States.) Nevertheless, our profs were certainly human. I recall the occasion when aging mathematics professor John Howie walked into his advanced calculus class and delivered the entire lecture he had prepared for the beginning class. The class reportedly sat there amused, but quiet, except to respond effectively to questions. When we gathered for chapel, members of the advanced class asked those of us in the beginning class to remain quiet and see if Howie would present the advanced lecture to us. We did, and sure enough, he did. No student was bold enough to tell Howie of his mistake; however, someone told Professor Jensen who relished the opportunity to tease, as he and Howie were wont to do. Professor Howie is reported to have said, "And I thought the beginning class was finally catching on to mathematics!" That classical "absentminded professor" served lemonade and cookies to each class at its next meeting.

A notable event in the career of Professor Enid Miller was writing with her newsman husband a play entitled *Manya*, about the life and work of Marie Curie. I served as student technical consultant and in that capacity read historical journals and questioned Professor Jensen about key points. "Doc" Miller (as she was known) and those of us who worked on that play had high hopes for its success. In fact, we envisioned it being produced on Broadway. Alas, a New York playwright wrote a play entitled *Madame Curie* and had it produced before we were ready. Nevertheless, *Manya* was well received when presented by the college players for the student body and community. Irene Pickering also was a major student contributor, and Hope Williams played the lead role.

Professor Marietta Snow taught German, and I was to regret that I did not take another year of her courses as a background for later cooperation with several German technical institutions.

John Christian Jensen at the NWU Observatory, 1942. Courtesy of NWU yearbook.

Although the classes were enjoyable, I resented the reason for studying the German language. It may be surprising that the concern was not related to the war in Europe of that time; rather, it was my resentment of the preeminence of German science. German society and its federal and state governments supported science through large institutions. To be a student of science or technology in America, the skill to read German literature in the chosen field was a necessity. Eventually, I began to wonder why the government of the United States did not support institutions of scientific learning to a similar degree. Perhaps it was through Professor Jensen's influence that I became aware of the need for our country to develop first-rate national research institutions.

Then there was Dean-Emeritus Francis Alabaster, who seemed appropriately named; he stood straight and stiff enough to resemble an alabaster statue. However, his teaching of Greek and Latin to student pastors was reportedly excellent. When punctuated with his dry humor, those classes were enjoyable as was every encounter with this distinguished gentleman, who in his youth had been an athlete and coach.

Professor John Christian Jensen was head of the physics department. He was of a remarkably gentle nature, but his gruff approach, tough standards, and rumpled appearance often concealed his compassion. Some students avoided his classes to their disadvantage. I shall never forget my encounter with him during freshman registration. He looked over my entrance exams and other records to remark, "If you do not get good grades, you ought to be shot." I was shocked, both by the threat and the compliment; fortunately, I was never in jeopardy of execution. Similar comments by Professor Jensen struck many students with awe and fear. In reality Jensen was as supportive as any teacher that I have ever known.

Professor Jensen had built the physics department from its beginning, having continued at the college in 1909 when he received his undergraduate degree. In 1938, he was finally finishing his Ph.D. from the University of Iowa. Jensen, who was born in a sod house, came to national prominence both for his own research and for the successes of his students—among them, all now retired, Harvard Hull of Sperry Gyroscope, Paul Copeland, head of the Department of Physics at the Illinois Institute of Technology, and Chris Keim and Dale Magnuson, both retired from the Oak Ridge National Laboratory. John Dunning, NWU 1927, now deceased, was the first scientist in the United States to determine that U-235 was the fissionable isotope of uranium. Dunning later was made dean of the School of Engineering at Columbia University. Other NWU graduates include heads of research and chief engineers for several major corporations, distinguished professors of universities, heads of small companies, and persons in other important positions of science, engineering, meteorology, and management.

Jensen's doctoral research also added intrigue to his image. The subject was lightning discharges. He photographed lightning at night with open shutters in cameras located at attic windows of Old Main, the tallest building on campus. Following lightning strikes, he closed shutters and reloaded film in the old Speed Graphic cameras. Thus, whenever a thunderstorm was predicted or in formation, Jensen would rush to the campus and tend the cameras, often at a very late hour. He also recorded meteorological information to accompany the visual and photographic observations. Jensen photographed ball lightning to prove that this infrequently observed phenomenon was real.

Professor Jensen told many stories of storms, one about conducting lightning research during a Minnesota vacation. The site was a lakeshore where he took photographs of clouds and electrical discharges with a camera partially covered for protection from rain. The procedure required measuring the relative humidity with an instrument called a sling psychrometer. This device—consisting of two thermometers, the bulb of one covered by a wet mantle—was twirled by a handle to increase contact with air to evaporate water and cool the mantel. The difference in temperature measured by the two thermometers provided a measure of the moisture content of the air. Resort neighbors viewed him as a very eccentric professor who obviously did not know enough to come in from the rain and who engaged in decidedly peculiar activities. Jensen, like Benjamin Franklin, was fortunate not to have been killed in this intense pursuit of lightning storm experimentation.

The lightning research paid off, not only with a doctoral degree but also with important contributions to the field. One concerned the protection of airplanes from lightning strikes. The small cables with frayed ends that trail from the wings of propeller-driven commercial aircraft to release static charges were derived from his studies. Professor Jensen was also employed by the newly formed Tennessee Valley Authority to design the lightning protection system for its electrical transmission lines. I found Jensen's work with the TVA particularly interesting, but his descriptions of the poor roads and poverty of East Tennessee were suggestive of a place to be avoided. (Only eight years later it became my home where I quickly came to love its natural beauty, the TVA lakes, and the friendly people of the Tennessee Valley.)

I earned half of my tuition by working jobs that proved to be rewarding beyond the twenty cents per hour in wages. During my first year, I worked as the janitor for the Physics Building, which was valuable both for the relationship with Professor Jensen and the considerable exercise gained through sweeping rough wooden floors. As a sophomore I was given responsibility for the stockroom equipment and supplies. This included checking instruments and experimental equipment for calibration and condition and making minor repairs. Those responsibilities reinforced first-year learning and provided insight for that yet to come; however, the third year was best. As a National Youth Administration worker, I not only commanded the princely wage of twenty-five cents per hour but worked at building and repairing equipment. The fourth year as a laboratory assistant was technically a promotion, but I lost a nickel per hour in wage.

That third year's work was both interesting and educational. One project was to build a heliostat, a device to focus sunlight continuously on one spot using mirrors and a clock motor to follow the solar position. The heliostat was to provide a steady source of intense sunlight for several student exercises, such as to analyze the colors and spectral lines of white light. Construction of the device involved fabricating a box to hold a small synchronous motor and gear box that rotated a mirror continuously to follow the sun's position. An attached framework supported a second adjustable mirror to receive the reflected light and direct it to the desired location. The finished product was to be attractive and of a quality consistent with commercially available laboratory equipment that was too expensive for NWU. We had materials suitable for the purpose but not the tools for the required fabrication and assembly. Professor Jensen arranged with the owner of a sheet-metal fabricating shop for use of their tools without charge.

I went to the sheet-metal shop, my first such visit, expecting Jensen's friend to show me the equipment and assist in its use. On arrival I was greeted cordially and shown to a table where I could

Don setting up a student lecture table, heliostat at right, 1942. Courtesy of NWU yearbook.

lay out my materials and the drawings previously made using the drafting skills learned as a freshman. As I set to work scribing patterns on the metal, the shop foreman casually came by to chat and complimented my drawings and technique. He then informed me that their entire staff was leaving immediately for field work and would not return until evening. The shop was mine to use for the day. They apparently assumed that one who had reached his third year of college must surely know how to work in their shop. I was terrified of making mistakes that would either spoil my material or damage their machines. At least there was a full day to experiment and learn; the restrictive rules of insurers had not yet been imposed.

Fortunately, the shop had a large bin of scrap metal in sizes suitable for practice in the use of the bend-break, metal shears, drill presses, lathes, and punches. After much careful experimentation and reduction of their scrap metal to further ruin, the time came to address the task at hand with my preciously small amount of aluminum sheet, tubing, screws, and blocks of brass. Much to my relief, the learning process had been adequate and the device, once assembled, worked well. I was able to leave the shop just before the staff returned. They would not know that the half-day job had taken so long unless the carefully covered debris in their bin was examined carefully. Over the years as an alumnus visitor to the physics department, I have proudly noted the heliostat still among the current equipment and free from dust.

During a field trip to the local natural gas-fired electric generating plant, Professor Jensen once again impressed us with his broad range of knowledge. The foreman apparently assumed that we were casual visitors and was explaining in excellent detail the mechanisms for controlling the plant, when Jensen appeared after taking a telephone call elsewhere in the plant. Realizing that we were from NWU, the foreman would say no more. Professor Jensen continued the lecture very well. In addition to teaching physics and some electrical engineering, he was responsible for the school's heating and electrical generating plant.

I became intimately acquainted with one of Professor Jensen's innovations on the campus. The auditorium for chapel and other events was in the old CC White Building that has long since yielded to the wrecking ball. The architect had created a large room and balcony designed for maximum seating but apparently had given little attention to acoustics. Persons located under the balcony could hear nothing directly from the stage. Professor Jensen, in pioneering with electronics, had installed speakers there that were powered by an amplifier fed from an early-model carbon microphone. Students seated in the "dead section" were not to be deprived of compulsory chapel programs, although I often would have preferred an undisturbed nap.

One of my junior-year tasks was to operate that chapel sound system. For some unknown reason, the amplifier was located in a small room with no view of the stage. If the person speaking stood directly in front of the mike, the sound as monitored through headsets was easily adjusted. Unfortunately, the carbon microphone's reception was quite directional. When the speaker chose to walk about the stage, I could only increase the amplifier gain in an attempt to compensate. Often the speaker would then return to the podium and deliver a loud, forceful point, the auditory impact of which would nearly knock my monitoring headset askew and startle students under the balcony who had been peacefully lulled to sleep.

Once, upon increasing the gain in an attempt to pick up the lecture, I discovered that a defect in the equipment caused it to receive the signal from nearby and powerful KFAB radio. Assuming that faculty members were safely seated in their places on the stage, I fed soft radio music to the students under the balcony. As it happened, a young professor of education, Lois Leavitt, had arrived late and found a seat in the rear of the auditorium. After chapel, she came to the amplifier room to inform me that she had heard the radio music and sternly warned that Professor Jensen would disapprove. I expressed surprise at her news and assured her that whatever was wrong with the equipment would promptly be repaired. After warning again on leaving that this should not happen again, she turned back and said quietly, "To tell you the truth, I rather enjoyed it."

An important task during my junior and senior years was to service the seismograph, located in the basement of the Rose Memorial Astronomical Observatory. This instrument, which measured signals from east-west motions, was mounted on a pedestal isolated from the building floor and set deep into the clay subsoil. The temperature was a constant 56 degrees Fahrenheit, the soil temperature at the deep basement level. The instrument previously had recorded several California quakes and received much local publicity as the only instrument of its kind in the area.

This seismograph met U.S. Geological Survey standards and was part of the national network. To coordinate the signals from many instruments, accurate time marks were required on the record. In those days before digital timepieces, time marks were provided from electrical contacts on a large grandfather-type clock having a temperature-compensated pendulum weighted with jars of mercury. Each day at 11:00 A.M., I checked the clock time to within a fraction of a second with a radio signal from the short wave station WWV of the National Bureau of Standards, now the National Institute of Standards and Technology. The clock was quite reliable and small deviations were to be noted on the seismic record, but occasionally it was necessary to stop the clock, adjust the pendulum, and restart it at the precise moment of the radio signal.

The seismic activity was recorded on a slowly moving photographic film by a tiny, focused light beam reflected by a mirror located on a hinged and weighted inertial arm that remained stationary as other equipment, including the recording film, moved with the earth tremors. Each week, it was exciting to remove and develop the film to see if an earthquake had been recorded. Occasionally, the signal was from a heavy truck that had strayed too close to the remotely located observatory or from some other disturbance, such as blasting for construction in the vicinity. These signals were clearly distinguishable from seismic events, and I soon learned to interpret the valid signals and determine a probable location of a seismic source even having only the unidirectional record.

Once, upon entering the very dark room to change the record, the light beam was oscillating wildly. To avoid disturbing the instrument, I stood motionless and cold in a short-sleeved shirt for half an hour until the motion ceased. I changed the photographic film, replacing it quickly to record aftershocks and developed the record with great interest. I placed the location in a remote area of the Pacific Ocean east of the Philippine Islands, which proved to be correct. It was thrilling as a student to realize that I may have been the first person in the world to observe evidence of that earthquake.

An interesting package marked FRAGILE arrived at the Physics Building in 1940, courtesy of the Sylvania Corporation. The box contained an experimental model of a fluorescent lamp. As I recall, the glass tube was nearly three inches in diameter and four feet long. It also had a large, heavy auto-transformer and separate starter. After some study, it was deemed suitable to illuminate the lecture-room blackboard. My task was to build the necessary supports and properly connect the wiring. The lamp served well and was used for several years as a special feature before fluorescent lamps were available on the market. A 1929 alumnus of the department, Roland Zabel, then chief engineer of the Sylvania Lamp Division, had sent the equipment to Professor Jensen for evaluation. Thus, I had the privilege of installing one of the very first fluorescent lamps. We also received an early model of an audio wire recorder, a forerunner of the tape recorder, but I no longer recall its source.

Association with those early devices was in keeping with the traditions of the NWU physics department. In the early 1920s it had been the first college in the nation to have a radio broadcasting station, WCAJ. In addition to broadcasts of news and music, it was a pioneer in education through radio. Several classes were developed for students who could not be on campus. Unfortunately, the station was sold later to reduce the college debt and to resolve a dispute with a commercial station. Professor Jensen also had experimented with a spark radio transmitter as early as 1905 and was recognized by President Hoover in 1933 with nomination to the Federal Radio Commission.

As a student, I was privileged to study with men who were to succeed in many areas of endeavor, although there were only three physics majors in my class. One of those, Keith Wycoff, became a successful developer and marketer of electronic equipment through his own business enterprise. The other completed the minimum course requirements but chose a career with the U.S. Postal Service. Since our group was small, advanced courses were taken with the class of 1941 that included many who would come to prominence. My friend Dale Magnuson, at Columbia University, and Homer Ibser, at the University of Chicago, would become early participants in the Manhattan District Project to produce nuclear weapons. Others in the class of 1941 included Claude Clements, who as a meteorologist had responsibility for the safe flights of President Roosevelt to the 1944 meeting with Winston Churchill in Malta. Dwight Hamilton became prominent in aeronautical engineering at the Cornell Aero Laboratory, and Ralph Clary retired as a colonel in the United States Air Force.

Thus the courses were challenging, both for content and competition. Professor Howie, on the final examination day for a course in advanced calculus and differential equations, surprised us again by serving punch and cake instead of a test. He explained that for the several periodic examinations of the term, this class had a numerical average higher than that of any previous class in this course. Also, the lowest individual grade was well above the previous average for "A" grades. Needless to say, we relaxed and enjoyed the occasion.

It was not possible to stay at the Barrett home during my sophomore year because the rooms were needed for family members. Willard and I then found a room at the Garner home that was popular for student housing. The entire basement and one room on the main floor had been furnished for students. The rooms were spacious and a great room was available for communal activities of the eight student residents. The commons area was shared with the gas furnace that heated the upper floor. The furnace was a monstrous unit originally designed to burn coal. The extensive duct work and fire box were heavily covered with exposed asbestos insulation. Even so, the waste heat sufficed to keep the room warm during cold Nebraska winter days, and none of the occupants is known to have suffered lung damage from the asbestos. The commons was equipped with a studio couch, a large, round table, small kitchen stove, small sink, cabinets, and a pantry. Thus, we could prepare meals and avoid the Hays house desserts.

Unfortunately, none of us really knew much about cooking, and the initial quality of food preparation was dubious at best.

Leavening or other ingredients sometimes were forgotten in attempts at baking, and the first preparation of pasta failed to anticipate the change in volume during cooking. Even after dividing the contents into several pots, the system overflowed.

Often, my dear Mom came to the rescue by sending prepared food (when transport was available) and by providing "foolproof" recipes. These were so simple and explicit that they usually proved true to their billing. We first worked in teams of two, one team to cook and the other to wash dishes, on a fixed schedule. It soon became apparent that the team that cooked had also to wash to improve efficiency in the use of pots and pans. We shopped primarily at a very small neighborhood grocery. Although prices there were higher than in the village store, the owner gave us a discount that kept each person's food cost to about four dollars per week. Four dollars per person also was the monthly price of the rooms, providing a very low living cost in a generally acceptable setting.

The commons room frequently was the site for "bull sessions," as discussions concerning college women and world events were known. The couch also offered occasional hospitality to a visiting male friend. On the one hand, we enjoyed good comradeship and fairly good privacy for study in the rooms. On the other hand, it was a spartan existence. There was a half bath for the six men in the basement plus a mirror over the kitchen sink. The separate shower was a design afterthought, poorly screened in its location near the basement entrance stairs. The floor had been raised to accommodate the drain, which meant that one could not stand erect while showering. Shaving was a problem because of poor lighting and awkward mirror locations. My Remington electric shaver was envied; so to help finances, I rented its use to several of the men. One, Paul Johnson, had two wrist watches, so my price to him was the use of a watch that I needed to meet the precise time demands for servicing the clock and mechanism at the seismograph. The dollar pocket watch Grandfather Trauger had given me years before had long since given up keeping time.

The basement rooms at Garners were damp in winter from the moisture produced by the exposed gas flame heater in each bed/study room. Those rooms also would not meet today's fire codes; direct escape was possible only by moving furniture to climb out of the small windows near the ceiling. The windows also provided a view of the commons room for two college women who lived on the second floor of the neighboring house. We not only knew that they watched us but took some pride in their interest. On one warm spring evening when windows were open, Jim, a preministerial student who lived on the main floor, heard one of the women say, "Move over and let me see too." Jim then raced downstairs to berate Max, who had emerged dripping wet from the shower, and was walking up and down toweling his back in the commons room while lecturing us on one of the many topics of his concerns.

Willard dropped out of school to return to the farm at the end of the first semester of the sophomore year, and I recruited Luther Powell's brother-in-law, Marvin Snyder, to share the room. Luther had married Marvin's sister during the previous summer. Marvin was a great roommate. He was as fastidious as I in housekeeping and spent little time in the room. He also became a pretty good cook as we all improved. Our cooking progressed so we could add three boarders from another house to share costs. Thus, eleven squeezed around the table when all were present. Now some fifty years later, it would be nice to know the fortunes of the men of that house. I can no longer recall all of the names, but Wilbur Bulkley, Carl Christianson, Jim Irwin, Paul Johnson, Max Kemling, Gerald Mattock, John McCallum, and Jim Varney come to mind.

In the spring of our sophomore year the college gave patch tests for tuberculosis, a disease of some prevalence at the time. I had a positive reaction to this, my first TB test. Marvin had such a severe reaction that he was bedridden for two days with a greatly swollen arm and much overall discomfort. We were frightened, thinking that Marvin and I were sick with that dreaded disease and had exposed our fellow housemates. When the rash and swellings subsided, the college sent a group of those with reactions to a hospital for chest X rays. As Marvin stepped in front of the fluoroscope, a common technique of the time, the doctors reacted excitedly. They spent much time studying the image. Those of us previously exposed or waiting moved into positions to see what was so interesting in Marvin's body. To our shock, a massive shadow dominated his chest region. We naturally thought that poor Marvin was about to be lost, and the next few days were tortured.

The report on my reaction was less dramatic. My positive response had been caused by a small tubercular incursion of some earlier year. The scar had calcified and no active disease was present. Marvin's eventual report was also a false alarm. His severe reaction was merely allergic in nature, and the report revealed that what we had seen on the screen was his lunch. Apparently, a birth defect had placed his stomach above the chest diaphragm. He was cautioned not to engage in overly strenuous exercise or overeat because of the reduced lung capacity caused by his stomach placement, but otherwise he was reported to be healthy.

Marvin, who later dropped out of college to study accounting and become a CPA, remained a close friend. He married Alatha, a girl I had known and dated in Fairmont, and they raised a fine family. Tragically, Marvin died of leukemia twenty years after the X-ray examination. I have always related his illness and death to the massive dose of radiation that he surely received during that extended examination with early, poorly designed X-ray equipment. It would have produced a high exposure to both patient and doctors and probably to those of us who watched in amazement and horror. I feel certain that my radiation exposure from that one occasion was greater than the total received while working with radioactive materials for over fifty years.

In my senior year (1941–42) with funds nearly assured to finish college, I joined the Delta Omega Phi fraternity. I probably had known more Phi Kappa Tau men, but this local fraternity was less expensive. Also, Loren Kottner, son of the rooming house "parents" where I lived in more comfortable and pleasant surroundings that year, was a close friend and persuasive "Delt" member. Being a senior plebe presented a problem both for me and the fraternity. How should a busy senior be treated with regard to duties and hazing? We concluded that I would not wash windows or have other duties at the house. That seemed reasonable since I did not live there. The Friday hazing night, however, was conducted with no compromise. Each initiate was stripped nearly naked, duly paddled, otherwise humiliated, and smeared with molasses and feathers. Then at 3:00 A.M., dressed in old clothes and blindfolded, each was taken to a different remote spot in the countryside without money, to find the way home.

Finding myself on a lonely dirt road, I was grateful that the cold of that October air partially alleviated the discomfort of the feathers. I walked for an hour or two in the direction of the city lights before finding a lighted barn. The farmer interrupted his milking to give directions to reach the nearest county road. He was surprisingly helpful, given that I was wearing a very old, ill-fitting coat, which at least covered up the feathers, left by my grandfather and surely looked quite degenerate. Once on the improved road, another farmer in a pickup truck soon stopped for me on his way to the Lincoln produce market. He was curious about my circumstance and responsive to my contrived tale of looking for work. (The story was only partially misleading because I was actually speaking of applying for "work" at graduate schools.) He was very sympathetic, told me of his problems as a young man looking for work, and offered money for breakfast and to tide me over until I could find a job. I refused, of course, although a few dollars would have been helpful. I asked him to drop me off at a point consistent with seeking a laboring job, even though it meant walking several extra blocks to the college. I did not want this kind man to know that I was a college student out on a foolish mission, and thus perhaps lessen his opinion of NWU, or to destroy his satisfaction in having assisted an unfortunate soul.

I arrived at the Kottners's just as the sun appeared. Bathing was a first priority using care to avoid clogging the bathroom drain with feathers. What a relief. I then spent the day working on a project in the laboratory and escorted a date to a college play that evening. The resilience of youth is great to remember.

I have treasured the experience of that night for its lesson about generosity. The farmer who helped me appeared to have very little of this world's resources but, nevertheless, was willing to give to a stranger in need.

My senior year was very demanding for many reasons. Work as a laboratory assistant took extra time for tutoring, study for honors exams was demanding (although often ignored), and I was obligated to spend much time at home. In August my brother had been diagnosed with early-onset Perthes disease, in which the rounded upper end of the thigh bone deteriorates, and was fitted with a plaster cast from waist to toes, to be so confined for most of a year in recovery. I spent many weekends and all holidays in assisting with his care, working in the Farm Program and helping on the farm. We had been concerned about Bob's discomfort in walking, and when our parents learned of the diagnosis and plan of therapy, they found it difficult to tell him of the year-long disablement. That task fell to me. Surprisingly, this proved almost a pleasure, for his response was, essentially, OK, I'll do it. This was a relief, almost good news for Bob; he had feared that walking might be lost or impaired for life.

A highlight early in my senior year was the arrival of a freshman woman named Betty Winquest. I met her after lunch in the cafeteria with several of her housemates. Our first date was to a football game on a cold September evening. We had many good times together much of that year before drifting apart. The friendship was briefly renewed each summer when I returned from New York on vacation, but the distances and time lapses were too great. I met her as Mrs. Cooper many years later when she worked in the U.S. Congress. This long-standing friendship has been helpful in both our careers. I provided technical information related to legislation, and her knowledge of the Congress and U.S. government was most helpful to me in anticipating legislative action that might affect my work.

My occasional trips to Exeter during college were mostly by thumb: hitchhiking. There was relatively little danger then, although caution was in order. On two occasions the driver was speeding beyond comfort, even for a young passenger; ninety and more miles per hour on narrow highways in cars of the 1930s was frightening. One trip was made in the back of a pickup truck among crates of chickens with feathers flying everywhere; another was on a pile of lumber in the back of a flatbed truck.

My only really disturbing hitchhiking experience had occurred years earlier when walking home from high school. A luxury model Desoto coupe stopped, and the driver invited me to ride. I declined, but this large man clad in a heavy, black leather jacket, insisted and assured me that I would be all right. My decision, foolish or not, was to ride with him rather than to refuse. Closing the door, I was gripped with fear on noticing a sawed-off shotgun in an auto holster and a pistol holster bulge under his jacket. The driver, sensing my concern, again assured me that I was safe and asked my name. On hearing my answer, he excitedly explained that he had worked for my father as a hired hand in the 1920s and wanted to know about my parents and the community. His questions were rapid fire as he drove a bit out of his way to the intersection with our road, where he stopped to let me out. The car then roared back to the highway and was quickly out of sight. Arriving home, I learned that this man was a Chicago gangster who indeed had worked on our farm in his quieter years. He was wanted by the FBI and the Chicago police.

Two serious farm accidents occurred during the summers of my college years. The first, like most accidents, resulted from lack of care and good judgment. The task was to eradicate a patch of bindweeds, a flowering, morning-glory-type vine that spreads both by root runners and by seed. Bindweeds thrive in wet or dry weather and if left untended will smother desirable plants. We had a small infestation on one side of the farm where it had spread from a neighbor's field across the road. Today, bindweeds are easily eradicated with herbicides, but then the only way to kill the plant was to till the soil frequently for three successive growing seasons.

I had taken the tractor and a disk harrow to the site for the scheduled tilling and was passing under fence wires that had been nailed to the telephone poles at a height to allow the tractor and driver to pass underneath. Characteristic of Nebraska weather, the wind was blowing hard. As I passed under the wire, the wind caused a loose wire end at a splice to swing down and catch in the shoulder buckle of my overalls. This dragged me from the seat as the tractor moved forward and prevented me from disengaging the clutch. I fell to the draw bar of the tractor, landing on my right thigh and hip, then to the ground on my right shoulder. The disk blades cutting deep into the soil were rapidly approaching only a few feet away. Fortunately, I had broken no bones, and I executed a rapid backward somersault to escape the disk, landing so close to it that I could feel and hear the soil movement induced by the outermost disk blade as it passed. Had the maneuver yielded a foot less, I would have been decapitated.

After assessing my condition and finding no apparent serious damage, the tractor already was some distance away moving rapidly toward the boundary fence. I was able to stand up, run to the tractor, climb over the disk and draw bar, resume a position on the seat, and stop the machine. After sitting there quietly and reflectively for some time, I decided against trying to complete the job. The accident had been too unnerving for that. I also considered walking home to leave the tractor and disk connected for further work, but chose to drive the tractor home.

After some rest, I felt ready for a different task; return to the bindweed site was unthinkable. I began to dig post holes for a new barnyard fence. Shortly after initiating that task, I tried to stand from a kneeling position and found my right leg inoperable. Examination showed that it was badly lacerated with much blood caked inside my overalls. The impact on the draw bar had severely broken the skin and damaged muscles, but an apparent state of shock had temporarily masked the trauma and pain. Several days passed before I could walk or work effectively. Fortunately, the injuries healed without any permanent scar or impairment.

The second accident, occurring in a later year, was different; this time I was mowing grass for hay in the beloved prairie. The grass was thick and as tall as the horses' backs as they pulled and powered a mowing machine of a type used for decades before the advent of the rubber-tired tractor. The mechanism transferred its operational power through the use of cast-iron drive wheels fitted with a series of gears and an eccentric to oscillate the shearing blades across a horizontal cutter bar. The wheels were prevented from slipping over the ground by equally spaced lugs that protruded nearly an inch into the soil. This shearing mechanism was essentially that invented by Cyrus McCormick in 1831, as improved by others prior to 1860. Since the cutting velocity was directly related to the speed of forward motion, the heavy grass stand could be cut properly only by urging the horses to move rapidly. This took advantage of the flywheel effect of the gears and heavily balanced eccentric.

At one point the cutter bar became jammed, and the thick standing grass stopped the machine. More than three thousand pounds of rapidly moving horse flesh was suddenly thrust hard against the collars, the powerful inertial force transmitted to the machine. The mower immediately buckled at the hinged cutter bar, throwing one side of the machine rapidly upward. I was thrust from the seat into the air to land on my back with my left leg just behind one of the lugged wheels. The natural reaction of the horses was to retreat from this impossible load. In reversing, they quickly backed the machine and the lugged wheel ran up my leg. Three lugs mashed into flesh before the movement stopped. Thus pinned to the ground under the weight of the machine (several hundred pounds on that wheel), I could not move that leg. There was no hope of escape unless I could induce the horses to move forward far enough to release me. After positioning arms and free leg to move away quickly upon release, I desperately called to the horses, "Gidap, gidap," and they responded. The three lugs again traversed my leg: crush, crush, crush. Suddenly I was free and rapidly moved far enough to watch the horses again reverse the mower, this time without harm.

Most fortunately, the lugs had not struck bone; the first badly bruised the Achilles tendon, the second impressed the flesh of the calf, and the third stressed tendons at the knee. I could not use my leg and painfully hopped around the mower to free it from the grass. Only with difficulty could I mount the mower seat for the ride to the house. Passing through gates also was a challenge with more hopping and difficult maneuvers. Despite it all, I was fortunate. The injuries were not permanent and recovery was rapid. Had the lugs struck and crushed the ankle and knee bones, the accident would have cost me my leg.

Farming is indeed a hazardous occupation; some accidents, such as this one, were hardly preventable in the past. Today's modern tractor-mowers are geared to adjust the cutting speed independently of forward motion, and a tractor does not respond like horses. These accidents served to make me particularly sensitive to safety and safe practices, a desirable attitude for one who later would work with large pieces of experimental equipment.

One summer my father and I were building a new wire fence, altering the shape of a grassland pasture so as to include some tilled land susceptible to erosion, as a soil conservation measure. On that hot afternoon we took a break in the shade of the wagon to rest and drink iced tea. Pop hesitantly initiated discussion of my plans beyond college, seemingly anticipating an unfavorable response. He asked, "Do you think that you might return to the farm after graduation?" My answer, painful as I perceived it to be for him, had to be negative. He then asked what I might do. I had become certain that graduate school was desirable, perhaps necessary and probably feasible, and I said so, but he wanted to know what occupation might follow schooling. I had given the matter some thought and told him of the options I envisioned, mostly in industry. Pop, who felt great security in the farm, had concern for the lack of job security in industry, suggesting government civil service. In response to my insistence on a more industrially oriented career, he advised that I should continuously develop new skills that were vitally important to the employer. It was sound advice.

Back at college, even in those times of the depression and World War II, I had many sociopolitical concerns beyond the purview of class assignments. One was over the nation's expanding use of fossil fuels. For instance, the rapid expansion of oil use for home heating, agriculture, and transportation would soon deplete the national petroleum resources as they were understood at the time. All fossil fuels, except coal, seemed limited with respect to the growing national need, and at that time the "greenhouse effect" was not a recognized problem.

The news in 1939 about the discovery of uranium fission and its large release of energy was both exciting and relevant to my studies. The news came as I was midway in a course on atomic physics, the first advanced course of my college studies. It was thrilling to read in the newspapers and scientific journals about the work of Niels Bohr, Rudolph Peierls, Herbert Anderson, Enrico Fermi, Eugene Wigner, Otto Hahn, Earnest O. Lawrence, J. Robert Oppenheimer, Fritz Strassman, Harold Urey, and others, many of whom had been featured in text books. The media reports of these people, who suddenly became more than figures in the technical literature, were stimulating to a budding scientist. I found myself eagerly awaiting the arrival at the library of magazines such as *Science* and *Science News Letter*. At that stage the more profound treatises, such as papers in the *Physical Review*, were often too mathematical for me to fully understand.

I read whatever was available on the subject, and an issue of the *Physical Review* particularly excited me. The issue contained a letter to the editor titled "The Fission of Uranium" (*Physical Review*, March 1, 1939, by H. L. Anderson, E. T. Booth, J. R. Dunning, E. Fermi, G. N. Glassoe, and F. G. Slack). This paper presented technical details of the process and indications of high energy release. The attraction for me was heightened because John Dunning had graduated from NWU only a dozen years earlier. I could not imagine that I would soon meet every one of these people and that three of them would have a direct and valued impact on my career.

Because of the fundamental importance to my life of those enormous events in physics that occurred in the late 1930s, I will roughly describe some of the early reported work in the order of discovery. Neutrons from a Ra-Be source and those from cyclotron bombardment were observed to cause uranium to fission, that is, to result in nuclear fragmentation accompanied by a very large release of energy. The amount of energy could be calculated by comparing the ionization of gasses by the fragments to that from the known energy of alpha particles in the natural decay of uranium. The energy release from the fission of a single uranium atom was estimated to be approximately 200 million electron volts, also consistent with that reported in the German work by Frish and Meitner. This release was orders of magnitude greater than that from any chemical reaction, such as the burning of coal. This news really piqued my already significant interest in energy sources.

The efficiency of neutrons to cause fission or to be captured by U-238 to form U-239 and plutonium was examined in various papers as a function of neutron velocity. Physicists also deduced that uranium fissions were from the isotope U-235, with the first confirmation in the United States credited to John Dunning. These publications were followed by reports from the American Physical Society winter meeting, including work by Bohr and Fermi.

Articles on details of the fission process followed. Fermi reported on the discovery of delayed neutrons and speculated on their importance in the control of the nuclear process. Work at the Carnegie Institute, Columbia University, the University of Chicago, and the University of California at Berkeley was reported in rapid succession. As knowledge of the number and nature of neutrons released per fission and their interaction with materials was gained, the possibility of a chain reaction became apparent. Early measurements indicated that between two and three neutrons were released in each fission event, but the probability of one striking another fissionable atom to cause a chain reaction leading to practical energy production was small.

It seemed more probable that enough neutrons would be captured by uranium or other materials or lost from the assembly to make a sustained reaction difficult. At best, this would require a large assembly; hence the word *pile* was coined to describe what is now known as a nuclear reactor. It was also observed that the occurrence of fission was highly dependent on neutron velocity. The neutrons must be slowed down but not captured by other atoms, and it was apparent that light atoms that recoil to slow the neutrons would be most effective. Hydrogen, beryllium, and carbon were proposed to moderate the neutron energies. All of this was interesting and exciting for me to learn as it was regularly published.

The cross-sections of materials for capture of neutrons had been studied extensively before fission was observed. Because uranium was recovered from ore containing many elements, some of these impurities would capture neutrons. Also, since the materials to slow the neutrons must have small cross sections, they too must be free of impurities. Refinement of materials to obtain a sustained reaction was one of the problems faced by the scientists who sought to produce useful energy from this new source. K. K. Darrow pointed out in 1940 that the presence of the absorbing impurities may have saved some early experimenters from a disastrous release of energy in the course of their work. Nature had provided a safeguard.

The achievement of practical nuclear energy, however, was to be even more difficult because only the isotope U-235 is readily fissionable by the slow neutrons. Thus, many physicists recognized quite early that the separation of uranium isotopes to enhance the lighter element, present in nature at only 0.7 percent, would be an important requirement. Isotope separation had already been studied for elements such as nitrogen and boron to obtain marked materials for studies of chemical reactions and biological processes. But the separation of uranium isotopes that differed in mass by less than one percent was a formidable task. Various methods under study for other isotopic separations employed centrifuges, thermal diffusion columns, mass spectrometers, and several chemical reactions, but none of these processes had been developed for more than experimental purposes. Gaseous diffusion through microscopic-sized pores in thin metal sheets and ceramic structures also had been studied for separating isotopes, but seemed not to be favored early because of the difficulty of building a large plant with specialized membranes. Some even delighted in calculating the number of years required to obtain a few grams of U-235 by the methods under consideration. These ranged up to tens of thousands of years and dampened enthusiasm for nuclear energy. Again, I could not conceive that this challenge would be the focus of my work for several important years.

However, as an impressionable student I focused on the positive and became increasingly convinced that science could find a way to resolve the problems. I was captured by the dream that nuclear fission could be the answer to my concerns for an adequate supply of energy. I had noted that the experimenters had identified the fission fragments by their radioactivity. Although I do not recall that the intensity of the radiation was reported in the literature, it was known for many of the elements. That intensity of radiation was to make difficult the practical use of nuclear energy. The vast amount of radioactivity that results from fission requires heavy shielding and unfailing containment of reactors to produce useful quantities of energy.

At meetings of the American Nuclear Society early in 1939, discussions also had turned to nuclear reactions that might occur in fractional millionths of a second with energy releases that could be equivalent to thousands of tons of TNT. The implications were enormous, particularly in view of the darkness descending on Europe. Although many believed that the difficulty of separating the uranium isotopes would negate all uses of nuclear energy for a long time, the discussions continued. Professor Jensen knew Enrico Fermi and others who had expressed concern for the weapon potential of nuclear fission and related some of those thoughts to me. Much later it was revealed that some researchers voluntarily withheld publication of their work because of this frightening potential.

Popular writers of 1940 speculated on limitless uses of this energy, not primarily for electric power production but for the more glamorous applications of power for ships, railroad locomotives, automobiles, and perhaps aircraft. Even with my limited experience, I knew that the presence of any radioactive material would require shielding and containment that would make the use of nuclear energy impractical for an automobile. It seemed feasible for a locomotive to carry the necessary heavy radiation shielding, but even I could see that an electrified railroad supplied with power from a nuclear-electric generating plant would be far preferable. (It seems a pity that even after these many intervening years, we have not electrified much of our railroad locomotive power.) Later, I actually worked on a military version of a nuclear powered aircraft. That somewhat absurd story comes later.

My interest and enthusiasm for nuclear energy (then called atomic energy) prompted presentations I made in two successive years on the status of nuclear technology in meetings of the Alpha Gamma Beta Club for science majors. These lectures no doubt were naive but seemed well received. My natural shyness in speaking before groups was overcome on two counts: I was speaking to an interested audience and speaking with enthusiasm about a subject I felt compelled to share. In the question period that followed, I recall embarrassment only from my lack of understanding about some of the chemical aspects of isotope separation methods. I recently reviewed with renewed excitement the journal references I had recorded in a college notebook between 1939 and 1941, but, unfortunately, I did not retain the notes used in the presentations.

A speech course in my senior year, 1941–42, was a challenge because shyness and stage fright were still problems when speaking before general audiences, and the selection of a suitable subject for presentation worried me. It was stipulated that the subject had to be of some social importance and not simply focused on personal experiences. When confronted with that final examination consisting of both a presentation to the class and a written paper, atomic energy was a natural choice. It was so interesting to me that stage fright did not occur.

The presentation dwelt first on the great potential value of this new source of energy for electric power generation and for ships. I then discussed the tremendous potential for destruction by a nuclear weapon that seemed possible as described briefly in some of the technical and popular literature. I felt that the world should consider the overall potential of this new energy as one of colossal proportions. I proposed

Enid Miller, Professor of Debate and Theater, 1941. Courtesy of NWU Yearbook.

that its attractiveness as a power source was so great that the technology would surely be developed. Once available, there would be great danger for its use as a weapon of destruction in war or for other evil. The basic concern of the speech was for nuclear weapon proliferation as the threat is now known. The presentation closed with the recommendation that an international organization should be created to control this formidable force. If that did not occur, I argued, nuclear materials might fall into the hands of irresponsible groups or governments with consequences too disastrous for contemplation. Of course, I had the order wrong; weapons were to come first. Although the international organizations formed soon after World War II—the United Nations and its subsidiary International Atomic Energy Agency—have served well, they are not as strong as that envisioned in my speech.

Professor Enid Miller gave me an A on the paper and the speech but wrote a qualifying note on the paper: "An excellent paper and presentation, but in the future please choose more timely and credible subjects."

On December 7, 1941, after completing studies at my desk in the Physics Building, I telephoned Glenn Downey, my cousin who was then enrolled at the University of Nebraska, to suggest that we meet for dinner. He responded with the news of the attack on Pearl Harbor. The plan for dinner was abandoned so that each of us could follow the news by radio. This greatly heightened my interest in the potentially destructive nature of nuclear energy. The Germans, who had been the first to discover fission, might have a major effort in the field under a cloak of secrecy.

In following the work of the principal American nuclear researchers, I had noted an increasing scarcity of papers early in 1941. The implication of an external inducement to stop reporting progress seemed likely, for previously most of these scientists had been consistently publishing. Further investigation revealed that many of the scientists whose work had been most interesting were now associated with either Columbia University or the University of Chicago. I wondered to myself if nuclear energy could have become a secret wartime program centered at these universities.

My senior year rapidly drew to a close. Before graduation I had received and accepted a working assistantship at the Illinois Institute of Technology in Chicago for the coming year. I graduated from NWU with honors, although not the top honor that was awarded. I had borrowed one hundred dollars from my father to finish the year, but by the last week of school all funds were totally exhausted. While I was finishing the tasks to pay for tuition, Professor Jensen's son, Robert, an NWU physics graduate, 1933, then the chief engineer for the National Broadcasting Company in Los Angeles, came by for a visit. On departing he commented that he would see me at the Alumni Banquet that evening. I explained that I still had work to do and would not be going. He sensed that I did not have money for the ticket, handed me three dollars, and repeated his expectation for our meeting at. the dinner. I enjoyed the banquet and have been ever grateful to Robert Jensen who had adopted the generous qualities of his father. Unfortunately, I never saw Robert again, and he died a few years ago.

After graduation, I returned to the farm for the summer. I had come under the draft deferment for university study and that status was to continue in the fall. My draft board, consisting of farmers, seemed content with a summer's labor to provide food for the war effort that was having an impact even in the remote Midwest. I also collected scrap iron, aluminum, and rubber tires for recycle, even removing the tires from the wheels of the old Jackson auto of my parents' courting days. Amazingly, one tire still had thirty pounds of air pressure after twenty-two years. There is not much else to report for the summer, except that crops were much better than in the thirties and afforded opportunity to earn needed money. The stipend for work at the Illinois Institute would provide no more than basic subsistence in Chicago.

In mid-August 1942 Mom called me to the telephone with the unbelievable word that John Dunning was waiting to speak with me. Picture the contrast: Dunning no doubt was speaking from his twelfth-floor office in the Pupin Physics Building of Columbia University, and I was standing at the mouthpiece of an early, wall-mounted phone equipped with a hand-cranked signal generator to alert the operator. After some pleasantries about Nebraska Wesleyan, he invited me to work and study at Columbia. The work was to be on a secret project of which he could tell me nothing more than that it would be challenging, interesting, and important to the war effort. I expressed interest, mentally presuming the project to be about nuclear energy but explained that I was committed to work and study at Illinois Institute of Technology. He countered that he had discussed the position and the IIT commitment with Professor Jensen and had telephoned me on Jensen's recommendation. Dunning then suggested that I might wish to discuss the opportunity with Jensen, although he also would not know more about the work than I had been told. Being of a cautious nature, I asked for a few days to see Jensen and decide. I made an appointment with the professor for the next day and drove to Lincoln.

I told Professor Jensen of my deductions concerning a war effort on nuclear energy. He was of the opinion that an undertaking described by my notions would be too large a task to undertake in wartime and that the secrecy I had noted was only to prevent further nuclear information reaching the Germans in the unlikely event that they were proceeding. He thought that the secret project of Dunning's call probably was to develop radar. That work at Columbia University was known to have a high priority and had been made secret. I restated my belief that the project was nuclear but that I would have no problem working on radar. On that basis I preferred to accept Dunning's offer, if I could be released from IIT. Jensen advised that there was little possibility of an amiable release and that I should send them a wire of resignation.

COLUMBIA UNIVERSITY
DIVISION OF WAR RESEARCH

August 6, 1942

REPLY TO UNDERSIGNED AT

PUPIN PHYSICS LABORATORIES
COLUMBIA UNIVERSITY
538 WEST 120TH STREET
NEW YORK, N. Y.
TEL. UNIVERSITY 4-3200

Mr. Donald Trauger
Exeter, Nebraska

Dear Mr. Trauger:

Professor Jensen spoke so highly of you in reply to a
recent telegram of mine inq uiring about new men that I decided to
write to you directly.

We are engaged in an extremely urgent war research project undertaken for the Army and Navy, under the Office of Scientific Research and Development in the Executive Office of the
President. This has worked out so well that the government has
asked us to expand our efforts greatly, in order to push this program through to production at the fastest possible rate. I am
sorry I cannot tell you more about it, since it is entirely SECRET,
but I can say that it may well be a major factor in ending the
war.

We are extremely anxious to find able men just out of
college to help. Dale Magnuson, whom you probably know, is with
us now. It is possible to take some graduate work here at the
same time, and so combine the two, to some extent, although
normally only about 1/4 full schedule can be maintained, as this
is a full-time job. However, the job itself is excellent physics
training.

The salary depends on how good you become, but we can
pay $145.00 per month to start, beginning immediately if possible,
and could probably cover some of your travelling expenses here.

I understand that you now have an assistantship at
Illinois Institute of Technology. If it were not for the serious
emergency which faces us I would not write you, but the situation
needs all the effort the country can bring to bear.

If you are interested, would you please let me know?

Sincerely yours,

J. R. Dunning
Associate Professor
of Physics
Official Investigator
O.S.R.D. in charge of
work under OEMsr-412.

JRD:kfm

While I was in Lincoln, the following wire arrived from Dunning: "Need you urgently for work. Trust substitute at Illinois Institute can be arranged. Reimbursement for traveling expense and living needs can be arranged upon arrival in New York. Looking forward to seeing you." I wired my acceptance to Dunning's office and set the date of September 1, 1942, to report for work. I was to receive the magnificent salary of $145 per month, nearly twice that of the IIT appointment. Professor Jensen was correct; the wire of resignation to IIT brought a curt, almost abusive reply.

Souvenir from the Mahattan District Project, 1945.

FIVE

The Manhattan
District Project

A fter arriving at Grand Central Station and mastering the IRT subway, I reached the Pupin Physics Laboratories of Columbia University with suitcase in hand. The twelve-story brick building was impressive, set on 120th Street in a beautifully wooded courtyard protected by a high and ornate wrought-iron fence. It seemed the right place for me to work and study. Columbia University Heights contrasted favorably with surroundings of the Illinois Institute of Technology (IIT), where I had stopped en route to New York. Nebraska Wesleyan graduates Francis Breeden and Dwight Hamilton, both in the graduate school of IIT, were enthusiastic about their studies and expressed disappointment that I would not be joining them. The security aspect of my new assignment prevented revealing the reason for moving to Columbia. They advised against seeing Dean Grinter, who remained furious about my resignation. I have regretted the inconvenience to IIT but never the decision favoring Columbia.

In Pupin I was directed to the basement, where the desk of John Dunning's secretary was strategically placed so that she served as receptionist and security guard. Although uniformed guards of sophisticated protection systems were to control the doors or gates of my work for the next fifty years, Miss Paul may have provided the best security. If she was not positive of identification and purpose, there would be no entry. Even though I arrived as expected, it was necessary to show my Nebraska driver's license and the letter from Dunning. After completing some paper work, Miss Paul sent me to a room at the end of the hall where my friend, Dale Magnuson, had been working since June. Almost immediately I saw a single piece of equipment that assured me of the nature of the project. The machine was surely for isotope separation research in the gaseous diffusion process to increase the concentration of the fissionable isotope uranium 235.

Dale greeted me warmly, assuring me that I could stay for a few days in the apartment he shared with three other men, until one returned from vacation. In apologizing profusely for not describing his work, he cited security and suggested that in due course the management would tell me about it. Dale then excused himself to complete a few tasks before showing me to the apartment some three blocks away.

While waiting, I watched a little pump working noisily, studied the configuration of the piping and equipment attached to it, and noted a thermometer reading—all serving to confirm my identification of the basic nature of the work at hand. However, many interesting aspects of the gaseous diffusion project would indeed prove challenging to learn over the next twelve years. The process is basically simple: in the gaseous compound uranium hexafluride, the fissionable isotope uranium 235 diffuses slightly more rapidly than the heavier isotope 238 in passing through a membrane (called barrier) having very fine pores. However, the equipment required to achieve this separation on a large scale is massive and complex. A great cascade of many repetitive stages is required to enrich the concentration of uranium 235 from its natural concentration of 0.7 percent to about the 95 percent required for a weapon.

Good fortune had placed me in a position where the work applied both to the weapons project and to future generations of nuclear electric power; uranium enrichment would be needed for both applications. The feeling of awesome responsibility for producing such fearsome weapons was not lessened by this dual role, but it offered a more satisfying outlook. As much as I wanted the project to succeed, I would not avoid hoping, against existing knowledge, that a weapon would not be possible. If a weapon was impossible, the concern over

German nuclear work would cease and the effort could be redirected to developing a major new source for electric power production. Realization of difficulties in the latter goal would come much later. These thoughts were to be troublesome throughout the project, would continue during the Cold War period, and extend to the present. People working on the project could hardly avoid ponderings such as mine, but such philosophical conflicts were seldom discussed, owing to pressures of the war effort.

On formally reporting for work on September 1, 1942, I completed a security questionnaire and other required documents. I soon met William W. Havens Jr., Willard F. Libby, Rex Pontius, Magnuson's supervisor, and James Rainwater of the cyclotron team. Both Rainwater and Libby were later awarded Nobel Prizes. I was to work under Francis G. Slack, who was on leave from his position as head of the physics department of Vanderbilt University. Professor Slack welcomed me and suggested adjourning to a park bench in the courtyard for his briefing—a pleasant beginning on a beautiful autumn day. I have often thought of providing a similar amenity for meeting with a new employee but have never been blessed with a comparable setting. As we discussed the project, I found few surprises, except for the magnitude of the effort on gaseous diffusion and the progress being made in research. Slack described the planning already underway for a large, cascaded plant, indicating the urgency for developing the barrier material, a major task of his department.

We discussed the matters of secrecy and draft status somewhat jointly. Deferment was expected, so long as a person's work was satisfactory. If dismissed for a security violation or poor performance, however, a military assignment in the South Pacific theater was assured. This was to prevent potential capture by Germans of a person with project knowledge. Safety was also emphasized, citing the pioneering nature of the work. I remember most vividly his discussion of personal responsibility. I was expected to know or learn the technology and safety precautions for each action and to proceed with dispatch in completing the task. The managers, "old gray heads" in their late thirties or early forties, were concerned with project planning and had little time for training or supervision. I was surprised with the assignment to work with E. O. Norris and Edward Adler, then principal developers of the barrier material, the principal key to project success. That seemed very interesting, but I expressed concern because their tasks involved chemistry and metallurgy, technologies where I had little schooling or experience. Dr. Slack assured me that they needed help most urgently. The interview lasted some thirty or forty minutes, after which I was considered indoctrinated and ready for work. Introduction to my new supervisors followed immediately.

Adler was an intense, intellectual, and somewhat impetuous individual who contrasted with the more laid-back, inventive, and innovative Norris. They seemed an unlikely team but obviously had made good progress in that difficult development of large areas of thin membranes having exceedingly fine pores. I struggled for several days to understand the process and avoid making mistakes in sifting and characterizing powders, mixing chemicals, and developing slurries. After two or three weeks, I felt more confident about working independently in the lab, except for the disorderly work benches. Nothing seemed assigned to a fixed place. I had continually to relearn the changing patterns for the location of laboratory equipment and supplies. However, the apparent confusion seemed not to adversely affect Adler and Norris's progress. More importantly, they really needed the assistance of an experienced Ph.D. in metallurgy or inorganic chemistry.

At about that point Adler asked me to report to Professor Slack, who informed me that new people had been hired for the barrier development; I was to move to barrier testing. This was simultaneously a disappointment and relief. On the one hand, even with my inexperience, the barrier development was innovative and involved tasks that were interesting and exciting. I had seen the barrier testing room where the work was much more routine and seemingly not as challenging. On the other hand, testing was conducted on two shifts, and I could therefore work evenings and attend some graduate classes in the mornings.

I had learned that the work at Columbia was known as the DBS project. The D stood for Dunning, the B for Eugene Booth, the first Rhodes Scholar I had met, and the S for Slack. That part of the project was known later by the letters SAM. The identification of those letters as an acronym was not announced, but it was rumored to mean Substitute Alloyed Materials. That connotation was a good secrecy cover, since many materials employed were unusual in commerce, especially in the large quantities that would be required. I was not informed until much later that the immense Manhattan District Project had been formed a month earlier, only a few days before John Dunning's call to me. General Leslie Groves, who would lead the project until completion, was placed in charge on September 17, 1942. That portion of the project in which I worked then had a staff of about seventy-five persons. A few weeks passed before I met John Dunning. He asked after Professor Jensen and then spent a few minutes sharing his enthusiasm about the project.

Robert T. Lagemann, also a Vanderbilt physics professor on leave, was in charge of the barrier testing group. The task was to perform quick evaluations of small samples of barrier in order to guide the research including that from my initial assignment. The screening test machines used a mixture of helium and carbon dioxide for which the theoretical (or ideal) separation factor was 3.3 compared to 1.004 for the isotopes in uranium hexafluoride. Thus, the work contrasted with the much more laborious and definitive testing that Pontius, Magnuson, and others were conducting with uranium hexafluoride in the room at the end of the hall. We were to provide quick tests to guide the staff struggling to develop thin metal membranes penetrated by as many tiny holes as possible while retaining mechanical strength. One person could test two membrane samples per hour, whereas the uranium testing required hours, days, or even weeks.

Room 109 was filled with six or seven screening test rigs. Each had a small rubber diaphragm pump used to circulate the gas mixture through a large tank and at specified measured rates and pressures as it flowed across the face of the barrier carefully mounted in a device called a holder. A separate vacuum pumping system maintained a low, fixed pressure on the back side of the barrier. Barrier effectiveness was determined by sampling each stream of gas, by analyzing the mixture by removing the carbon dioxide with potassium hydroxide in a burette, and thus by determining the increase of helium concentration in the stream that passed through the membrane. The operator was to mount the barrier sample carefully in the precisely made holder to avoid breaking the delicate material. It was then necessary to operate valves to establish pressure and flow equilibrium, as measured by orifice meters and displayed on oil and mercury manometers. The final step was to operate the burettes. This was a routine and almost boring task, except for developing the skills required to obtain precise measurements. It was also interesting to see the progress, or lack thereof, for different barrier-manufacturing formulations as samples were tested. The brief experience with barrier preparation enhanced my interest in testing; by visual examination I could often identify the variable under study.

The barriers we received were either of a copper-manganese alloy or those created by Norris and Adler (N-A). The former consisted of thin flat sheets of brass-type alloys that had been made porous by acid etching to remove one component, such as manganese, and to leave small, irregular holes through the thin metal sheet. These specimens were either of low porosity or had been etched more extensively to have larger pores and little strength. If the porosity was adequate, often the pores were too large, and poor separation capability resulted. If they functioned well, the product was often extremely fragile. Norris and Adler had succeeded in obtaining better barrier strength through their process in which nickel powder was applied as a slurry to form a coherent structure. When the particles were precisely sintered together (melded without melting) in a furnace, an acceptably strong product was formed. The spaces between the powder particles formed the passages through which the lighter element of a gaseous mixture would diffuse more rapidly than the heavy component. Acceptable separation was achieved for either barrier type only when the holes were so small that most molecular collisions were with the wall rather than between gas molecules. This was on the order of one tenth of a micron or 0.000004 of an inch.

The N-A barriers clearly performed better than the etched types, but both were brittle. However, there were more variables to explore and improve in the complex process of N-A formation, options that had made my learning both interesting and difficult. Cutting and mounting either type barrier sheet into flat, one-centimeter-diameter test specimens, without breakage or leakage, was an exacting challenge. Clean gloves and tweezers were required; oil from one's fingers would plug the barrier, and tweezers helped to minimize breakage in handling. These barrier samples and their methods of fabrication seemed far from a manufacturing process that could provide the quantities needed. A uranium isotope separation plant would require a total barrier area that could be specified in acres. I wondered: Could that be achieved in time to affect the war?

Room 109 fell far short of modern industrial health and safety standards. It was crowded, had a high noise level from the multitude of pumps, only one door, no windows, no forced ventilation, and a bare concrete floor. Once, as I passed a test rig, a stream of mercury suddenly shot to the ceiling to descend as heavy rain on my head and shoulders. An operator had inadvertently applied excessive pressure from a gas storage system to her test system to propel mercury upward from an open-top glass manometer tube. After recovering from the shock, I checked my clothing for deposits of mercury, even pouring some of the shiny liquid metal from a pocket and from each shoe. In cleaning up that spill, we discovered quite a bit of mercury from earlier events residing on the floor. I worried about inhalation of toxic mercury vapor in that enclosed room. However, because dirt and oil cause mercury surfaces to have very low vapor pressure, the dirty concrete floor was actually protective, saving the occupants from mercury vapor inhalation and poisoning. I know of several people who worked in that room much longer than I, and none has shown evidence of injury attributable to mercury. Room 109 was in several ways an undesirable working space, but there were no complaints. The team in that room was dedicated to the task as part of the war effort.

One member of the test group was from Georgia Tech. He was the only southerner in the group and was often teased about his speech. He seemed not to mind and frequently drew attention to himself with stories during breaks from work. His tales were humorous, his antics sometimes amusing, and both provided relief from the monotony of repetitive actions. However, his horseplay was once nearly tragic. Contrary to strict rules requiring that gas cylinders be securely tied to a rigid frame, his was not. (These cylinders are about four feet tall with thick steel walls and are filled with gas pressurized to a thousand pounds per square inch or higher.) In an enthusiastic, arm-waving gesture to emphasize a statement, he toppled a nitrogen gas storage cylinder that crashed against a test rig and then to the floor. The cylinder closure valve was open (another safety violation), and when the regulator connection severed upon impact with the floor, the roar of the issuing gas was deafening and terrifying. The heavy cylinder, weighing perhaps a hundred pounds, propelled by the roaring gas jet issuing perpendicular to its axis, oscillated wildly on the floor. Each time the broken pipe pointed downward, the force of the gas jet flipped the whole cylinder, even from end to end, to again crash resoundingly. Everyone scrambled furiously to the top of the rigs to escape the erratic path of the flailing cylinder as it repeatedly slammed against metal frames and floor until all the gas had escaped. The period seemed interminable but was probably less than a minute. We assessed the status, first from awkward positions atop the testing benches, to find that no one was hurt and, surprisingly, that the lab had suffered little damage.

That event caused all of us to examine the room closely for other compromises to safety, and sure enough, we noted a hydrogen—cylinder standing beside the one that had toppled—also unchained with its valve open and regulator attached. It could just as easily have been dislodged, falling similarly but with terrible consequences—intense fire or explosion. Beyond the probable loss of life, the explosion of that hydrogen cylinder could have delayed or even canceled the gaseous diffusion effort, because the prospects for that effort at the time were somewhat tenuous.

After a few months of testing, Bob Lagemann, in conducting quality control analyses of results, noted that my data consistently did not quite agree with the theoretical curve for pressure dependence of the separation factor. The other operators were consistent in producing results that fit the expected pattern. We traded samples and exchanged test rigs with samples mounted in search of the discrepancy, but in all of this, my data did not agree. It differed from that of the dozen or so operators from both shifts, as well as from the theoretical curve. Although the difference was within experimental error for each data point, it was statistically significant for the large number of tests on record. I was distressed, even mortified. The whole issue raised in me a cloud of self-doubt: could I not succeed at anything on this important and desired project in which I had great interest? Rather than tolerate an inconsistency, Lagemann assigned me to build a new test rig to be located in the one remaining open corner of the room. The work was expanding with the project growth, and new people were to be hired for testing. Except for the cause of reassignment, this work for which I was fully prepared, was a pleasure.

Midway through the first task, another member of the test team, a Columbia University graduate, stopped me as I was drilling a hole in a panel board. He cautioned that university rules allowed such work only on the specific approval of a member of the university's senior staff. I responded, "The instrument needs to be at this place; we have much to do and it will be placed here." The former student had not realized that we, in fact, were members of the university staff. My immediate goal was to complete tasks so quickly that Lagemann would have to scramble to keep me busy. This went very well until he assigned the fabrication of a rather complex glass apparatus. Glass blowing is a challenge for me; the tubing seems consistently to bend, bulge, or break at undesired places. I worked through the evening shift as unobtrusively as possible to avoid revealing incompetence to colleagues and even harder after they had gone home. It was daylight when I finally emerged from the building; the "masterpiece in glass" awaited Lagemann's arrival.

After some further days of this frenzied, satisfying work, one of the analysts burst into the room to explain that an error had been found in the theory and that they would provide a revised equation for validating the testing. My data fit the new curve exactly—what a relief! I was ecstatic, even though I have never found a fully satisfying explanation for the variance in test results. The fact remains, however, that the analysts found a well-established difference between the earlier model and the new equation. The new theoretical curve proved to be correct, and so was my test data.

This turn of events soon worked to my advantage. Having gained recognition for accuracy in measurement and a demonstrated capability to build equipment, I received a new assignment. Sidney Visner, a more senior project member, had been constructing a new apparatus for testing the performance of larger barrier specimens. Because of that experience he had been chosen to assist with a small pilot plant to test features of a gaseous diffusion plant cascade. I was to complete his former project.

It was exhilarating to have my own laboratory space as large as Room 109, equipped with a desk and table, a work bench, a set of tools, and storage space for supplies. This was great luxury, for the building had become severely crowded by the project's rapid expansion. My new daytime task involved research to determine the sensitivity of the separation factor to gas flow patterns on the high pressure side of the barrier. As the light component of the gas is depleted, an increased concentration of the heavy element occurs at the surface, the result being separation reduction. Thus, flow conditions that continuously replenish the mixture at the barrier surface become a necessity. This was a particularly serious problem for the flat barrier structures originally envisioned for the plant. Although the use of barrier in tubular form was under consideration, flat barrier was planned for at least the higher enrichment stages at the top of the cascade. The uranium hexafluoride volume at the top must be small, to minimize holdup of the valuable separation achieved and to avoid a nuclear criticality accident with the enriched gas. Flat barrier assemblies were preferred to minimize these problems.

At first glance, this equipment resembled a "Rube Goldberg" cartoon image of apparatus assembled by a mad scientist or an overly zealous engineer. A maze of plastic and copper tubes connected gauges and manometers to key points in the equipment, all necessary to measure small differences in pressure and to convey information needed to confirm or supplement theory. Viewed from the door to the lab, the scene might be described as jungle-like. The odd appearance was emphasized when the newly appointed project leader, Harold Urey, entered the room on one of his rare visits to the basement floor. He exclaimed brusquely, "*What* is going on in here?" A few moments of explanation seemed to satisfy. Advising Urey that the work was guided by Clark Williams was probably the most reassuring statement. I soon gained a high regard for Clark, who was a senior project member and my new boss. (A few years ago I visited with Sidney Visner, shortly before his death. We reminisced about the project, including the curious array of apparatus he had constructed. Sid was shocked to hear of my and Urey's first impressions of that equipment, but I reassured him that it proved to be rather well designed.)

I studied the system, completed rigorous leak testing, gained comfort in operating the machine and learned to mount the larger, still brittle, copper-manganese barrier samples chosen as most suitable for the initial tests. I then carefully explored other features of the room. Careful swishing of a broom over the floor revealed mercury in amount and appearance similar to that in Room 109. Although this room also had no window to open, I was pleased to observe that it had both an exhaust blower and another to bring in fresh air. Surprisingly, the incoming air was not particularly cold, even though it was winter and snow covered the ground. Further investigation showed that the exhaust air eventually returned through the inlet by traversing a closed window well. With proper revision of the ventilation and a thorough removal of mercury from the floor, the room seemed a safe and efficient place for work.

Properly mounting larger samples of the N-A material proved more difficult than for the etched barriers. The problem was solvable for the test rigs but foretold problems when installing barrier in the plant. Soon after I had gotten measurements underway for flat barrier, a decision was made to use barrier tubes in most, and probably all, of the cascade. The next task was to convert this equipment to test short sections of tubular barrier; it would be the first machine with that capability. I worked with a designer, A. A. Abbatiello, to prepare a device suitable to support the tubular barriers for testing. Shop service was excellent, and I was ready to put tubes to the test on the requested date. However, the engineers designing the fabrication process for forming fragile barrier sheets into tubes had encountered problems; no tubes were available, not even for evaluation of the test equipment.

I chose to make a barrier tube suitable for that purpose myself. Even after selecting material known from previous experience to be relatively ductile, fabrication of a flat sheet into a tube was difficult. I had to prepare a mandrel for bending, devise innovative clamps to hold the curved sheet, and experiment with cleaning techniques to prepare the edges for closure. The whole process made me appreciate the reasons for the development group's delay in forming tubular barrier.

Finally a barrier sheet was formed to the proper diameter, and after much experimentation I succeeded in tin soldering the seam without a cleaning flux. The flux would have plugged the barrier pores and rendered it useless, even for my purpose. Thus, I was the first to produce a workable barrier tube and to obtain test data for that configuration, but the distinction received little recognition because the tube was not prototypic. Soldering was an impractical closure for larger-scale manufacturing, but it served my purpose for several weeks of experimentation until the project's more practical tube fabrication was successful.

I also continued to work for some time to complete a program of flat barrier tests and to refine the equipment for holding the tubular barrier for testing, without leakage at the ends of the test piece. We called the device a holder, the name also used for flat barrier tests. This work was of great importance because for some time this would be the only test for barrier as tubes. Even though flat barrier samples would continue to be tested as produced, those tests would not address concern for possible damage in the subsequent bending and sealing of closures and in the overall performance of barrier as tubes. Convenience in installation of specimens and effective closures would be important for routine and rapid testing of many specimens as their development progressed. In addition, much remained to assist the theory group in their analysis of the data from both flat and tubular barrier tests.

The test equipment required devices to measure gas flows with some precision. An orifice-type flow meter was a simple device to make, but the small sizes required for our laboratory use did not follow the well-established rules for the rate of flow as determined from the pressure drop measured across the orifice. Calibration of each flow meter by liquid displacement, a commonly used technique, was a distraction that required a good deal of time. I examined one of the small orifices under a microscope to observe that the tiny hole was neither round nor sharp edged, as were larger orifices made commercially that have predictable characteristics. The Columbia University specialty shop had a high speed precision drill press and exceptionally good drills in small sizes. By very careful use of that equipment, it was possible to obtain round holes having diameters of only a few thousandths of an inch. Even with that equipment, the edges were rough with burrs. It was possible to remove the burrs by working with sharp tools under a microscope, and when the face of the plate was polished with a very fine emery stone, the edges were quite sharp. These small "homemade" orifices followed the established rules, yielding a relatively high degree of precision and did not require calibration if the diameter was measured precisely using a microscope. I became a supplier of orifices for the laboratory as a minor sideline.

On January 18, 1943 the Carbide & Carbon Chemicals Company, a division of Union Carbide, signed a contract with the Manhattan Project to operate a gaseous diffusion plant to enrich uranium to a high concentration of the U-235 isotope. It was designated as K-25 and to be located in Oak Ridge, Tennessee. This important milestone indicated that barrier development and testing were considered promising and that our effort was important and urgent. At that stage, however, barrier quality seemed far from a product that could be made in quantities and of a quality to meet the needs of the large plant about to be constructed.

An organization chart of February 15, 1943, shows 185 persons in the SAM Division-One staff. (This document was found among the possessions of its creator, A. A. Abbatiello, following his untimely death a few years ago.) The chart had no secrecy markings but should have been classified; knowledge of the staff size and membership was closely guarded. Evidence found after the war indicated that German and Russian spies had monitored persons known to have worked on nuclear processes even before security was imposed. Their purposes were not to satisfy curiosity as in my college observances but to find persons who might inadvertently, or intentionally, reveal information.

Very few civilian members of the staff had problems with their draft status and, hence, with their positions on the project. My draft board was a notable exception. I had taken no steps to obtain deferment, but the Project did so repeatedly as required. In the summer of 1943, I was called to a preinduction physical examination at the

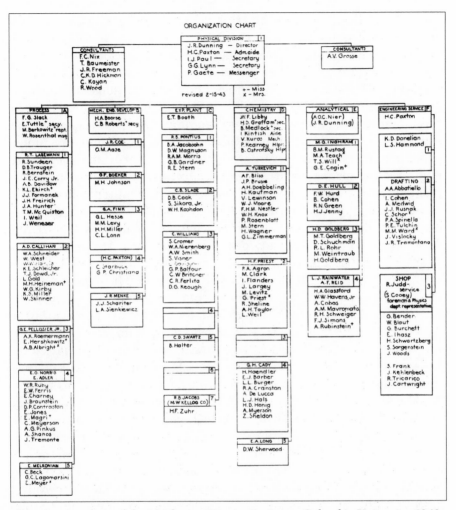

Organization chart of the Manhattan District Project at Columbia University, 1943. Note "D. B.Trauger" in column one, group two. Chart drawn by A. A. Abbatiello.

New York Armory before the Project could press an appeal. It was a mass production where, after blood and urine samples were obtained, inductees removed all clothing and proceeded to the armory court. We followed a prescribed path marked on the floor to specialized doctor stations. Because it was a hot summer day, our nakedness did not make us physically uncomfortable, although some, including me, were a bit shy. That feeling was amplified when I noticed the high windows wide open for ventilation and comfort of the medical staff. Through those windows we could see a tall office building whose windows were crowded with young women enjoying a remarkable view of masculine flesh.

After passing stations where one "assumed the position" for rectal or stood for "short arm" examinations, I came to the otolaryngology specialist. He was talkative while perfunctorily glancing at ears, nostrils, and throat. He then suddenly jabbed the tongue depressor to choking intensity and peered for a long time, calling colleagues to see something unusual. The other doctors first ignored him, but upon being urged came and were followed by curious inductees. He reminded the medics of his previous tutorials on tonsillectomy then pointed to the thin scar line at the back of my throat reporting, "That is the way the incision should be closed." I jerked my head away to say, "Doc, my tonsils have not been removed." The observers roared with laughter as I was quickly dismissed to the next station. I passed and, after several appeals by the Project, was notified that I was to report for induction. Dr. Slack advised that I would be returned to the Project immediately following basic military training.

In late August, about three days before the reporting day, I received a Presidential Deferment, a new draft status that reportedly was created at the meeting of Winston Churchill and President Roosevelt in Quebec on August 18–19, 1943. It was part of the decision to join the British and American nuclear projects and to pursue the U.S. plan for weapons; the British program had been less focused and included work on electricity production. In a few weeks we began working with some very competent British scientists, Professor Kurti for instance, whom I remember as quite helpful in the testing program.

Barrier development was both the most unusual and difficult of many challenges faced in producing weapons-grade uranium through gaseous diffusion. Although I was involved primarily with barrier testing, other problems were discussed among the staff. For example, new instrumentation and control systems were required. Mass spectrometers to measure the isotopic enrichment needed improvement in precision, in standardization of design, and in the detailed procedures required for newly trained operators. Thousands of measurements would be necessary for routine monitoring of plant performance. Development of shaft seals to separate the process stream in the gas compressors from drive motors presented one of the more demanding tasks. Leakage of the uranium hexafluoride into the atmosphere was unacceptable, and leakage of air, moisture, or lubricant into the system could irreversibly plug the barrier pores.

Paralleling seal development, compressors with sealed-inside motors were built, and a strange reciprocating unit with flexible bellows seals had been designed and partially constructed by Newton Underwood, who had preceded Lagemann as supervisor of barrier testing in Room 109. It fell my lot to assemble and test this unusual unit, which we called the "Underwood Flapper." The long "cylinder" was trapezoidal in vertical cross section, wide at the rounded bottom and narrow at the top, from where the "piston" was supported by arms attached to a heavy horizontal shaft. The bellows were welded around the support arms to seal the pump interior from the eccentric drive. These bellows flexed, allowing the piston to rock back and forth with no interior or exterior leakage. A series of reed valves mounted in the hollow and ducted sides of the sloping outer housing walls allowed the gas to enter and leave the cylinder. The ducts led to piping connections that would fit into the plant cascade. The unit could provide a large displacement with a minimum movement of the bellows and a small volume of uranium hexafluoride. This pump had been conceived when flat barrier was favored and system size was to be severely minimized at the top of the cascade. The machine was subject to severe vibration, despite the use of counter-balancing weights to offset the momentum of the "flapping" piston; thus, we did not operate it long enough to determine the bellows' reliability or lifetime. I mention this artifact only to illustrate the depths of redundancy that the project employed to assure a solution for each major problem.

The development of valves, instruments, and seals would provide another story, but that would be second-hand reporting. I must note, however, that Eugene Booth of the Columbia University senior staff made several contributions to the design and improvement of many components of the plant and laboratory. These included the invention of two small diaphragm pumps, one used for uranium hexafluoride and the other in the helium-carbon dioxide testing; a widely used pressure gauge, which was developed cooperatively with Sylvan Cromer; and other laboratory devices. Eugene Booth has not received proper recognition for his many significant contributions to the Manhattan Project.

With tubular-barrier manufacture approaching, my testing operation needed to expand. More suitable machines were required to accelerate the type of testing that started with my initial sample using the research equipment built by Visner. Also, as barrier was manufactured at the plant, new and more rapid barrier-quality test devices would be needed for each tube as manufactured, or at least for many obtained by random sampling. It was necessary to move to a new facility and recruit engineers to meet the expected workload. Over the period from late 1943 to early 1944, much of the project moved to the Nash Building, an eight-story parking garage on Broadway at 132nd Street. Once again I marveled, as that building was quickly transformed into laboratories and offices, at the capabilities the Project commanded. At the new address, my work was again relegated to the basement. The testing of full-sized barrier tubes required heavy gas compressors that needed firm support. Their vibration, had they been placed at an upper level, would have adversely affected sensitive equipment used by others.

Security badges were first issued while my work was located at Columbia, and guards were positioned in the lobby of Pupin Hall. We were unaccustomed to encountering armed guards routinely, and we joked that the circular badges might be targets. Testing security became a game, although none of us were as ingenious as Richard P. Feynman described in his book *Surely You're Joking, Mr. Feynman!* He reported daring and surreptitious opening of safes and violation of fences at Los Alamos. At the Nash Building security systems became more sophisticated, and we were issued exchange badges. A pocket picture badge was provided to carry concealed while out of the building and to be exchanged for another to pin on clothing while at work. The photographs were to assure proper identification by the guard. Nevertheless, security did not seem to improve much. One evening at dinner a colleague and I decided to exchange pocket badges as a test. In case either was challenged, we had conjured an elaborate explanation to account for the mix up as accidental. His picture showed much black hair and thick glasses that contrasted with my thinning brown hair and good eyesight of those days. After a week of unchallenged entry, we tired of the game and thereafter used our own badges.

The Nash Building basement had advantages. We were near the extensive and experienced shop services required in constructing new equipment. These support people were remarkably capable and experienced. For example, the person in charge of welding, for both modification of the building and equipment fabrication, had been the welding supervisor for construction of the Empire State Building. As an accomplished welder, he delighted in showing me how to determine quickly the competence of an electric arc welder by observing the pattern of sparks produced in the initial striking of an arc. Other craft needs were served by people of comparable capability and experience. One machinist had operated lathes and milling machines on British warships for many years before coming to America. He could produce larger chips from a turning lathe, while holding close tolerances, than anyone I have seen before or since. When it was decided that hourly workers were to punch time clocks, he absolutely refused.

Manhattan Project security badge.

Because of the Project's recognition of quality work, he remained the only craftsman who came and left at will.

My team had built slowly at first. George Toal, an English major and bookkeeper, joined early to collect and correlate data, and a secretary was assigned to assist and type reports. Jack Hunter, an inventive engineer, worked effectively to develop simplified test equipment for the production plant. Our more elaborate tests were practical only for statistical sampling and to guide research. One civilian, assigned from the Kellex Company and whose name I recall only as Bernie, and five other men who came from the Army—Lewis Carpenter, Larry O'Rourke, Stan Stancko, Bill Tewes, and Bob Vogel—were to build and operate the new test rigs. Each was a mechanical, chemical, or civil engineer. These men were from the Special Engineering Detachment created to fill technical staffing needs, not only in the Nash Building laboratory but also for many roles in Oak Ridge, Los Alamos, and other Project sites. We also employed a female mechanical engineer; this enterprising young woman was small of stature but fiercely independent. We had built the test equipment for larger people, but she was undaunted; where not tall enough, she found a ladder; if the valve was too stiff to turn, she found a wrench. We lost an excellent team member when she chose to join her husband who was an early transfer from the European Theater to the Panama Canal Zone. Regrettably, I cannot recall her name.

My desk was in a triangular office shared with Dixon Callihan and William G. Pollard. The unusual office configuration was in accommodation to the helical ramp of the former garage. Dixon, a physicist, was then in charge of the group in which Dale Magnuson worked before transferring to Oak Ridge. They continued to evaluate barrier performance and chemical stability with uranium hexafluoride under plant operating pressures and temperatures. While a professor at the University of Tennessee before the war, Bill Pollard had published papers on the interaction of gases with metal surfaces and contributed greatly to the theory of isotope separation. The theory group was housed on the eighth floor, but Bill preferred to be near the sources of experimental data.

The several people of each testing group at times created considerable traffic and discussion in the office. None of this seemed to bother Pollard; unperturbed by the noise, his pen turned out equations at a steady rate. On one hectic occasion, Dixon turned to Bill and asked, "How can you concentrate so intently in the midst of all our disturbances?" Pollard answered, "I have four sons at home. If I had not learned to concentrate, I long ago would have gone mad."

Pollard worked closely with Richard Present in developing barrier theory. Both returned to the University of Tennessee after the war and continued to consult with the gaseous diffusion program. I took graduate courses under Present in Oak Ridge, and Bill Pollard moved there to found the Oak Ridge Institute of Nuclear Studies, now Oak Ridge Associated Universities and Oak Ridge Institute of Science and Energy. Bill Pollard was agnostic at the time but later joined the Episcopal Church, eventually studying for and entering the priesthood. He remained a close friend until his death in 1989.

The Project was highly compartmentalized for security, and we were therefore not always informed about sources for barriers submitted for testing. Each sample was identified only by a code. I had continuing familiarity with the laboratory source of the N-A samples and once had reason to visit the restricted part of the Schermerhorn building at Columbia. Equipment there appeared appropriate for (and was) the pilot-plant source for N-A material. Later, we were expecting test specimens from the plant operated by the Houdaille-Hershey Corporation in Decatur, Illinois, where the barrier was to be produced for the gaseous diffusion plant designated as K-25. Construction of both the barrier plant and K-25 had started in the spring of 1943. We grew increasingly disturbed that our testing of the N-A barrier from the pilot plant still showed wide variations in quality, even by the time the manufacturing plant was nearing completion.

From time to time in 1943 small samples of a different barrier material had arrived from an unidentified source. It was not as porous as the N-A barrier, but was consistently improving. We learned later that this barrier was developed by a combination of Bell Laboratories, the Bakelite Division of Union Carbide, and the Kellex Company. Although the best of the N-A barrier was better, we frequently found and reported it to have defects that indicated it would never be satisfactory. In January of 1944 the project managers and General Groves arrived at the same conclusion. The decision had been difficult, since it was necessary to strip the production plant of the N-A process equipment and replace it with the process for the new barrier. Testing of this better, sintered-type nickel barrier suddenly had the highest priority, but it was to be several months before large quantities could be expected. This delay was gravely serious since construction of K-25 was proceeding well in Tennessee. Some cell units of the diffusion plant were assembled for testing purposes and later dismantled for barrier installation. However, the value of having redundant effort for key components was clearly demonstrated; without a readily available alternative barrier, the K-25 plant could not have produced enriched uranium by August of 1945.

The waiting time was used to construct new equipment and to refine our test rigs as well as our data analysis to meet anticipated rush demands. The high priority of this wartime project became clear during construction of one test machine. On a Friday morning we discovered that a control valve to be installed on the following Monday would not perform as expected. I perused catalogs and data sheets in the purchasing office for more suitable valves and identified one with the proper characteristics. The young woman who handled our work then undertook to find it, a difficult task in that time of war-created shortages. However, at about four that afternoon she reported having located the required valve in an Ottumwa, Iowa, warehouse. Thirty years before the advent of Federal Express, the part was flown to LaGuardia Airport by Sunday afternoon. We had it picked up and installed the valve as scheduled on Monday.

On another occasion we had difficulty finding a particular type of gas compressor. After one was delivered, we learned that it had been removed from a naval vessel partially disabled in the Pacific. We were able to obtain the ship's compressor because the time needed for the vessel's repairs also provided the time required to order and receive a new one for the ship. Minor hardware items could be picked up and charged through a contract at the large Weil Brothers hardware store on 125th Street. This eliminated the need for an extensive stockroom and the attendant purchasing procedures. Such efficiency and priority measures contributed greatly to the completion of the Manhattan Project objectives in record time and at a reasonable cost. Even so, bureaucratic paperwork was much in evidence: seven invoices were required for each purchase, three for the Army and four for Columbia University, and all were typed with multiple carbons.

As indicated earlier, test equipment had to be absolutely leakproof to protect the barrier and to assure the integrity of the chemical analyses. By the time the project moved to the Nash Building, mass-spectrometer-type helium leak detectors had been developed by A. O. Neir of the University of Minnesota for ensuring that the K-25 plant was made leakproof. Our laboratory also received at least two of these machines. Although the early leak detector machines were slow to respond and hard to keep in adjustment, even they were much more sensitive and far easier to use than the previous method of painting surfaces of a pressurized system with a soap and water solution, which meant watching for bubbles before the soap solution dried.

Having been able to hear sound above 22,000 cycles per second while working with audio sound generators in college, I wondered if I could hear the high frequency hiss of helium leaking from a pressurized system. I tried this for a large test rig where the helium leak detector operator had experienced difficulty. By listening in the late-night quiet, after the nearby Broadway Line of the New York subway had ceased operation, I could effectively locate even very small leaks. I could also hear air leaking into an evacuated system. This saved valuable time but did not eliminate the need for the helium detector. Today, leak detectors are compact and sensitive beyond anything imagined in 1944.

The official work week was six days, Monday through Saturday, eight hours a day. Actual hours for most were from eight o'clock in the morning until five, with a break for lunch, and many workers often returned after supper. Some also worked on Sunday, but I felt it desirable to get away for a full day each week and was present on Sunday only to meet very special requirements.

Project management apparently had heard of efficiency experts and engaged a person named George to inspire us. The tedious process required assembling groups of the staff, reminding us that we were at war, challenging us to meet our great responsibilities, and encouraging us to work hard. This was all terribly annoying. The meeting time subtracted from the available hours in the day, and knowledge that the Germans might achieve nuclear weapons first was incentive far beyond any efficiency expert's words. The "pep talks" and the time they required were so irritating that we sometimes took off from work for an evening movie, for the relief it provided, but newsreel reports of the war pricked consciences and forced return to late working hours.

Given the rather extensive knowledge we had about other aspects of the project, it was surprising that we did not know more about the sources of barrier from the development efforts outside the Columbia laboratory during 1943. The "grapevine" information system had worked well to circumvent much of the compartmentalizing imposed by security. For example, we received frequent reports of construction progress for the electromagnetic plant for uranium isotope separation at the Y-12 site, and later about the S-50 plant for the thermal diffusion separation process, both in Oak Ridge, Tennessee. Also, the centrifuge work under Jesse Beams at the University of Virginia and by the Standard Oil Company were known. All of these were alternative to or backup for the gaseous diffusion plant.

Beyond the isotope separation program, we received news of the graphite reactor construction at X-10 in Oak Ridge. Los Alamos was a familiar name, and reports from there were of great interest, even though sketchy at best. I knew of the project at Hanford, Washington, but I could not even imagine its magnitude. All of these projects were well concealed from most Americans, but occasionally one would meet a citizen who was obviously aware of the Project. I tried never to show any recognition in response to statements by those who seemed to know. I understood that I might do harm by giving any information at all; in addition, we all knew that such individuals might be FBI agents testing one's integrity. I was contacted frequently by an FBI agent who requested that any suspicious activity be reported to him, but I never observed an infraction of security.

As the war progressed in Europe, Adolph Hitler repeatedly made threats of new weapons. Although these proved to be of new missiles, we remained keenly cognizant of the German nuclear weapon potential. Members of the public often scoffed at Hitler's pronouncements, and those of us involved in the Project just as often wanted to counter with the warning that the possibilities were not only imaginable but very serious. Of course, the correct posture was to join in the skepticism, and I did so. We later learned that our government and military did not have accurate intelligence about the decline of German nuclear work, and our concern as Project workers was, therefore, understandable and appropriate.

After the war ended and security restrictions gradually relaxed, it became apparent that the German program had not prospered. Destruction of the Norwegian heavy water plant by patriots and British Commandos, the ineptness of the German high command, and Hitler's priorities had combined to prevent significant progress. Allied bombers also had taken their toll on German facilities that could have contributed. Project members were not alone in their continuing concern over the German capabilities, however, as confirmed by the fascinating story of the ALSOS project. There, our government engaged qualified scientists to follow closely behind U.S. and British frontline soldiers invading Germany. They were to learn of the nuclear work, obtain nuclear materials, and interview technicians. Another ALSOS purpose was to reach the German nuclear scientists early to prevent capture of experienced nuclear staff by, or transfer of their nuclear information to, the USSR.

The revised barrier plant and its equipment were still under construction in mid-1944, and we were increasingly anxious to see and test full-length tubes of the new barrier material. Our large helium–carbon dioxide separation test machine for full-length barrier tubes was completed, a second was under construction, and the screening equipment to be used at the barrier manufacturing plant was ready for evaluation. Both "in" and "out" boxes had been prepared, but the in-box was notably oversized for the short samples we received from the development laboratories.

One day, the electrical shop left several pieces of new conduit standing against a garage pillar near our equipment. These were approximately of the diameter and length of the anticipated new barrier, but of course they lacked clearly identifying features. As a joke, I moved three pieces of the conduit to the in-box. Nearby colleagues recognized the ruse, but someone from another group saw and misinterpreted them; a rumor spread like wildfire that barrier tubes had finally arrived. Apparently, the word reached high levels before the hoax was discovered. I was thoroughly surprised by the success of the joke because virtually everyone knew of the incomplete status of the barrier plant modification. Embarrassed parties who had fallen for the joke were understandably unhappy and advised me in sharp language that it was not funny. Although joking about problems was an important part of maintaining morale, I had crossed the line. In due course the long barrier tubes arrived, and testing shifted into high gear.

The last separation test rig constructed was the most interesting to build and operate. We had been provided two centrifugal-type compressors with "canned" rotors and impellers supported by newly developed gas bearings. The motor windings of these units were isolated from the helium–carbon dioxide mixture by a thin membrane of stainless steel. These units had been intended for use in the upper end of the plant where, as mentioned previously, gas volume, leak tightness, and containment of the uranium hexafluoride were especially important. Developed in parallel with the "Underwood Flapper," these units were yet another example of the Project's policy of providing alternative equipment.

Our challenge was to determine quickly whether the novel bearings designed for the heavy uranium hexafluoride would support the rotors when operated with lighter gases. The gas bearings were welcome because conventionally lubricated compressors required elaborate filters to remove traces of oil from the system. Even with this filtering applied to standard compressors, a slow plugging of the barrier would occur to limit the testing time. The gas bearings worked remarkably well; the motor brought the machine to speed quickly, resulting in minimal rubbing of the metal surfaces before the grooves in the thrust and journal bearings could provide enough pumping action to support the load. However, little braking force was available to stop the compressors because the gas mixture offered little frictional resistance to turning at the slower speeds. Rubbing between the metal surfaces in the dry gasses and with no lubrication could damage the bearings. To listen for the noise of internal metallic rubbing, we fashioned sticks to conduct sound from the housing to our ears, a technique of old-time auto mechanics who similarly listened for vibrations to guide their repair plan for an ailing motor. It took about twenty minutes for the rotors to stop. The bearings had been machined precisely and were so highly polished that the actual contact was barely detectable.

Since my group was working with relatively large equipment, it was natural that we had overlapping interests with the cascade pilot plant built on an upper floor by Eugene Booth, Sylvan Cromer, and Clark Williams. I felt that this six-stage unit was inadequate to model a plant having thousands of units with much larger equipment, but it was more representative in size and design than the earlier, very small twelve-stage cascade operated in Pupin by Jack Slade. The scaling in size from pilot test to the working plant was perhaps a hundred times greater than that usually considered appropriate, especially before the advent of computer modeling. That the enormous working facility at Oak Ridge was created successfully from experience with such small units is a tribute to the genius of Manson Benedict, Karl Cohen, Arthur Squires, and others of the design group.

Once barrier tubes arrived in great numbers, pressure mounted to provide prompt test results, and we initiated shift operation. The barrier plant was rushing to meet the requirements for the final assembly of K-25 startup cells. (A cell consisted of six stages.) It was most encouraging to find the barrier product improving steadily and to be free of the difficulties encountered with the N-A product. Even so, tube closures continued to present some problems. I communicated frequently by telephone with Bob Carter, our liaison at the barrier plant. To maintain some semblance of security, we tried to use language that might have protected information from the casual listener (intrusions were not unusual for the telephone system of the time), but our amateur efforts would not likely have troubled an expert.

All applicable resources were utilized in solving problems with the barrier materials. For example, early electron microscopes assembled and operated by L. T. Newman and Warren Harris were used to understand the porous structures better than was possible with optical units. The new quality tests we developed for barrier tubes at the production plant coordinated well with the more complex machines for measuring the separation of helium and carbon dioxide. Tests with uranium hexafluoride both on a small scale and in the six-stage pilot plant also were consistent with our findings.

It was highly satisfying to know finally that the barrier, the diffusion plant component that had required the most development, would be successful. We knew that the porosity and separation factors were adequate. Tests by others had shown that the barrier was stable when exposed to uranium hexafluoride for extended times in laboratory systems. This was particularly reassuring because Willard F. Libby had been quite concerned about chemical corrosion and plugging of the pores. Since barrier stability was dependent on keeping impurities out of the plant, the development of satisfactory shaft seals for the gas compressors was a major accomplishment. Such components as excellent valves, unique instrumentation, leakproof welded joints, and protective plating of surfaces exposed to uranium hexafluoride all required features not previously available. Although development of the barrier was the single most difficult technical challenge of the Project, achieving each of those components should be considered an outstanding accomplishment. And the plant perhaps could not have been made free of leaks on schedule without development of the helium leak detector.

On March 1, 1945, Union Carbide assumed the contract for the laboratory in New York. That day marked the end of my relationship with Columbia University, except as a student, and the beginning of my employment with Carbide that lasted nearly forty years. We noted little change in paychecks: only the paper color, but not an increase in the amounts printed. The Project, however, continued to function under the same technical leadership of Nobel prize winners and others of comparable talent and the management genius of General Groves.

The SAM Laboratory not only served to conduct the research and testing necessary to carry out the Project mission, but also was a training facility for K-25. As the facility emerged from the construction stage, experienced people were required for planning of operations, for laboratory duties there, and eventually to operate the plant. An early need in Oak Ridge was for operators of the helium-leak detectors, and a few were sent to train people hired from the Tennessee area. Later, persons with experience in handling uranium hexafluoride were required, as were workers having skills with special equipment, such as instrumentation closely related to plant needs. Thus, one by one, many of our colleagues came to say good-bye as they departed for "Dogpatch," as Oak Ridge was jokingly known. Reports from the site concerning muddy streets, primitive housing for many, and commodity shortages often confirmed similarities to the Dogpatch community of Al Capp's then popular comic strip, "Li'l Abner"—a far cry from what Oak Ridge would soon become, one of the leading centers in the world in the field of nuclear research.

Staffing in New York was substantially smaller in mid-1945, well below the peak of about twenty-five hundred reached in the previous year. Some people had come only for short periods of training before moving to Oak Ridge. Thus, the laboratory served the important function of training people in this entirely new technology as well as for research and development. My group was not raided early because of the urgency for barrier-plant product-quality control. Barrier manufacture and testing would not be located in Oak Ridge until after the war, when the Decatur plant function was moved there.

During 1945 we had gradually learned more about progress in starting up sections of the K-25 plant, the completion and operation of the electromagnetic separation units of Y-12, and of the thermal diffusion plant construction. It seemed that enriched uranium might soon be available for weapons. Although news from Los Alamos was sparse, the apparent size of the facility suggested that they could soon be ready with a weapon design. Speculation increased on when, rather than if, a nuclear weapon might be tested or used. This expectation fostered discussion concerning the appropriate deployment of these fearsome devices that we were working to produce: Should a demonstration test be shown to the enemy or should the bomb be deployed destructively? Ideas at our level, of course, were only conjecture. After the war, it was reported that Dunning, Fermi, Teller, Urey, Wigner, and others who were in positions to influence the decision had corresponding concerns. Whatever the choice was to be, everyone wanted the terrible Pacific conflict to end quickly. Even after resolution of the question of a German nuclear threat on "VE Day," May 7, 1945, the pace of work did not diminish.

On August 6, word flashed through the SAM Laboratory that Hiroshima, Japan, had been destroyed by an "atomic" bomb. Gossip sessions, coffee breaks, and meetings at the water fountain had not been the norm of the laboratory; this day was a notable exception. Everyone was eager to know as much as possible. Despite misgivings about use of the weapon to kill so many people, Project members were rejoicing that this would surely end the war. I shared that feeling—the Manhattan District Project had succeeded! We were aware that portions of the K-25 plant were operating and assumed correctly that it had contributed to the first weapon. However, that enrichment could only have slightly lightened the burden of the Thermal Diffusion Plant and Y-12 Calutrons in producing the high enrichment required for the weapon.

The long awaited news had come earlier than expected. My own guess had been that a test might be made in September and the use or demonstration of a weapon might occur a few weeks later. We were surprised not to have learned of the test at Alamogordo in July. That security system had obviously been beyond the capabilities of the amateur intelligence gathering systems within the Project staff.

After work I purchased a copy of the *New York Times* and proceeded to the apartment of Enid Miller, my college speech professor, who was teaching that summer at Columbia University. As a popular and respected teacher Professor Miller was often the center of focus for the several NWU graduates who were in New York. We even occasionally met for dinner as a group. After proper greetings, I placed the paper with its massive headline "ATOMIC BOMB DROPPED ON JAPAN" on her table and asked if she recalled my final examination speech of 1942. She remembered both the speech and the critical note about timeliness she had penned on the paper. We had a lively discussion that evening concerning the news of the new weapon and its significance to the world. She easily shared both my concerns about nuclear weapons and aspirations to see peaceful applications. I was told much later that Professor Miller related the story of my atomic energy speech and the revelation of the weapon to her classes as long as she taught.

The intensity of work on the project was unabated through August, as the Nagasaki plutonium bomb was reported and the Japanese surrender occurred. The K-25 plant was to be completed, and much work remained to be done. It was a period of important effort for our small group. The barrier plant continued to produce at wartime speed, and our testing of barrier tubes had priority. The transfer of staff to Oak Ridge continued unabated, and consequently the SAM Laboratory population declined steadily. Speculation among the group was heightened. How long would the Project last? Would it eventually pursue nuclear energy for peacetime use? Discussion groups formed to debate the issues inherent in a future with nuclear weapons. On an individual level, I wondered if I should look for other employment. The time of year was right for younger members to resume graduate school and for university professors to reassume their positions. It was a period of continued important project work for our small group and a time of great anticipation and stress for me.

Don and Elaine below Riverside Church, 1945. Photograph by Richard Bernstein.

SIX

Relaxing in
New York City

For a Nebraska farm boy like me, life in New York City in the 1940s was a thrill. Not only was New York the place where I began my career; it was where I lived when I met, fell in love with, and married my wife of now fifty-six years, Elaine Causey Trauger. New York was also where I first encountered some of the early enriching experiences of my life outside of work. I recall life in New York fondly as Sunday respites from work between September 1, 1942, and August 31, 1945, followed by a six month "honeymoon" period with Elaine until March of 1946. The latter months, although busy, were encumbered only by a declining work load and one graduate school course. The Sundays off from work and occasional evenings of three years had afforded exploration and enjoyment of the great city with its cultural and entertainment offerings. Each Sunday was a relaxing change from the six work days, when sleeping was often shortchanged. The stimulating tasks, the mission's urgency, and the opportunities to learn had combined to make the effort satisfying, certainly not a burden. Nevertheless, the 150 wartime Sundays and last months in New York City contrast so greatly with my work on

Pioneer Zephyr at Exeter depot.

the Project that I describe them separately in this chapter to emphasize their thrills for a youth who had come there straight from the farm.

The initial trip to New York from Nebraska by rail coach had been pleasant; in my naiveté, however, I never anticipated reimbursement for sleeping accommodations; to save money, I rode coach all the way—but loved it. The first night on the Burlington's Denver-to-Chicago Zephyr was as exciting as my short ride on the Pioneer Zephyr in 1934. In Chicago I admired massive, ornate Union Station, then, after visiting friends at the Illinois Institute of Technology, boarded the New York Central Limited. Both trains were at the peak of rail luxury, and in those youthful days sleeping in a coach seat was no problem. I especially recall the thrill of awakening to see the beautiful Hudson River cradled in mountains that I had known through the stories of Washington Irving and other authors. The water appeared clear and deceptively clean, in contrast to the muddy Platte and Missouri Rivers.

Initially settled in New York in Dale Magnuson's temporarily vacant apartment room, I explored the surroundings of Columbia University. On Sunday Dale and I attended services at Riverside Church, the great cathedral-like structure on Columbia Heights above the Hudson River. Entranced by the grandeur and beauty of the nave, with its ornate stonework and stained glass windows, I paid little attention to the service. But when the sermon began, I was startled to hear a familiar voice. The guest speaker was Rev. Harold Bosley, who three months earlier had given the commencement address at Nebraska Wesleyan. In stunning contrast, that afternoon we visited Coney Island, my first and only visit, and rode the giant roller coaster. What an experience—slowly reaching the top to marvel at seeing the Atlantic Ocean for the first time, then falling suddenly, shaking in the wild and terrifying gyrations of that machine. Almost as amazing was to see that mass of bathing-suited humanity crowded onto that small stretch of sand.

Dale then lived with three talented men. André DeBethune impressed me by quickly working Sunday *New York Times* crossword puzzles using only the horizontal or the vertical clues. After obtaining his doctorate at Columbia, André moved to a lifetime of teaching at Boston College. A second, Alan Reid, had been runner-up in the Minnesota chess tournament the previous year. He once enticed me to play with him, and, amazingly, I won. Although I had previously played some chess, the win was a fluke. He interpreted an inept move on my part as an unusual strategy to entice him into a fatal mistake himself. The next time we played, I was impressed with how few moves he required to place my board at checkmate. The third was Ed Messervy, a social scientist of note at Columbia.

I found it difficult leaving that environment for the single room I had found through the University Placement Service. At first I was pleased about the ten dollar cost of the small room, until realizing that it was priced per week; my last college room had cost five dollars per month. What a contrast from the room in the Kottners's house, where three windows offered views of lawns and towering trees. The sole window of this New York room opened to a narrow, vine-covered courtyard. Kitchen privileges were available for breakfast, but the accommodation offered little beyond a place to sleep.

Almost immediately after moving my few belongings into the room, I learned of an opening at the Morningside Residence Club. That room was one of four in a corridor where three other men shared kitchen and bath. The view was of a larger, but even less interesting, courtyard, but the overall amenities were attractive. Also, my former landlady, sympathetic to a "freshman" in the city, was kind enough to release me from her contract and even refunded the advance payment. New York seemed not as severe as I had feared.

The Morningside apartment building was a residence club only in name, except that it did have a large, well-furnished commons room that could be used for parties. Unfortunately, most residents were working on the Manhattan Project or were in other war-related tasks and had no time to plan or hold large parties. Building occupants were congenial, and this pleasant place was made even more friendly by Julian, the doorman and daytime telephone operator who came daily from Harlem. Julian knew and befriended everyone. The building corridors were not locked, and ours, 5F, connected with 5E at slightly offset kitchens. The only security was at the building outer door, alternately protected by Julian and a night watchman. Even so, in over three years the only loss from the accessible kitchen was one small bottle of maple syrup.

I spent many Sunday afternoons exploring New York City. When walking in Morningside Heights on a beautiful October afternoon of 1942, I came upon the towering, ornate facade of an uncompleted building. Upon entry, I was amazed to discover the immense nave of the Cathedral of Saint John the Divine. Such ventures also brought me to the Empire State Building, the Statue of Liberty, the harbor ferry boats, the Cloisters, and eventually to most of the city's museums. Favorites were the Frick, because it was housed in the industrialist's former home, beautiful beyond my imagination, and the Museum of the City of New York, for its history.

For one autumn Sunday outing, I boarded a bus labeled "Hackensack, Patterson and Passaic." It seemed that places with such names might be interesting. They were much like other cities, but that afternoon provided my first acquaintance with New Jersey. Later, picnics with friends on the heights of the Palisades were more rewarding for the beauty of the Hudson River and the George Washington Bridge. As I became increasingly familiar with the metropolitan area, it was fun to escort New York City natives who worked on the Project to see some of its sights. We would eat Sunday dinner at a modest restaurant, then visit parts of the city they had never seen. The success of these events was indicated by the readiness of the local residents to participate in new adventures. Sunday afternoons in New York were by far the most intense and rewarding cultural hours of my life to that time.

Sunday mornings also were enlightening. When Rev. Harry Emerson Fosdick was preaching at Riverside Church, I was there if at all possible. Fosdick is recognized as one of the most profound preachers and religious writers of that time, perhaps of the twentieth century. Fosdick had retired from preaching before the war, but he returned to Riverside to minister to those in the military and others, like myself, who were displaced from familiar surroundings. I since have collected and read many of his books. My favorite is *On Being a Real Person* (Harper & Brothers, 1943). (I recall vividly our conversation with Fosdick upon leaving New York. Knowing that our destination was Oak Ridge, he charged us to help make nuclear energy a peacetime benefit and not a continuing instrument of war, which reinforced my earlier resolve.)

On Sundays when Fosdick was not preaching, I frequently visited the magnificent Christ Church of Rev. Ralph Sockman. Even after coming to know and admire Sockman, that Methodist church was never a comfortable place. Its ornate walls and ceilings covered with gold leaf and expensive furnishings were lavish beyond my appreciation. Also the wealth and apparent stuffiness of the parishioners often left one of modest means with an unwelcome feeling. Having arrived early one Sunday morning, two women in furs advised me in firm language to move to the rear of the church. I had inadvertently chosen to sit in "their pew."

Signs of the war were much in evidence in 1942 as many service people came to New York for embarkation to Europe. The port facilities were always busy and interesting to watch. In mid-October we were alerted to a larger than usual number of ships anchored in the Hudson River. In the late evening, Dale Magnuson and I walked to Riverside Park along the Hudson and saw merchant ships anchored closely, extending as far as one could see. The ships had no lights and the decks were shrouded in canvas coverings, but outlines of large guns, armored tanks, and aircraft were discernible. We took the subway to the southern tip of Manhattan and the ferry across the Upper Bay to Staten Island and back, finding it also filled with ships, all seemingly positioned to move out. By daylight, they were gone. We thought the massive convoy was for an invasion in the European Theater, but radio reports in early November described a large procession of ships passing Gibraltar into the Mediterranean Sea. The objective was North Africa.

Air-raid sirens gave further evidence of the war. Daytime drills were hardly noticed in the basement environs of Pupin and the Nash Building. Late evening siren tests often brought Morningside residents to the roof for a view of the city in darkness. Blackouts were largely effective, although occasionally lighted windows attracted the wardens' attention. However, this city of skyscrapers presented a definitive silhouette that was enhanced on a moonlit night; it seemed vulnerable except for the vastness of the Atlantic Ocean. The weekly appearances of the two great British ships *Queen Mary* and *Queen Elizabeth* were exciting. They alternated five-day crossings at full speed, outrunning German submarines as they carried troops to Britain and later to France. As I recall, the *Queen Elizabeth* was damaged once and did not appear for some time, but their regularity was reassuring.

The transition from autumn to winter brought another surprise. I had already noticed that my shirts were soiling more rapidly than on dusty days in Nebraska. As coal-fired furnaces for heat and steam-electric generation met the seasons' challenges, a white shirt was not acceptable in a few hours, and all shirts required daily laundering. Smog slowly became the norm, except for briskly cool days of refreshing winds from the north to clear the atmosphere.

The turnover of residents in the Morningside Residence Club was rapid throughout the wartime years. By late 1942 Paul Doty, later a prominent chemistry professor at Harvard, and his suitemate moved away. Karl Schleicher and I then occupied the two front rooms connected with French doors. Each room was larger than my previous space. Karl's seniority in the corridor gave him the front room on 120th Street, from which, by leaning out the window, one could see the Hudson River. Because by this time I had been moved to barrier testing and an evening work schedule and Karl's project schedule was during the day, we had little occasion to be present at the same time. But these were lonely months as I struggled to pursue graduate classes and

New York skies were often laden with smog, circa 1945.

studies during the day and to work at night. After the first year, the work schedule became too intense to allow for formal study too. Although I had to give up graduate classes, life became more interesting. I gained time to breathe, and even a Saturday night was occasionally available for leisure or a movie.

To save a bit of money, Karl and I, after my work schedule eventually shifted to days, sometimes chose to prepare dinner in the 5F kitchen. We had a treasured cookbook of simple recipes entitled *The Working Girl Must Eat*. Since I could rely on my college cooking experience and Karl had some skill with specialty dishes, we did pretty well. However, our schedules often did not permit eating the food as planned. When we chose to cook an eggplant, for instance, our trusted book cautioned, "First soak it overnight in salt water." We did, but then did not eat there for a few days. Osmosis did its thing; the eggplant swelled until as tight as an over-inflated basketball.

What should one do with a potential eggplant bomb? The solution: We waited until late evening and carefully carried the offending ovoid to a courtyard window and, with only slight hesitation, dropped the bomb. The resounding explosion resembled that of the shotgun once accidentally fired by my young cousin in the family farm kitchen. The eggplant's shocking thunder reverberated, then echoed in the chasms of the court and street. It seemed prudent to close our window quickly and turn out the lights. We could hear the sound of rising windows throughout the court. After waiting for our neighbors' speculation to subside slightly, we turned on the lights, raised our own window, and joined the questioning. Whatever could that frightening noise have been? As morning light revealed the aftermath, we marveled at how extensively the eggplant had plastered itself on the courtyard floor and lower walls, an unfortunate byproduct surely overwhelmed by the overall success of the exercise.

The vacated rooms of 5F were soon occupied by Mal Clark and Jack Largey, recent Williams College graduates working as chemists on the Manhattan Project. We found little time in common; schedules varied even for breakfast, although occasionally a party was planned for late Saturday night or Sunday evening. Many of our friends across the country came to New York, some on their way to European war theaters. As an accommodation we always left the key to our suite in a cup located in the cabinet of the common kitchen. It was not unusual for either Karl or me to come in late from evening work to find several men sleeping soundly on the floor as guests of the other, or of Mal and Jack. This, of course, violated the house rules, but we worked things out with Julian, the doorman, and the night watchman. However, on one evening when a group of young women slipped past him into the building, the night watchman, assuming that the women were headed in our direction, called us, sharply accused us of harboring them, and ordered that the women were to leave. We had not been so favored, and my angry response generated a new respect from the watchman but did not improve our relationship.

Loneliness was broken by weekly correspondence with my mother and telephone calls on special occasions such as birthdays and Christmas. Even this use of the telephone was a luxury in those days of difficult routing and poor connections to that rural line. Pop did not like to use the telephone, so I seldom heard his voice. I often recall my conversation with him the day he helped pack my typewriter in the trunk for shipment to New York; he naively supposed that I would see him at Christmas, an impossible schedule for me. In compensation I spent my annual two-week vacation each autumn with them at the farm by repeating the long trip by rail coach. Mom also helped alleviate my loneliness by an occasional shipment of cookies that could be shared with acquaintances, although it was difficult not to hoard them for evening snacks.

New York's principal characteristic, for visitor as well as for resident, is its abundance of interesting people, sometimes found in surprising places. The building superintendent was one with whom I occasionally enjoyed a chat in his basement quarters. The surroundings, including the asbestos covered furnace, were not unlike those of the Garners's commons room at Nebraska Wesleyan. However, these walls were decorated with news clippings and other memorabilia from many parts of the world. Mr. Henderson regaled me with stories of his adventures as a newsman, businessman, and writer. His greatest pride was in having fulfilled a longtime ambition to ride in a parade with the Lord Mayor of London. On the one hand, the story hardly seemed plausible for this man living in the modest accommodations of that Morningside basement. On the other hand, Henderson seemed educated, was articulate, and told the stories credibly, when sober. Evidence of his former prominence was displayed in treasured news photographs showing the Lord Mayor in full dress regalia with someone that appeared to be a younger Mr. Henderson.

Henderson was sorely needed, but absent, one noon as I returned home to prepare a sandwich for lunch. As I approached our building, the window sill of the 5E kitchen seemed to be the crest of a waterfall cascading to the courtyard in a torrent both audible and visible from the street. Hurrying to the fifth floor, I discovered that excessive steam from a neighbor's cooking had activated the overhead fire sprinkler, and water was flowing down the walls and spraying out the open kitchen window. The poor woman's hair and clothing were drenched, and the floor was rapidly flooding. Calls to the management only confirmed the superintendent's absence, and no one knew the location of the sprinkler system shutoff valve. A young assistant to Mr. Henderson was persuaded to stand on a ladder and hold the offending sprinkler head shut, but he soon tired, then relaxed, and was immediately soaked; the flood resumed. I built a dam of towels across the opening with the 5F corridor to little avail and retreated to the doorway of our rooms, which was easier to fortify. Stacking dampened towels alternated with layers of newspaper worked fairly effectively. In due course the fire department arrived to break down a locked door, close the valve, and replace the sprinkler unit. By that time water had penetrated floors to the basement. Doubts concerning the effectiveness of sprinklers to douse a fire would have been erased from the mind of even the most hardened skeptic witnessing that scene.

On occasion, the ordeals of life in New York had a military connection. A Naval lieutenant drilled cadets on 120th Street in early mornings. His remarkably strong voice issued orders at full volume. This 5:00 A.M. awakening, with windows open in summer, often after we had managed only a few hours of sleep, was most annoying. We and others who were engaged in war-related work complained to the Naval Commander at Columbia to no avail, and shouts from various windows were equally ineffective. A remedy eventually arose from an initially unrelated diversion. We occasionally amused ourselves by dropping a small paper bag of water from the fifth floor on an unsuspecting colleague trudging to or from work. The resulting shock was always entertaining, but as we all dressed simply and informally no harm was done. Practice in this art finally proved beneficial. Early one morning, as the lieutenant made a military turn at our sidewalk, I landed a large water bag on his shoulder. The trainees broke ranks to roar with laughter at his distress and were soon dismissed. Our complaints then seemed to be understood, his voice was lowered, and early morning sleep improved.

The Morningside Residence Club was a social institution primarily by virtue of having living quarters arranged with connecting corridors; meeting people was relatively easy, sometimes unavoidable. The obsolescent elevator produced some dramatic encounters by frequently stalling between floors and entrapping occupants. Screams of a claustrophobic victim were frequently heard throughout the building, and because Mr.

Henderson was not always available we learned to effect rescues. Because my life in New York was largely sedentary except for walking to work, I seldom rode the elevator, preferring to climb the stairs for exercise and, for the first year, to determine how much stamina I had lost since my summer of work on the farm. Besides, even when the elevator functioned at its best, climbing the stairs saved time.

Eventually we came to know several young women residents. It seemed logical to invite them to join us for dinner, with the expectation that they would feel sorry for us as amateur cooks and return the favor with a meal of better quality. Alas, it was soon apparent that they either knew less than we did about cooking or perceived our motive and concealed their talents. Our encounters led to several friendships that were pleasant but not romantic. Two of the women had interests far removed from our scientific fields. Ruthanne Huff and Janet Henthorn, nicknamed Jinx, had come to New York from Columbus, Ohio, to find fame and fortune as actresses on Broadway. Reality set in early for Jinx, who took an office job and seemed to fit easily into the hectic city pace. In contrast, Ruthanne lived an alternating life tied to the theater: haunting stage-door offices until funds ran out, then temporarily taking other theater jobs, such as ushering, until she again could pursue tryouts.

Sunday night parties that gathered in our two-room suite included men with whom we worked and other men and women from apartments in the immediate Columbia area. The evenings often started with a snack supper of sandwiches and soft drinks from a nearby deli and then progressed to parlor games of some sophistication. A favorite was charades. As familiarity grew, the quotations, titles, and other subjects to be silently acted out necessarily became increasingly obscure. Ruthanne was popular on any team because of her creative acting skills. On one occasion, when assigned to act out Admiral Farragut's "Damn the torpedoes, full speed ahead," she assumed a commanding posture, raised her arm, and brought it down with a snap of fingers, to which Larry responded with the quote. That record of five seconds was never matched or bettered. It would have been difficult to believe they had not cheated if her performances had not been so consistently effective. This kind of simple socialization is not what most people think of as entertainment in New York with its glittering night life, but it was relaxing, involved good company, and was affordable. Over the years I have noted with interest the similarities between our wartime lives in the big city and those reported from the Oak Ridge and Los Alamos communities, where fewer public entertainment options were available.

Somehow, the Episcopal Cathedral of Saint John the Divine, even with its marvelous choir and large attendance, had not been included on my church attendance agenda. On Palm Sunday of 1944 Ruthanne, desperate for someone to accompany her to this service of her denomination, persuaded me to go. The service closed with Communion served by priests on the altar steps. Ruthanne understandably wanted to take Communion but refused to go forward alone. Being unfamiliar with the Episcopal service, I stalled, urged her to go, and stalled further. When I finally yielded, we joined a queue about midway down a side aisle, and I was the last person in our line, which proved longer than the others. By chance or because I missed a subtle signal from one of the priests, I was finally alone, the last communicant, attempting to act confident in ascending the steps, arranging palm leaves and taking communion before some ten thousand people. The lesson—procrastination does not pay—is well remembered. At least I had not fallen over the accumulated palm leaves or stumbled on the steps.

One weekend, Larry O'Rourke invited me to his home in Buffalo for a Sunday visit. We set out late Saturday to hitchhike, starting at the George Washington bridge over the Hudson River. Traveling with a man in uniform was usually an advantage; drivers would frequently stop for Larry and let me come along, but we had not done well on this occasion. Some time after dark we were in Scranton, Pennsylvania, a long way from Buffalo. Our prospects were discouraging. Fortunately, after some further exercising of thumbs, a driver who was going to Buffalo invited us to ride. Our benefactor advised that he was really too tired to drive that far and had planned to spend the night somewhere, but if we would chauffeur while he slept on the back seat, we could get to Buffalo early Sunday morning. We alternated reading maps and driving, reaching Larry's home about 3:00 A.M. That Sunday afternoon I visited briefly with Professor Jensen, then working during the summer with the Hughes Company on lightning protection for aircraft. After a pleasant evening with the O'Rourke family, we returned to New York on the overnight train and were at work on time Monday morning. Larry and I made similar excursions to Providence, Rhode Island, and Waterbury, Connecticut.

Annual trips in the autumn of 1943 and 1944 were to Manchester, Vermont, to climb Big Bromley Mountain during the peak foliage season. That was a real thrill for a Nebraska boy. Today, the Midwest benefits from many cultivated trees displaying fall's colored leaves, but the autumn beauty I had known was mainly from sumac shrubs and poison ivy. Each year a different group assembled for this pilgrimage that had become traditional for Columbia graduate students. We stayed at rooming houses in Manchester and ate at the Quality Cafe, something of a misnomer but staffed with very friendly people who offered acceptable fare and who also packed lunches to be enjoyed on the mountain when we were ravenously hungry. Since we were familiar with the Union Carbide Corporation, a mansion on a nearby mountain built by an executive of the company was of interest. It was said to display, somewhere in its furnishings and structure, every plastic, fabric, and finish material made from a company product. The house was particularly noticeable as a break in the brilliantly wooded slope, a seeming fire-storm of red and yellow sugar-maple leaves. After each trip's Sunday night supper, we boarded the train to Albany, where the night coach returned us to New York and work.

———•◆•———

On November 2, 1944, I was invited to an engagement party for Dale Magnuson and Doris Lent at a friend's apartment. While attending that affair, I met Elaine Causey, a Columbia graduate student in the Foods and Nutrition program. Doris and another mutual friend, Barbara Lewis (now Heising), had told Elaine that I was an interesting person, someone she would enjoy getting to know. With that expectation, this North Carolinian approached me with her naturally outgoing enthusiasm. I completely botched the encounter, coming off as rude and aloof and very nearly ruining the opportunity before us. Elaine later consulted with her friends, thinking she must have spoken to the wrong person.

Fortunately, we met again quite by accident the following Sunday on 120th Street. Hank (then called Harry) McKown and I encountered Elaine and her friends returning from Riverside Church services. This time the interaction was different. We were invited to the women's apartment after lunch and, on that remarkably agreeable November afternoon, had an impromptu party on the building roof. Elaine was immediately my favorite, even though she remained cautious, based on our earlier encounter.

That pleasant occasion was followed by a date to a New York City Ballet performance. Dick Bernstein, with whom I occasionally enjoyed dinner and a brief, brisk walk that included challenging each other in memory or mathematical games, and I had purchased two tickets each. I invited Elaine to go and was pleased when she accepted. After the performance, she introduced me to persimmon pudding, which her mother had sent. Elaine and I then increasingly spent time together until our dates included essentially all of the free hours and minutes permitted by my work and her graduate studies.

Just before Christmas Elaine and I enjoyed a delightful dinner at a restaurant on Thirty-Fourth Street and viewed the decorations of Macy's store windows. I then escorted Elaine to Pennsylvania Station, where she was to board the train for the trip to her North Carolina home. At my urging she agreed to return early so that we could celebrate New Year's Eve together in Times Square. Early on the evening of 1944's last day, I eagerly awaited Elaine's arrival, her train scheduled to arrive well before midnight. But the train was very late, its status was unknown to the skeleton staff, and I spent hours alone in cavernous, nearly deserted Penn Station. Most of the festivities of Times Square had subsided by Elaine's 3:00 A.M. arrival, so only the Hawaiians could help us by radio to celebrate the arrival of 1945.

In February a group of our friends spontaneously planned a ski trip to Vermont. Most of us were novices, though we practiced a little on the snow-covered slopes of Riverside Park. We took the train to Manchester and found what would now be called a large bed-and-breakfast for lodging. Looking back on that trip now, it seems to me a wonder that people could actually survive such impromptu and disorganized expeditions.

Early Sunday afternoon, Barbara Lewis, whose skiing experience was limited, lost control and after displaying seemingly comic gyrations buried herself in loose snow at the side of the run. Unfortunately, she experienced a painful shoulder injury. The ski patrol quickly effected her rescue and provided transport to a nearby doctor's office, a simple affair reminiscent of an earlier time and located in his home. Hank, Elaine, and I went along. The doctor was an elderly fellow who confirmed the rescue team's diagnosis of a separated shoulder. He tried a quick, but unsuccessful, maneuver to relocate the shoulder but lacked the strength to accomplish the task. He asked me to help, which sent the squeamish Hank and Elaine to the waiting room. When our combined efforts failed, the doctor decided to use an anesthetic, preparing a mask with ether, which I was to apply. I shudder to think of my agreement and of the consequences that could have resulted. At that point the doctor's wife returned home. Seeing that Barbara was exposed in the upper half of her body to facilitate manipulation of the shoulder, she ordered me out of the

Vermont ski trip in 1945: Don Trauger, Elaine Causey, Hank McKown, and Barbara Lewis.

room and commandeered Elaine as anesthesiologist. Elaine was sensitive to strange circumstances, and I feared that the ether might make her ill. Sure enough, a thud was soon heard, and I returned to revive Elaine who had fainted: Barbara remained wide awake.

At that point, we decided it was time to escape the tender mercies of the country doctor and to take Barbara to the hospital in Albany. Unfortunately, the bus to Albany was crowded with only standing room, and no seated passenger was willing to yield for our injured Barbara. After a considerable wait at the hospital, the doctors quickly restored Barbara's shoulder to its normal position, apparently without great discomfort to her. We took the evening train to New York arriving at about the same time as the others. As the luggage was sorted out, we discovered that one suitcase carried for the round trip belonged to someone's roommate and not to anyone of our party. Such is the confusion of a group moving on an unorganized expedition. Except for Barbara's discomfort, it was a fun trip.

When Bill, a former boyfriend of Elaine's from Greensboro, arrived to see Elaine, I was frustrated by her insistence on entertaining him, though she welcomed me to come along. We spent an unusual Sunday afternoon as a foursome: I was identified as escort for one of her girl-friends. Since I knew New York and Bill did not, it was easy for me to place him in awkward seating situations in taxis, theaters, and restaurants. The day proved both interesting and amusing, with Elaine and "my date" cooperating at key points. Of course, I paid for my fun when Bill reported dire words of my behavior to Elaine's parents upon his return to North Carolina, which was followed shortly by a visit from Elaine's father on a supposed business trip to the city. I was unsure about having passed that inspection, but we seem to have gotten along well.

By spring Elaine and I were secretly engaged. We did not announce it because of a serious understanding she had with a soldier serving in the European Theater. I understood the importance of her not sending a "Dear John" letter to a man under fire in World War II. After the European armistice, it seemed time for announcement to family and friends, and I purchased a diamond ring. On the last Sunday of June we took the Hudson Day Line boat for an excursion with friends to Bear Mountain State Park and set forth to climb the mountain. With the ring safely pinned in a pocket, my intent was to present it in a romantic setting at the summit. But Elaine tired, perhaps three hundred yards short of the top and refused to proceed even with my persistent encouragement. It was not until after dinner on the following Saturday night in a restaurant setting overlooking Riverside Park and the Hudson River that I succeeded in slipping the ring onto her finger.

In retrospect Elaine was taking a risk in agreeing to marry one who worked at a secret place and could not explain his employment. Even worse, she could not visit my home or family because of wartime restraints, including travel restrictions. However, we had much in common: we both were of the Methodist faith, had graduated from small colleges, and had grown up on farms. Because of our unusual schedules, we had met at a variety of odd times, from early morning to late evenings, and found comfort together in many settings. On Memorial Day weekend of that year, 1945, I visited her beautiful family home and cattle farm in North Carolina. Elaine introduced me to many of her family members at their country church before we rushed back to New York. The Causeys were successful people with whom I could easily converse, although their farms in the Piedmont region bore little resemblance to my home.

As a home economics graduate of Greensboro College and with her masters degree in foods and nutrition to be awarded in June by Columbia University, Elaine often took pity on the nutritional situation in 5F. Also, she and her apartment mates, Tot Malone (now Pearson), Jeanne Sheets (now Johnston), and Yvonne Thompson (now Errickson) frequently invited Hank McKown and me for Sunday lunch. We were expected to bring a gallon of milk and sometimes other groceries, but overall Hank and I were the beneficiaries.

One August evening when Elaine had come early to the 5F kitchen, she discovered how much I disliked washing socks, often purchasing a new pair rather than facing the ordeal of sock washing. On that day, however, I had finally faced up to the task, washed a tub full of socks, and hung them to dry on several lines strung in the kitchen. As Elaine cooked alone in the common kitchen before Hank or I had arrived home, someone passing by in the "E" corridor and noticing Elaine working around all the sock lines remarked, "Had a busy day?" Upon my arrival from work, Elaine firmly demanded immediate removal of the sock laundry, dry or not.

The last weeks of August were intense. The marriage date and festivities of September 2 were looming as a happy time but with many prior demands. It was to be in Elaine's home community, at Mount Pleasant Methodist Church near Greensboro. Elaine's telephone charges in planning the elaborate wedding must have been enormous. There was much to do: buy a suit, plan my limited role in the festivities, and arrange for coverage at work during the previously scheduled two-week annual vacation that would accommodate our honeymoon trip. In addition, the armistice on August 14 introduced new uncertainty about the Manhattan Project's continuation and the ensuing prospects for employment.

We also needed a place to live in New York. Had I been able to predict accurately the completion date of the nuclear weapons effort, we might not have set our marriage date for this tumultuous time, but as events unfolded, the choice was fortuitous. With the war ending, Karl decided to leave as early as possible, solving the housing problem. We would occupy the two front rooms of 5F and share the kitchen and bath with Hank McKown. He had moved to the middle room some months earlier (after returning from an extended assignment on the Manhattan Project's heavy-water research and production facility at Chalk River, Ontario, Canada). Because I maintained rental possession of the apartment, the rent could not be increased under New York City controls. A nurse had rented the small room that I had first occupied, but she worked evenings and did not use the kitchen.

On Thursday, August 30, Hank McKown, Tot Malone, and I took the overnight train to Greensboro. Hank was to be my best man and Tot one of Elaine's attendants. Wartime travel discomfort prevailed,

Don Trauger, Hank McKown, and Karl Schleicher in apartment 5F of 100 Morningside Drive, 1945.

and we suffered in the unair-conditioned train on that very hot summer night. South of Washington, D.C. the air in the cars became thick with smoke from the steam-powered locomotive pouring in through open windows. To make matters worse, the car had a flat spot on a wheel rim directly below our seats in the fully occupied train, and we had to endure a constant, bumpy shaking. We arrived after 3:00 A.M., tired, sooty, and disheveled.

On Friday morning, Elaine and I went to purchase our marriage license at the Guilford County Courthouse. The clerk spent some time examining our health-test reports as completed by the state of New York on papers provided by her office a month earlier. After concluding her examination, she somberly stated that they were not the proper forms; we could not be granted a license. We protested: the information was correct, the tests were valid, and the elapsed times met requirements, but she remained adamant. After considerable argument, we made clear that we were leaving immediately to visit the *Greensboro Daily News* to report that the publicized Causey wedding would be postponed because of a mistake made in the County Clerk's office. We further threatened that if the paper would not print this as news, we would pay for an advertisement to that effect. She then found it quite possible to issue the license.

The church was beautifully, though modestly, decorated as appropriate during wartime shortages and with restrictions still in effect. Guests numbered more than four hundred, crowding the nave and every corner of an adjoining space on that hot September afternoon. Reassuring glances exchanged with Elaine's mother on the front row eased my tension, and seeing Elaine especially beautiful and smiling as she proceeded down the aisle overcame all my anxiety. Photography following the ceremony was limited to four poses of the wedding party, and one shot of us in the aisle upon leaving the altar. The pictures were made by an obliging news photographer who could obtain film. All guests were invited to a reception for punch and cake formally served from the beautifully appointed dining room table of the Causey home. I had met many of Elaine's immediate family including her sister Eloise and several cousins, but most guests were new acquaintances. For the next two decades, I frequently encountered people who seemed to expect remembrance from our having met at the reception. Not only were the numbers overwhelming, but by reception time my mind was nearly blank from the pleasantries and tensions of the day.

The honeymoon trip was limited by time and rationing to a few days in neighboring High Point, an afternoon train to Washington, D.C., and a series of flights to Lincoln, where Elaine would meet my family. On this, our first commercial airline travel, the American Airlines DC-3 winged its way to Wheeling, Cincinnati, Indianapolis, and Chicago. Finally, United Airlines carried us to Des Moines, Omaha, and Lincoln. We were exhausted and relaxed for two days in Lincoln, including being dinner guests of the Jensens at the Cornhusker Hotel. Professor Jensen, with his dry sense of humor and a slight smile remarked, "Well, Don, if you couldn't get a Nebraska girl, I think that you did rather well." Elaine, of course, took feigned offense, but she and Jensen got along nicely on that and other visits of later years. We were warmly received by my family, and Elaine was immediately popular on their first meeting.

Visiting family is not a widely recommended way to spend a honeymoon, but circumstances and travel limitations had been determining factors. The honeymoon was severely marred at about 2:00 A.M. one night, when I was awakened by a flickering light through our bedroom window: the horse barn loft was on fire. I dashed to the farmyard, returned to call my parents, and disconnected electricity to the barn; the fuse was already blown.

Remembering that two horses were tied in the barn, I attempted their rescue, but the large tongue of flames from the open hayloft door did not permit entry through the door normally used. Also, because horses sense a barn as a safe place, they might have refused to brave the towering flames visible above that door. The fire was leaping across the surfaces of

Just married at Mount Pleasant Methodist Church, September 2, 1945.

loose hay, growing in intensity. Although the barn structure seemed intact, the front wall could soon collapse. I tried the side door away from the flames, but it had been nailed shut. As I searched for a pry bar, Mom ran to the barn reporting that Pop was nowhere to be found. He was prone to insomnia and at times had been found relaxing or asleep in the barn. The prospect was horrifying; I abandoned efforts to rescue the horses to search for Pop, soon found napping on a cot in the workshop attic. I returned to the barn, but it was too late; the horses died in the fire. I still occasionally awaken suddenly in the night envisioning their suffering and demise. Perhaps in some small way that experience is like the trauma of veterans troubled by the horrors of war.

The Exeter Volunteer Fire Department arrived soon in response to Mother's call on the hand-cranked, party-line telephone, and that ring also alerted neighbors. We soon had ample help, but the barn was beyond salvage by any fire-fighting unit. The volunteers sprayed the nearby buildings with water from the livestock watering tanks and contained the fire to the barn. Elaine volunteered to prepare coffee and sandwiches for the many workers who were assisting. At daylight an overly-friendly cat slipped into the kitchen to interfere with her mobility. Elaine finally ejected the offending feline by use of the most convenient handle—its tail—just as Mom approached to meet her favorite cat suddenly tumbling in air. Despite this, Mom always dearly loved Elaine.

The blown fuse and pattern of the fire, as first observed, suggested that it had originated at the wiring entry to the barn. I recalled having used the improper entrance conduit and in early daylight searched for remains of the wiring to confirm the unpleasant prospect. Even though I knew where it should have fallen from the pattern of the barn wall collapse, it had disappeared; apparently someone observed the electrical damage and rather than distress me for having caused the fire had carried it away. That, alone, was evidence enough for remorse. The whole sad matter was a hard lesson for one who was to spend many years working in the nuclear field. Every decision and action must be made with due concern for safety: no code violations allowed!

I felt no better about things when, on reflection, I realized that the fire might have been avoided the previous evening. As Elaine and I were touring the barn and stroking the ill-fated horses, I considered and then rejected the urge to investigate the loft. Had I done so, the overheated components might have been sensed and the fire avoided. After the fire we worked to salvage grain stored in bins on one side of the barn. Damage to the grain had been limited by neighbors and firemen who pulled the burning hay and loft structure away from the grain with cables pulled by tractors. I later worked hard to salvage the undamaged grain and transport it to other storage. During this work, contact with a partially burned and damp top layer of wheat caused my shiny new platinum wedding band to turn black. Vigorous polishing, even with cleaning compounds, proved useless; it remained black. However, as a few weeks passed, the ring slowly regained its original sheen without further attention.

After returning to New York through Chicago on the Burlington Zephyr and the Pennsylvania Railroad, setting up our household in the Morningside suite was the first order of the arrival day. We had cooking and table utensils, but the dishes and sharp knives had belonged to Karl. Our immediately available resources totaled some $13.85; ten dollars of that had been a wedding present from Aunt Pearl. We spent twenty cents for the subway to Macy's Department Store and return. We found a four-place setting of dishes for ten dollars, a paring knife for thirty five cents, and we returned to Mr. Babson's neighborhood store to select, very carefully, $3.30 worth of groceries. My next pay check was only a few days away; we were okay. Actually, Elaine had funds remaining from her graduate allowance and I had purchased war bonds, but we were unwilling to spend those savings.

As expected, much had happened following the signing of the peace treaty on the USS *Missouri* on our wedding day. Some friends quipped that one war ended and another started. Nevertheless, the marriage has now prospered for over fifty years, even though it started out with a traumatic honeymoon event and severe financial shock. The Project had cut back from the official forty-eight hour week to forty hours, with a 30 percent reduction in my pay; Elaine's job at the New York State Emergency Food Commission for the summer had no provision for a leave. To circumvent this rule and retain an effective employee, her boss had arranged for Elaine Causey to resign; two weeks later she would hire Mrs. Donald Trauger as the replacement. However, effective September 1, Governor Dewey had terminated the commission. Elaine had no job. Fortunately, we had agreed to live on my salary and to save Elaine's earnings; thus, the impact was on savings, not on living expenses. It then took some time to achieve our first financial goal: to set aside one hundred dollars, not to be touched except for an emergency or very unusual opportunity.

With the shorter hours of work, I again registered in the Columbia University Graduate School as we settled into a delightful routine. With a bit more leisure time, New York was a fun place in which to live as newlyweds. Elaine was soon offered a temporary teaching position in clothing construction for home economics students at Hunter College. My work at the Nash Building continued unabated, as the Decatur plant completed manufacturing barrier for K-25 and K-27; the latter adjoining facility was for additional uranium enrichment capacity ordered late in the Project. Work intensified somewhat as individuals moved to Oak Ridge to participate in the continuing startup of the plant, returned to school or sought other occupations leaving more for the remaining team to do. We, of course, knew that this employment was only temporary, but we tried not to worry about things over which we had no control.

The total silence about nuclear technology was broken by the surprise release of Manhattan Project information in the *Smyth Report* of late September 1945 and by the distribution of souvenirs to participants. This permitted open discussion of issues created by the Project's success. Groups of scientists organized at Chicago, Los Alamos, Oak Ridge, Berkeley, and New York assembled to discuss key questions about controlling this new force. Peacetime applications were included but were greatly overshadowed by concerns for the control of nuclear weapons. Positions taken ranged from maintaining total secrecy beyond that already released to sharing the technology with the Russians, but in meetings at Columbia University those extremes were held by small minorities. Most scientists favored some kind of effective international control and focused on the potential for the United Nations, then in the process of formation. Some even favored a stronger United Nations, similar to that of my college projection. The May-Johnson bill in Congress that would have continued all nuclear research under strict secrecy was disliked as restricting scientific endeavor, but it seemed difficult to develop a consensus on anything else.

Since the meetings were open to the public, others gradually joined the biweekly gatherings until as many as fifty to seventy-five were present. As the group grew to this size, it became even less capable of focusing on actions or positions to be taken. Also, those from outside the Project became increasingly vocal. Project members who had lived in New York before the outbreak of the war recognized some of the outsiders as having been affiliated with a Communist organization. On one occasion, the outside group dominated the discussion. They appeared to have conspired to render the meeting ineffective or possibly to introduce a new position. Signals passed from one to another to gain the floor in succession. One member spoke at great length, exceeding time limits imposed earlier and ignoring requests for conclusion. Clarke Williams, who was the chairman, stood up to his full and impressive height to bring a meter stick down on the lecture bench with a resounding crash, followed by the words, "Shut Up!" Silence fell on the group and the meeting soon adjourned.

Subsequent meetings were orderly, but I lost interest as the likelihood of a useful outcome diminished even though the issues were close to my heart. Gradually, the Association of Atomic Scientists was formed from the groups at the different sites; its bulletin, noted for the clock hand moving toward midnight signaling a possible nuclear Armageddon, is still published.

New York can be beautiful in the autumn, and a mid-November Sunday of 1945 displayed a perfect, cloudless sky. By chance, Elaine and I also were dressed to match the beauty of the day. She was wearing a fine fur coat her father had given her as a buffer against New York winters adorned by a corsage of roses received from the Hunter College students in appreciation for her having taught until a WAVE returned to claim the job. I was wearing my wedding suit. Perhaps this attire gave credibility to the ruse we were unexpectedly to pursue that day.

As Elaine and I emerged from Fosdick's church, the shimmering Hudson River beckoned to direct our footsteps into Riverside Park. As we leisurely traversed pathways, a battleship came into view, and as it drew nearer we confirmed its identity by reading the name on the prow, USS *North Carolina*. We walked on to see the formidable outline and weaponry of the great ship, named for Elaine's home state. Coming essentially abreast of the vessel at the Seventy-Second Street dock, we were blocked from further progress by a fence and a no-admittance sign.

Led by our adventurous spirits, we slipped through an opening in the fence and greeted and chatted with a group of sailors aboard a launch and about to return to the ship. We asked if they could take us with them to see the ship and were told they could but only if we were friends of someone on board who would vouch for us. When Lieutenant McKiver, newly assigned to the *Carolina*, entered the craft, we told the sailors that we knew someone on the ship. We thought it quite possible that a crew of that size might well include someone we knew. Besides, we had confidence that during travel to the ship we could persuade the lieutenant to invite us on board. Though skeptical, they agreed, advising us that if not allowed on deck we would be left to sit on the access stair until their next excursion, two hours later.

Undaunted, we climbed the stairs just behind the lieutenant, met at the deck by the officer of the day, Lt. Comdr. Randolph Klipple. As Lieutenant McKiver parted to find his quarters, he thoughtfully said, "Here are some people who would like to see the ship."

Much to our surprise, the commander greeted us warmly, and invited us to join his mother, aunt, and uncle on a tour he was conducting for their benefit around the ship. He even insisted on showing us the ship's features that his party had already visited. As we observed an anti-aircraft gun station on the starboard side, Klipple suddenly asked about Lieutenant McKiver and his family. Klipple liked to learn about his new subordinates and their backgrounds. Elaine and I each separately panicked, recognizing that we were soon to be revealed as the impostors

Battleship USS North Carolina *in the Hudson River, November 1945. Photograph by Elaine Trauger.*

we had become. The nearby stair where the uninvited had to wait appeared hard and uninviting.

As we stood there speechless, each hoping for the other to devise a face-saving response, the ship's bells rang three times and our escort excused himself; they were calling him as officer of the day. Alone, we tried to figure out what tack should be taken. Fortunately, when Klipple returned he had forgotten his question for the moment and continued the tour. As we approached the stern, he again asked about the lieutenant, and for a second time we were saved by the bells. Standing on the open deck in the shade of sixteen-inch guns, we were desperately searching for a strategy when Lieutenant McKiver appeared to inspect the two pontoon-type aircraft that were his major responsibility. I met him at the rail and politely inquired after his experience, family, and plans for temporarily living in New York. Armed with a wealth of personal information on the lieutenant, we rejoined Klipple and his family.

Klipple again asked questions about the lieutenant. This time we were prepared, but, incredibly, the chiming of ship's bells interfered, and we were left with the family: Randy Klipple's mother, a nearly deaf uncle seemingly not much interested in conversation, and a maiden aunt, who was noticeably suspicious of us. Mrs. Klipple, in contrast, was pleased to meet New York residents who could befriend her son, expected to remain in port for some while. By the time we reached the officers' lounge at the end of the tour, Elaine had invited

Randy Klipple to dinner at our home on the coming Friday evening, much to the delight of his mother. We also asked him to extend the invitation to Lieutenant McKiver. We sipped tea and munched cookies from the beautiful dishes inscribed with the ship's insignia and again were asked about our friend; I confidently provided the information so recently obtained. Klipple was satisfied, the maiden aunt relaxed, and we had a jolly time until the hour for the launch to return.

As we were about to depart unscathed, Lt. Comdr. Klipple appeared with the ship's log, intent on entering the new officer's arrival. He asked me how to spell the McKiver name, was it "Mac" or "Mc" or spelled in some other way? When I finally admitted that we had only just met the lieutenant and had no idea which was the correct spelling, body reactions around the table were noticeable. But Mrs. Klipple was unflappable: her son now had friends and a dinner invitation for the next weekend; so who cared how this came about? Randy Klipple appeared for dinner promptly as invited, but the lieutenant did not come and was not mentioned that evening or ever again when we were with Randy. We learned that Klipple came from a long line of officers in the German Navy, including his father, who had emigrated to the United States following a naval career.

On Navy Day, we were invited to join Klipple on the battleship USS *New York*, a veteran of both World Wars. While Elaine and Klipple's date enjoyed the hospitality of the officers' lounge, I was taken on a tour of the engine rooms and other working parts of the ship, during which we came upon a sailor sound asleep at the boiler watch station, a potentially disastrous infraction. The initial, vocal stage of Navy discipline was loud and harsh; no doubt the poor fellow was dealt with severely after he was replaced at the watch and led away by a junior officer. Otherwise, the tour was pleasant and instructive, and we have always been grateful for our chance meeting and warm friendship with the lieutenant commander. Elaine and I spent many pleasant evenings with Randy before we left New York. Although eventually losing contact, we corresponded for some years and know that he was promoted to commander.

Near the end of 1945, a decision was made to close the laboratory in the Nash Building and to move it, or at least much of its function, to Oak Ridge. I was to rebuild the laboratory and improve equipment we had in the Nash Building. Although barrier tubes for testing were still coming from Decatur, they were samples from production of spare or replacement units for the completed diffusion plants. We were also to select the equipment to be moved to K-25 and to identify material to be sold as surplus. Thus, there was work to do, but the pace was slowing noticeably. Some groups in the building had little work, and small private enterprises began to develop. One was to bend and shape silver solder wire into costume jewelry. By etching copper from the surface, the wire took on a fine silver patina. A few artisans became quite skilled. One of the best was a young woman named Katherine, who, I was told, later established a jewelry shop.

There was no extracurricular activity in the basement, as my team continued to test barrier for separation performance and Dixon Callihan's group produced more data on the extended performance of barrier in uranium hexafluoride. It was sad to see that fine laboratory organization disintegrate and to tell friends good-bye. Many were moving to Oak Ridge, and my joining them there seemed likely. Nevertheless, it was great to live in New York while working only a forty-hour week. Since we were scheduled to leave before the end of the semester, I had not registered for additional courses in the winter/spring term. This left every evening and weekend free to enjoy New York and its surroundings.

One goal, largely to be fulfilled, was to see all of the plays on and off Broadway. Our modest income made this unrealistic at regular box office prices, but standing room was cheap. The strategy was to select two or three plays having starting times at twenty or thirty minute intervals. If tickets were not available for the first, perhaps a second or third choice would be available, and a movie was always a last resort. Usually we stood only through the first act until empty seats were spotted, often in excellent orchestra locations. We enjoyed opera and other musical events as pleasant alternatives.

In ventures beyond the city, the availability of Hank and his stored-nearby automobile often brightened weekends. The 1945 autumn trip to Vermont was made by automobile and was made less hectic because we could leave on Friday evening instead of on Saturday, as during the war. We also ventured to Hank's home in Rochester, New York, in January when the piles of snow beside the road for most of the way were as high as the Ford. A trip during this period to northern New Jersey in search of my grandfather's boyhood home was described earlier.

In February of 1946 Jack Craven and I made a trip to Oak Ridge to see the location chosen for our new laboratory facilities at K-25 and to engage in some advance planning. After passing through security at the Oak Ridge City perimeter fence, we found a rather spartan community that resembled a military base. Indeed, it had been constructed under the auspices of the Army Corps of Engineers. In being shown about the residential community, we noted the beauty of the setting and the attention that had been given to preserving trees and orienting houses to provide for the best views through picture windows. The multitude of shopping centers located within convenient walking distances of the homes was further evidence of the excellent war-time planning of this new city. The houses were of simple but practical construction.

It is to the credit of the Corps and other leadership that an effective effort had been made to provide attractive surroundings to retain the very talented people who were essential to this high-technology enterprise. We were spared the legendary mud and dust of Oak Ridge's construction phase, since most streets had been paved some months prior to our visit. Still, there were no street curbs, and pedestrians were accommodated either by boardwalks that provided diagonal shortcuts to stores or by street sidewalks of packed cinders held in place by boards at either side.

Our new laboratory was to be located in one corner of a large industrial-type structure of twenty acres on one floor called the Conditioning Building, K-1401. The building had been constructed to house operations for preparing equipment prior to installation in the K-25 plant and for a large machine shop and other services. We were to share the area where internal surfaces of plant equipment were treated to stabilize the metal against the highly corrosive uranium hexafluoride gas. It would be necessary to install laboratory rooms, offices, and other logistical facilities. The task was much like that of converting the Nash Building garage to a laboratory, except that this building had been constructed for technical uses and included major utility services and overhead cranes. Much urgent planning would be required because the initial installation of laboratory facilities was scheduled to occur in about one month.

Secrecy prevailed, and no reasons were given for continuing the work at the proposed urgent pace after the war had been won. However, several incentives seemed probable for continuation of a priority status. When the gaseous diffusion plant reached full operation, it proved more economical for producing enriched uranium than the labor-intensive process of the electromagnetic plant, Y-12. This was contrary to previous expectations, indicated by the more-permanent-type buildings of Y-12. In addition, our testing showed that barrier design and quality were contin-ually improving; thus, it might be economical to replace some barrier already in the plant. Furthermore, because Soviet adventurism in Eastern Europe was of increasing concern, more weapons-grade uranium might be requested. I continued to hope that non-military nuclear power would be developed that would need much enriched uranium.

As we returned to New York, Jack and I were quite excited about the prospect of these new developments. We even liked the thought of living in the unusual and interesting town of Oak Ridge. I also found pleasure in developing a close acquaintance with Jack Craven. He had taken over the test group (in which I had started) when Bob Lagemann returned to Vanderbilt University. The flight to New York, in contrast to the gray and rainy days in Oak Ridge, was on a beautiful, sunny day, and Jack's disser-tation on the geology of East Tennessee contributed to my appreciation of the interesting and beautiful features of the Tennessee River Valley. As we came to know Jack better in later years, we discovered him as versed not only in physics, his primary field, but in geology, astronomy, ornithology, and music. In the new Oak Ridge organization, I was to report to Jack as he assumed responsibility for both my group and that of Dixon Callihan, who was to have a new role.

The remaining weeks of work in New York were busy as we continued to test the new barriers, to plan for dismantling and shipping the equipment deemed suitable for the new laboratories, and to consider improvements in the test machines. We were also busy with restructuring the staff to fill the gaps as, one by one, colleagues opted to find new employment or schooling. Jack Hunter and Larry O'Rourke had moved to Decatur, Illinois, where they assisted with barrier testing using the machines Jack had designed.

At the apartment, Elaine and I studied the drawings of the promised Oak Ridge "B" house (one of five designations indicating house size) and, because Oak Ridge offered few options for shopping beyond the essentials of life, purchased a few pieces of furniture to be moved there. Finally, the date for moving was set for Saturday, March 22. We arranged for movers to pick up our few possessions on Friday and obtained flight reservations for Saturday morning. Hank was to drive us to LaGuardia Airport; he would move to Oak Ridge a bit later.

On Monday afternoon of our New York departure week, Al Tenney, who was in charge of transfers to Oak Ridge, telephoned to inform me that plans had been accelerated and I was to be in Oak Ridge for work on Thursday. Protests that there was more to be done in New York than was possible in one day were to no avail; it was imperative that I be in Oak Ridge as requested. Tuesday was incredibly busy with arranging for others to do uncompleted tasks, closing bank accounts, settling with the landlady and arranging transportation. We could not get two tickets for any single flight, so I was to fly to Knoxville on Wednesday morning, and Elaine would follow that evening. She needed additional time to separate our possessions from the furniture furnished by the apartment and to provide more detailed instructions to the movers. She also had a few minutes available to bid friends good-bye. I was obliged to forgo that pleasure.

The weather was clear as the plane left LaGuardia Airport that morning, and I was thrilled to see Manhattan Island from the air. Familiar landmarks were identified as each came into view: the Morningside Residence Club facing the Park and Harlem, Columbia University, and Riverside Church, all beautifully illuminated by the morning sun. We crossed the Hudson River while the DC-3 aircraft still strained for altitude. My thoughts were many. Life in New York had been exciting and rewarding beyond anything I had imagined. I had come to love the city with its grandeur, beautiful parks, and cultural offerings that I found difficult to leave. Even so, except for the sad fact that Elaine was not sharing with me that aerial view of the city, I felt not the slightest remorse in the departure. Soon, I was wondering what the next day would offer.

K-25 (U-shaped building in the background), the massive gaseous diffusion uranium isotope separation plant as completed in 1946 and within what is now the East Tennessee Technology Park. The building is seven stories high with each side half a mile long. Note hutments and trailer housing for construction workers in lower right corner. Approximately 10,000 of the 75,000 Oak Ridge residents lived in this "happy valley" housing. Photograph courtesy of the American Museum of Science and Energy.

SEVEN

Uranium Isotope Separation in Oak Ridge

———·—◦—·———

My first evening alone in Oak Ridge was hectic. With a population of about thirty-five thousand people, this was a strange town full of unusual street paterns, utilitarian buildings, few telephones, a government taxi service, and a formidable surrounding fence. The taxi from Knoxville's airport had taken me to the Rutherford Hotel in Oak Ridge to register for our assigned lodging, until arrangements could be made for a house. The term *hotel* was a misnomer. The amenities of this barracks-type dormitory were a registration desk, maid service, and one pay phone near the entrance. I then rode through the city on a public bus to find "Townsite," and the Central Cafeteria for dinner. The army headquarters, locally called the "Castle," where I could arrange for a taxi so I could pick up Elaine, was on a low hill above Townsite. Although a car and driver could be ordered for her alone, I wanted to meet Elaine on arrival and attempt to soften the shock of entering through a security gate to see this unusual city where we would live and work.

In asking the way to the taxi desk, I met by chance Bob Carter who had been my contact at the Houdaille-Hershey barrier plant. He had arrived earlier with a mission to assist in the transfer of barrier manufacturing to Oak Ridge. Bob and his friend Julie, later his wife, were to become longtime friends. I accepted Bob's offer to provide the airport transportation. He was among the very few people in Oak Ridge who had a private automobile. Elaine arrived tired but excited. We reviewed our diverse activities of the day, trying not to bore Bob and Julie. After passing through the security center at the Solway Gate, we arrived at the Rutherford exhausted; sleep was sound despite the hotel's inconveniences.

Thursday morning I left Elaine to rest and explore as I traveled by work bus to K-25, excited and eager to start work on the urgent assignment. Jack Craven, having come earlier, was there to greet me, and we hurried to meet Al Tenney. We found him in one of the many corridors of the rambling administration building, K-1000. Al greeted us warmly but soon reported on new orders from Washington to place the barrier laboratory planning on hold for a few days. We would be assigned office space in which to wait and perhaps to contemplate the new equipment design, but we were not to develop specific plans. Having worked for more than three years under the pressures and efficient management of the wartime project, we were unprepared for this seemingly unnecessary bureaucratic pause.

Perhaps the delay should not have been surprising. We were quite aware of the national and international concerns impacting national program planning: Were nuclear materials to have civilian or military control? What scale and urgency was prudent for continued development of weapons? Would those requirements expand because of adventurism by the USSR in Eastern Europe? Only two weeks had passed since Winston Churchill had introduced the term the "Iron Curtain" in his famous speech at Fulton, Missouri. Under the prevailing secrecy, we were not to know the reason for delay. Our role was to be on standby. The "pause," through extensions announced from day to day and later week to week, continued through the six-month period for which I had agreed to work. If the extent of delay had been known at the outset, we and several others might not have stayed long enough to appreciate fully the amenities of Oak Ridge, the beautiful East Tennessee countryside or the challenging work that would follow.

Waiting was frustrating. Nevertheless, we used the initial working days effectively to describe the test equipment and techniques developed but not formally documented during the frenzied war effort. We had settled into a long, narrow room of K-1000 with sash-type windows affording a view across the bus-parking lot to a wooded hill. Each person had a desk, slide rule or mechanical calculator, and access to a lockable file for protecting classified documents. Security for technical information was to continue at the wartime level, but identification of the project and employees was no longer restricted.

Those employees included Dorothy Weissberg, soon joined by her husband, Hal, who had been recently released from the military. Antony Grefig, Bill Kirby, Libby Johnson, Jim O'Brien, Dick Snyder, and others from Jack Craven's SAM group were there. Libby's husband, Ned Johnson, came later. My group had dissolved, except for those left to dispose of equipment and supplies not needed in Oak Ridge and to assist in closing the Nash Building. George Toal then moved to Oak Ridge. Bill Tewes and Roger DeLoor had come earlier but were temporarily engaged in other assignments. Some men in uniform did not transfer to Oak Ridge; rather, they chose other military roles or discharge.

Each evening, Elaine and I used the excellent bus system to explore the hillside streets in this city of contrasts. The four-lane Oak Ridge Turnpike with its traffic lights was much like any major city street, except for its paucity of traffic. Few employees owned cars but, instead, relied on crowded buses at shift-change times. Tennessee Avenue and Outer Drive respectively ran windingly parallel to the Turnpike near the base and top of 300-foot Black Oak Ridge. The city street grid was formed by other avenues that connected Outer Drive with the lower elevation streets. Labeled alphabetically by state names from Alabama to Pennsylvania from east to west, the avenues followed contours to accommodate side streets. These also were named alphabetically as the ridge ascended and each name began with the same first letter as its avenue.

Neat little houses, all owned by the Army, were nestled along the wooded streets with most separated from neighbors or the next street or avenue by naturally green areas. The houses were labeled A, B, C, D, and F according to size. "E" was reserved for buildings housing one and two-bedroom apartments. This heart of the original city was planned before the project magnitude was fully envisioned. The setting was a beautiful hillside on which the native forest trees had been preserved to the extent practicable between streets, with wider green-belts spanning ravines. Project leaders had insisted on an attractive city center to help retain highly capable technical staff who could easily find other wartime work.

When viewed from the bus on Tennessee Avenue, stores of the main shopping center, called Townsite, formally named Jackson Square, were aligned, oddly enough, with their service entrances facing the street, and we would learn why when winter came. The Square was attractive, with access to stores along covered sidewalks, although the shops seemed plain and temporary. The city with its "permanent" houses was said to have been designed for a five-year expected lifetime. However, the basic house construction of concrete footings, seasoned timber, oak floors, and walls of asbestos cement boards was sound. (The black iron piping, unusual roofing, and limited wiring often required replacement after one or more decades, but the foundations and basic structures still serve well more than a half century later.) There were few lawns and little landscaping of these rental units. We viewed houses with special interest; the project's promise of a 960-square-foot "B" house as an inducement for our move to Oak Ridge had been negated, and we were disappointed that new housing rules permitted only a smaller "A" house with 768 square feet of floor space. Houses of varying size were awarded according to family size.

The city had been built in segments to accommodate the ever-increasing project work force. Thus, housing types retrograded from the simple, but attractive, cement board units for professional workers and managers to less permanent structures. Prefabricated houses, called flat tops, set on posts were interspersed with Cemesto units, as space permitted. Others were clustered in ridge-top locations that provided spectacular views of the Cumberland Mountains about twelve miles to the west and north. There were many apartment units of different designs for families, and singles had rooms in dormitories patterned after army barracks. Camping-type trailers with accompanying wash houses for toilets and showers accommodated construction workers. Many of these had been removed before we arrived because that phase of the project phase had been completed, leaving many abandoned streets.

"Hutments," provided for black families and singles in this segregated city, were the most distressing to observe. These square structures with a single door and no windows had screened openings without glass and only wooden closures for privacy. Heat was provided by a coal stove in the center of the single room otherwise fitted with rudimentary furniture for sleeping, eating, and relaxing.

Toilet and washing facilities were like those of the trailer camps. Some white men of the Special Engineering Detachment also were temporarily placed in hutments: the only racial integration in the housing of early Oak Ridge. Many of the most temporary structures had been moved away before we arrived, but remnants (including the hutments) remained as reminders of the Oak Ridge wartime stories we had heard in New York.

Sunday morning brought a surprise. We had expected to worship in army-type chapels seen from the buses, but the Methodist Church services were held in the Ridge Theater of Townsite. Entering under a marquee with a movie billing such as "Chicken Every Sunday" hardly inspired religious thoughts. We had left the resplendent, Gothic nave of Riverside Church with Harry Emerson Fosdick in the pulpit on one Sunday, to encounter on the next a windowless movie theater of cinder-block walls. Arriving early, we found the ushers busily removing hymnals from apple crates packed at the close of the previous Sunday's service. Each parishioner received a book as a ritual of the usher's greeting. We edged our way from the aisle to theater seats, gingerly stepping on a floor sticky with discarded chewing gum. However, Pastor Bob Lundy, who would later become a bishop, was a talented young man, and the choir was good, though not of the Riverside's professional level. As we left the "church," the popcorn machines were ready for theater patrons of the early afternoon. Later, we climbed a ridge near the Jefferson shopping area and the Rutherford Hotel. I had hoped to show Elaine one of the plant areas, but my geography was wrong. Instead, we gained a view of the hutments of black laborers.

We also wanted to visit friends, for even as a newcomer I knew more people in Oak Ridge than in New York and Lincoln combined. They had worked and trained in the SAM Laboratories before transferring, but without easy access to a telephone at the Rutherford, and few phones in the community anyway, we had no convenient way to make contact. Thus, when I was at work, Elaine was alone with two choices: to endure the unpleasant lodging at the Rutherford or to travel around town by bus. On Monday, when our housing assignment was also temporarily deferred, she flew to Greensboro to spend two weeks with her parents.

That next week in Oak Ridge brought extended frustration. The days at work were confusing, but not beyond expectations for a new environment. However, nights in the Rutherford became intolerable. There was no sound insulation, and space below the doors accommodated rodents and cockroaches that scurried about with impunity. Human occupants arrived from various shifts and engaged in noisy activities, including entertainment of "ladies of the evening." Thursday night was so bad that it nearly precluded sleep. On Friday morning I packed my belongings in the green metal suitcase I had earlier carried to NWU and New York and checked out of the hotel. At the office of Tom Lane, head of personnel, I deposited the suitcase for the day and rashly informed him that I demanded better housing that day or he and the project would not see me again. The gamble was that they needed me, for I had no other specific plan. As the day progressed into afternoon, the future seemed a bit cloudy, but about 2:00 P.M. I was invited to see an "A" house. It was not nestled in the woods as were many, but had a bit of lawn seeded by the Army colonel who had lived there during the war. Most important, it had a telephone! Even knowing that my job probably did not qualify for that luxury, the possibility of retaining the telephone was persuasive in accepting the house at 328 East Fairview Road. We were to be among the very few with the convenience of telephone communication.

The house was to be available in a week, so the Magnusons accommodated me in their small "N" apartment, allowing me to sleep on a couch much as I had imposed on Dale upon arrival in New York. On Friday evening, with the house key pocketed, I flew to Greensboro to spend Saturday with Elaine and her family on the farm. Mr. Causey loaned us his pickup truck into which we loaded a bed, its spring and mattress set, linens, a few dishes, four breakfast-room chairs, a card table, and miscellaneous kitchen items to sustain life until our shipment would arrive from New York.

We carefully drove the truck over the Blue Ridge Mountains and through Knoxville. The countryside was much as Professor Jensen had described, sparsely populated beside poor roads, but it was beautiful on that pleasant spring day with the roadsides adorned by redbud, shadblow, and other flowering trees. In Knoxville, the pavement of Magnolia Avenue was so rough that I kept the pickup tires on the abandoned trolley tracks to avoid damage to our precious acquisitions carefully packed and covered in the back. On arrival at our new home, a next door neighbor, Ed Granek, immediately volunteered to help carry items into the house. In the harsh reality of the next morning's daylight, Elaine was disappointed that the house location was not as she had hoped, but that day initiated the seven delightful years we would spend living there.

Housekeeping then was rudimentary. Because the bedstead needed minor repair, we slept with the bedding on the floor. Our kitchen utensils and dishes were not to arrive for many days, but Elaine resourcefully cooked meat in a small iron skillet and vegetables in a large aluminum pitcher brought from the Causey home. With our paucity of furniture, even that small house had much open space. Despite the simplicity we continued our tradition of entertaining friends. If we expected more than two, they provided chairs and we transported both people and seating to our house via the truck. In due course shipments arrived from New York and from my family, who sent a new dinette set, an antique chestnut library table, a steel cot with folding sides that could be lifted and locked to make a double bed, and other pieces, including a rocking chair that I had refinished some years earlier.

The bed was a hazard; if you sat on an outer edge of the unoccupied bed, it would tip over and deposit you on the floor. We converted the cot into a sofa by altering one folding side to form a back support, cushioning the seat and back with thick horsehair pads scavenged from a scrap yard, and, with Elaine's skilled upholstering, making it attractive and almost comfortable. The needlework was facilitated by a new sewing machine purchased with Elaine's remaining graduate school funds. I later used a community workshop in the nearby TVA city of Norris to make and finish wooden arms for each end of the "new couch." These improvisations initiated a policy of making do with what we had until good quality merchandise could be purchased for the longer term. An early example was a fine sofa purchased in New York a few years later that still serves well.

The University of Tennessee offered graduate courses in Oak Ridge, but we arrived too late to register for the spring quarter. Elaine and I would study the new surroundings and enjoy an extended period of eight-hour work days. We explored the Tennessee countryside on weekends, usually with Dale and Doris Magnuson crammed into the truck seat with us and occasionally with others riding in the back on short trips. We also made longer excursions to Cumberland Falls in Kentucky and Fontana Dam and Village in Western North Carolina. We discovered that most roads away from the improvements made for Oak Ridge were in poor condition; this, indeed, was Appalachia. However, the TVA lakes and the beautiful Cumberland and Great Smoky Mountains were enchanting.

On one occasion the truck enabled us to traverse a fire-lane road to the top of Frozen Head Mountain for a spectacular view of the valley, including Oak Ridge cradled on three sides by the Clinch River. Smog from energy systems had not then seriously invaded the Tennessee Valley. The road had been so narrow that Dale sometimes walked backward in front of the truck for guidance because ledges cut in the mountainside were only inches wider than the wheel span. That area is now a Tennessee state park, but one must hike to the mountain top.

In the enviable position of having a pickup truck, we received many requests for transportation, particularly to obtain lumber available from wartime construction scrap. On one occasion, I accommodated Jack Craven and Al Tenney. The venture was successful with each having carefully placed and identified his "loot" in the truck bed. On the return I ran out of gas in a pleasant area near the ridge top. Fortunately, the location of my embarrassment left only a short walk to their homes. I sent them on ahead and told them I would join them after finding some gas. Left alone to deal with the truck, I saw that the road contour toward the community favored me; it was possible through maneuver and the absence of traffic to coast more than a mile to one of the city's few gasoline service stations. Jack and Al were pleasantly surprised when the scavenged lumber was delivered in ten minutes. After three weeks of this luxury, the truck was returned, and we reverted to riding buses. That was not difficult; the Oak Ridge bus system had continued service at near the wartime scale with quality that exceeded that of many major cities.

Elaine and I soon found that bus travel in areas outside the city contrasted negatively with that of Oak Ridge. The first occasion was for a visit with friends in Norris who had invited us to dinner. Near Clinton, our bus and a truck approached opposite ends of a narrow, aging bridge across the Clinch River. Space between rusted members of the supporting superstructure appeared inadequate to accommodate both vehicles. Undaunted, each driver proceeded, expecting the other to retreat. We met at the center of the river span, each vehicle contacting the other and touching a bridge support. The drivers exchanged verbal threats, increasing the tension of the moment. Finally, our driver shifted to a low-low gear and proceeded to pass with much crunching of rivets and scraping of paint on each side. We feared that the old bridge would give way, but the bus, scratched and dented, delivered us to Norris on time. After a delicious dinner and pleasant evening, we were pleased when George Gerhardt offered to drive us home. That fall Mr. Causey was able to purchase a 1946 model Chevrolet and gave us his 1942 Chevy, already driven 140,000

Our "A" house at 328 East Fairview Road, Oak Ridge, in August 1948.

miles. I soon joined a car pool with Roger DeLoor and Charley Allen, members of the Barrier Department, for more convenient transportation to K-25. As automobiles became more plentiful, the Oak Ridge bus system deteriorated toward eventual abandonment.

A young secretary, Mary Pemberton (now Strobbe), and I learned the techniques of dictation to expedite report writing. I was apprehensive about sounding like the amateur I was and suspect she may have been equally afraid of making mistakes. It was good to learn how valuable a secretary can be in extending one's capabilities. Our secretary in New York had been fully occupied in compiling, filing, and distributing data and typing weekly progress reports from handwritten copy. Work gradually shifted from reporting prior findings to evaluating data statistically for better options to improve barrier manufacture and testing.

Elaine used the summer months to decorate the house interior attractively, and we worked together preparing flower beds and planting a few shrubs, even though we could only rent the house. I made a small vegetable garden but quickly learned that East Tennessee red clay was not easily tilled. Its properties of sticky mud when wet and rock-like texture when dry require introducing heavy amounts of organic materials and fertilizer. As our friend John Hill is fond of saying, "The first job in East Tennessee gardening is to make dirt." We used seasoned sawdust from the great deposits created during city and plant construction as well as prior-year sludge from the city sewage treatment plant. The latter also provided a superb crop of volunteer tomatoes!

It was apparent that flower beds were more to our liking than the vegetable garden, although savings from two summers of vegetable gardening paid for some of the fertilizer and tools. We even established a fairly respectable lawn on the government-owned property after learning to cope with crabgrass. Some of the trees and shrubbery we planted in the 1940s still thrive at that address.

When authorization came through to plan the laboratory, Jack Craven and I reviewed drawings for the proposed space and refined designs for new equipment. Authorization for construction of the barrier-fabricating plant and the associated research laboratories finally came in September of 1946. The time commitment I had made for rebuilding the laboratory had nearly passed, but we were enjoying life in Oak Ridge, and it was stimulating to resume work under pressure. Our project's position afforded the assistance of Plant Engineering for the final design of equipment, and we could engage excellent shop services for the construction and installation of the new machines. Oak Ridge was favored with services comparable to those we had enjoyed in New York.

Our offices and laboratories would occupy part of a sizable enclosed area of building K-1401, which also housed furnaces to prepare equipment for installation in the diffusion plant. We thus learned the safety and health rules for work in that environment. Diffusion plant equipment and the inner surfaces of piping were treated with fluorine or fluorine trifluoride to resist the effects of the very corrosive uranium hexafluoride. These hazardous materials were familiar since small quantities were used in our laboratory equipment, but provisions such as area evacuation drills were now included in safety programs. Fortunately, such emergency actions were never required.

Research and test equipment in each laboratory room was arranged around a glass-enclosed office space so that the operators could monitor the rigs from a quiet area conducive to thought and analysis. The machines were either of improved designs or extensively remodeled units moved from New York. We were to explore the capabilities of advanced barrier materials thoroughly as they were developed, and to refine the machines used for testing all tubular barriers manufactured at the new plant. Thus we participated in developing features of the manufacturing process, which was managed by Hank Stoner and John Blair.

Although we had been authorized to proceed, those days continued to be a time of uncertainty. The future of Oak Ridge remained difficult to project because of the complexities in establishing national objectives. The massive world conflict was a year in the past, but the Cold War had become at least partially foreseeable—much as World War II had been in the thirties. The uncertainty itself made weapons-related programs seem likely to continue. The issue about civilian versus military control of nuclear weapon manufacture was resolved. The Atomic Energy Commission (AEC) was authorized but not then completely staffed, its functions not yet fully known. A few scientists and policy makers were urging peacetime application of this new energy source for electricity-generating plants. Our own family members and friends in Nebraska and North Carolina seemed impatient that nuclear power was not developing rapidly; the difficulties in building reliable and cost-effective nuclear power plants were still greatly underestimated. The future of Clinton Laboratories (now Oak Ridge National Laboratory), which had produced the first plutonium for research, was in jeopardy. The value of large, government-sponsored laboratories had been made clear by success of the Manhattan Project, but some officials favored a location in the Northeast. The long-term future of Oak Ridge indeed was unclear, but the gaseous diffusion process seemed assured as vital to both civilian and military programs. This provided a measure of confidence in the stability of Oak Ridge and our work.

In the fall of 1946 our lives reached an intensity resembling the war period, but with a much broader focus. The new testing laboratory was under construction. Elaine had taken a position as food service director for the Oak Ridge High School, and I registered for graduate courses at the University of Tennessee. We became active members of the First Methodist Church of Oak Ridge despite the theater environment and enjoyed entertaining friends in our home. With winter approaching, we observed the reason for Oak Ridge houses and businesses' being positioned with service entrances and coal bins on the street side for convenience in deliveries. With no lock on a small exterior door, coal delivery was made at the supplier's convenience; a lockable, ill-fitting inner door provided security against unwanted entry. Evidence of delivery was a drift of coal dust across the utility room floor, but if we had neglected closing that door, the drift extended throughout the kitchen.

With the garbage cans also exposed to view, this "back door" orientation made the house somewhat unattractive. To conceal the offending containers, I visited the scrap lumber yard and found square posts and pallets that resembled units of picket fence to form an enclosure. As it was nearing completion, Ed Granek stopped on his way home from work to suggest that I should have a building permit for the addition. I told him that it was only a very small project and chose not to bother with such red tape. He rather insisted that I should, but after some discussion of the issue went on to his home. The next evening as I was painting the fence, Ed appeared with a proper permit fully completed for me to sign. At the top of the bureaucratic form was the printed name EDWARD GRANEK, DIRECTOR OF HOUSING. I signed the papers.

Life under this partially socialistic system was economical and interesting but often frustrating. The housing landlord, the Roane-Anderson Company, managed all city functions, first under the Army, then with AEC direction. It was great paying only thirty-eight dollars per month for rent with water, electricity, and coal included. But smoke from burning bituminous coal was so invasive that cleaning interior walls became a spring ritual, even if the furnace had not been used. Much neighborhood discussion focused on the problems and advantages of different wall-washing techniques and cleaning agents. A compound resembling modeling clay would remove and retain most of the deposited soot. The walls and ceiling then could be washed with less difficulty. Unfortunately, about three washings also effectively removed most of the paint, which, in turn, brought on a new problem—to persuade the housing department to repaint the cleaned interior. Paint colors were limited, but any fresh surface was welcome.

Exterior painting was determined bureaucratically. Wood house-trim that also held the wall panels of Cemesto board in place was caulked and painted in accord with a city-wide schedule. I can recall only gray, green, blue, and brown as the available colors, although the white trim originally used for all houses was retained for some. All porch floors were painted gray, and paint crews were narrowly focused, presumably for efficiency, specializing even in front porches or back porches. Thus, it was possible for one crew to paint the front porch and another the back porch on the same day. This actually happened to us; we returned home one evening to find wet, slow drying, oil paint on both porch floors, virtually barring us from entry. Hastily built bridges of wooden blocks and boards of scrap lumber made passage possible but quite hazardous.

Our "A" house heating system was especially difficult to control. Because no furnaces of proper size for our small house were available during that early phase of Oak Ridge housing construction, a furnace suitable for the larger D and F size houses had been installed. It crowded the utility room and proved impossible to regulate; the house was either too hot or too cold. We tried banking the fire, but that often caused combustible gases to accumulate and explode, blowing open the furnace doors to spew out soot and cinders. We finally abandoned the furnace and bought an electric heater for the bathroom, opened the oven door to heat the kitchen, and shivered readying for work on cold mornings. Although we did not pay for the electric power, the amount we could use for heaters was restricted by the limited house wiring and fuse sizes. The cost to the city of providing unmetered electricity was remarkably small under the low rates of TVA. Use of the living room fireplace on cold evenings provided my first experience with the warmth and charm of an open flame.

Although happy with both work and life in Oak Ridge, the bureaucratic system controlled many activities in a way I found absurdly intrusive. For instance, special passes were required to admit friends and family from other places. We wore badges at all times, and I often felt more constrained than we really were. Nevertheless, I have to admit that the system had advantages beyond the economic. The house and property were secure and protected, so that we never locked the exterior doors. Even the car keys, when not in use, were conveniently placed on a shelf inside the back door. Milk was delivered by a fine person from the Norris Dairy who quickly learned the amount normally used, brought full bottles all the way to the refrigerator, and picked up empties left on the porch. In the first years in Oak Ridge, I carried keys only when we locked the house, perhaps unnecessarily, while away on vacation.

The Oak Ridge community was turbulent in 1947. Uranium isotope separation would continue, but its expansion was only conjecture. The Materials Test Reactor, conceived and designed in Oak Ridge, was to be constructed at a new nuclear reservation in Idaho. Transfer of nuclear reactor development to Argonne, the first site to be called a National Laboratory, seemed even more threatening. The Clinton Laboratories' status remained uncertain throughout the year, and in late December transfer of its operation from the University of Chicago to Union Carbide was also of concern to the staff. Most scientists seemed more accustomed to university environments than to those of industry.

The Y-12 Electromagnetic Separation Plant had been shut down, and the city population declined accordingly. Most of the decrease from a wartime population peak of 75,000 to about 30,000 had occurred before we arrived. The construction forces had moved elsewhere leaving large areas of vacant housing. January 1948 brought new hope as Clinton Laboratories also gained status as the Oak Ridge National Laboratory (ORNL), and it became apparent that some nuclear reactor work as well as nuclear fuel reprocessing development would continue. This chemical separation of plutonium from irradiated uranium-reactor fuel had been the original mission of Clinton Laboratories, where the X-10 graphite moderated reactor had been built to produce plutonium in gram-size quantities for this purpose.

Uncertainty had also somewhat clouded the gaseous diffusion program, but only concerning its ultimate size since preliminary plans for expansion were being considered. If additional and larger plants were to be built, it seemed that even our new and improved testing equipment would be inadequate. I proposed with others, including the late Dick Korsemeyer, that a pilot plant would be needed. This seemed important, both to evaluate new barrier properly and to optimize operating conditions, such as temperature and pressure, for the production plants. The counter argument by management was that pilot plants were costly and K-25 had been built successfully with only very small pilot units. These discussions continued, but a decision was long in coming.

Even though civilian control of nuclear energy had been assured by creation of the AEC, it was difficult to see a bright, long-term future in Oak Ridge. We had come to love the city, its remarkable people, and its setting in East Tennessee, but we wondered from time to time where our future should lie. Elaine's father was making changes in his business plans and invited me to join him to build and operate a new commercial cold-storage locker facility to couple with his wholesale meat business. That work could not rival interest in nuclear development, but because I had once envisioned a later career in commerce or industry, the opportunity seemed appropriate for evaluation.

I sought counsel from the University of Tennessee personal consulting staff and took a series of tests designed to help in making difficult career decisions. Those tests, as interpreted by the university, were hardly conclusive, indicating only that a business enterprise might be desirable. I discussed the choices with Jack Craven, who thought I should stay with technical work. Additional interviews with the psychologist, who was also considering a career change, produced further uncertainty. After several sessions I began to feel that our discussions were more about his problem than mine. Elaine and I struggled over this decision for weeks.

Before we could reach a conclusion, Jack Craven announced his pending move to the University of Tennessee as the director of an expanded research facility for improving cotton fiber textiles. Jack invited me to join him at the university, but he further proposed that I might be considered for his position, if I were not leaving K-25. I replied immediately, "If that promotion is possible, I have absolutely no intention to leave." I got the job as head of the Barrier Testing Department. With the diffusion plant expansion also having been assured, this was a good opportunity and a heavy responsibility. As a new manager I was helped greatly by the experience and cooperation of the department staff. I approached my new responsibilities recalling how I had wisely relied on experienced help in the farm harvest of 1937. In this case, however, some of the largest production plants ever built would be significantly dependent on our services.

Jack Craven had assembled a great group of people doing interesting and important work. The assistance of an experienced secretary, Betty Jeffers, was of inestimable value. The department consisted of my former group, placed under Bill Kirby, and three more: those testing barrier with uranium hexafluoride, a theory group led by David Justice, and a service unit. Department personnel enjoyed a good rapport, perhaps derived from Jack's relaxed but effective management style, and we augmented that work rapport with weekend social events. We visited picturesque Tennessee state parks in summer, transport provided by those having automobiles, and occasionally met socially in a staff member's home—usually a "D" house. I first reported to Dick Wiswall, who also had worked in New York, later to Dixon Callihan, my boss shortly before leaving SAM, and then to Clifford Beck as the plant expanded and individuals moved to new roles.

Elaine's position at the Oak Ridge High School was rewarding both to her and the school. Although only a few years older than the seniors, she maintained strict order and used her food and nutrition training to provide well-balanced diets, attractively served to students and faculty. The school cafeteria became an inviting place for lunch; both Atomic Energy Commission employees and merchants of the Jackson Square area frequently chose to eat there. They paid somewhat more than students, thus augmenting cafeteria finances. I sometimes teased Elaine, saying that she had balanced the budget by

Ernest Jones, Dixon Callihan, Hank and Helen McKown, Elaine Trauger, Jack Craven, Alva Callihan, and Frances Jones on Myrtle Point, Mount LeConte, Great Smoky Mountains National Park, circa 1947.

slicing ham so thin that the plate's pattern could be enjoyed while eating. The abundance of servings, however, satisfied ravenous appetites of teenage customers, even members of the football team.

On rare occasions when I had official business in town, I too had lunch at the school. Attendance policies for K-25 plant workers had not been well defined but were illustrated following one of my visits to the school cafeteria. Upon return to the plant I was confronted by a central management representative following through on a complaint. Apparently, someone from the AEC, who also had eaten at the high school, had reported that I was not at work, even though I routinely informed the Laboratory Division Office of my location and plans for lunch to assure availability when away from the plant. As AEC members became more engaged in important agency duties, they no longer had time for such trivia.

Shortly after achieving some confidence in my role in the Barrier Testing Department, another change occurred. We were to receive less funding with inadequate support for the staff. Two members, Carl Bauman and Radford Carroll, easily transferred to X-10, and two others, Arnold Lansche and Jack Feyerbacher, left Oak Ridge. In addition, it fell to me to release Libby Johnson because of a hard-and-fast rule of upper management. Because her husband, Ned, was also employed in the department, the rules required that she leave. In those days before affirmative action, we lost a valuable employee. Fortunately, Dixon Callihan had formed a criticality group to ensure, by testing, that the new enrichment plant facilities would be fully safe from an accidental nuclear event. Libby applied with Callihan and soon joined that group for an extended and successful career.

From the low period of 1947 our ranks grew slowly as the new plant expansion activities unfolded. Others who continued to support or were added to the research effort included Bob Ackley, Jerry Klobe, C. R. Lay, Loren Lund, Jim O'Brien, Larry O'Rourke, Danny Palmer, Dick Quinn, Alice Rohr, Bob Siebert, Jack Thompson, and Herbert Trammell. The support group was managed by Brooks Bowen. It was an effective staff of engineers and technicians. They contributed steadily to barrier improvement, in cooperation with the development group, also located nearby inside building K-1401.

The barrier plant soon began production and required a heavy testing load as new equipment was employed and processes improved. Soon after startup, this plant was also to apply new findings from research; the project's interest in and need for barrier testing and evaluation were increasing. Providing close coordination with the barrier plant was much easier after the whole operation was moved to Oak Ridge, where only a short walk was required to meet with their staff and plan the manufacturing steps and testing requirements. That contrasted with the long distance telephone discussions between New York and Decatur with language structured to maintain wartime secrecy. In Oak Ridge we could select samples for our studies instead of having to use whatever materials the Decatur plant chose to send. Similarly, the gaseous diffusion separation plant staff was on hand to identify test needs and to suggest new features to be included in research. Thus, we worked closely with both the plant and the nearby research group that was developing improved barrier under Sam Flanders's guidance.

Newton Underwood came to work at K-25 in 1948. I had known of him as designer of the "Underwood Flapper" and was impressed by reports of his early role in the development and testing of barrier. Underwood was not amused by the irreverent name "flapper" for his invention, but he was as innovative in this new role as he had been earlier. Although Underwood was in barrier development rather than testing, we found common interests and worked well together on an informal basis. Underwood devised and conducted experiments to improve the manufacture of barrier. This development was resisted by management both in the barrier plant and in K-25 because it appeared difficult to implement. I was impressed by Underwood's slow and careful persistence in the work until his technique was adopted and proved successful.

The need for a pilot plant was finally recognized, and a twenty-stage unit was built and managed by John Michel. Its program was appropriately and closely aligned with needs for the barrier plant and the new cascade plant designs. Although costly, the pilot plant provided a more precise measure of each new development or discovery in barrier design, manufacture, and operation than was possible with our testing methods. This small pilot plant proved so effective that a thirty-stage pilot unit was built later with larger equipment that could better represent cascade operation. The need for precise measurement and effective modeling arose from the cost for operation of the three large installations finally built at Oak Ridge, Paducah, Kentucky, and Portsmouth, Ohio. They required 6,000,000 kilowatts of electrical energy combined. An improvement in performance of only 1 percent represented 60,000 kW(e). Consider the importance: even such small improvements at 5 cents per kilowatt hour could save $3,000 per hour or up to $26,280,000 per year.

With the pilot plant providing a more precise measure of barrier performance, we reduced the number of measurements for statistical determinations of separation efficiency in order to concentrate on the temperature and pressure dependence of barrier performance. An important objective was to improve the theory of diffusion separation for gases for a wide range of operating conditions. The small theory group under David Justice included Murray Lyon, John Keyes, Mario Goglia of the Georgia Institute of Technology, part time, and George Somers. Bill Pollard occasionally visited as a consultant. These studies served to guide the more costly and definitive pilot plant tests as well as our experimental work. We also studied the behavior of other porous materials such as zeolites and filters.

After two years of managing the high school cafeteria, Elaine resigned to start our family. Our savings fund allowed us to purchase a Baldwin Acrosonic piano so that she could resume practice, abandoned since high school. She remained active in her field by writing a weekly food column of recipes, cooking advice, and household tips, until Thursday's *Oak Ridge Journal* was replaced by the daily the *Oak Ridger*. Our first son, Byron, was born in 1949. Elaine was then well known in the community for her work with the schools, church programs, and newspaper articles. On the day she and Byron arrived home from the hospital, twenty-six people came to extend congratulations and to visit. At the end of the day, both mother and child were exhausted.

A few days later, the editors of the company paper, the *Carbide Courier*, took note of the event. The article read, "Born to Mr. and Mrs. Don Trauger, a son Byron, August 20, 1949," which was immediately followed by a safety slogan filler: "If you need to know the safe way, ask your supervisor."

The K-25 Laboratory consisted of a large chemistry department, a metallurgy group, a mass spectrometry group, a physics department under Gus Cameron, and the barrier testing and development departments. This was under the dual management of Clifford Beck and Frank Hurd, both formerly at SAM Laboratories. Because of the prominence and sometimes crossed purposes of both men, the staff dubbed the operation the "Beckhurd" laboratory. It later functioned fairly well with Beck taking primary interest in the physical programs and Hurd in the chemical. Cliff Beck soon moved to North Carolina State College, now North Carolina State University at Raleigh, to set up a nuclear engineering department, the first in the country. He offered me a position helping to plan the new reactor facility while pursuing a related graduate program, eventually to be in charge of the reactor operation. This possibility for work in North Carolina was attractive, particularly since the reactor was designed for research. However, because the project did not seem to lead to nuclear energy production, and because the barrier development of the time was particularly exciting, I declined.

As the designs of the new and much larger separation plants progressed, the importance of even small improvements in barrier performance became more fully understood. The huge TVA grid would eventually be stressed and expanded as would other power suppliers at Paducah, Kentucky, and Portsmouth, Ohio. Some of this foresight came from the realistic test conditions and increased precision of the pilot plant measurements, guided by laboratory findings. As a result, plant management concluded that more effort in barrier development was desirable, and their confidence in the expansion effort was also stimulated by the innovative and successful work of Newton Underwood. Charles Mahoney, a senior but newly employed metallurgist, was put in charge of barrier development. Mahoney was a personable and ambitious man, and after some months he persuaded management to combine barrier development and testing into one large department. He assumed the position of department head in late 1950, and I was to assist Frank Hurd, then the sole laboratory director. The consolidation did offer an increase in efficiency, but evaluation more independent of development has its own merits as well. Regardless, an impressive overall improvement in barrier production and performance was made over the period 1946 through 1953, the period when most of the new separation-plant expansion needs had been met. Most of the basic advances can be credited to the laboratory programs and pilot plant. Those carefully refined tools were innovatively applied with excellence to manufacture barrier and in effective design, management, and operation of the cascades.

Frank Hurd was a man from whose creative mind ideas issued rapid fire; most had merit. As technical assistant, I accumulated, on average, about one new assignment each day. Many would require several days or even weeks to research and report, so that the work soon was overwhelming. When first swamped, I went to my former staff and others for help. Although the response was always cooperative, that avenue offered limited support. After that I more carefully documented the date, content, estimated effort, and required completion time for each assigned task, then chose to do whatever seemed most interesting or potentially important. Hurd seldom asked for the finished work, and on several occasions when I proudly presented a completed product, I was forced to show Hurd my records to prove that he had requested the work. Most assignments were never mentioned again. This role became untenable, and after a few months I transferred to the Physics Department under Gus Cameron.

Gus was an altogether different boss who suggested searching for tasks of interest to the K-25 project, presenting him with proposals, and then mutually selecting an assignment. I was pleased to return to experimental work and first chose to study the concentration gradients that were established on the barrier's high pressure side by the separation process. This followed from my first independent studies with Sid Visner's equipment at Columbia. A directly coupled mass spectrometer was used for the analysis of tiny streams sampled through fine probes that caused a minimal disturbance of the gas mixture. Although the gradients were measurable, the information did not seem to add much, if anything, to the existing theory. In the second study I chose to investigate separating isotopes with a Hilsch tube, named for its inventor. In this device, a large, high-velocity gas flow is caused to swirl, creating a centrifugal force. The process was to function much as a mechanical centrifuge, which, through its spinning action, thrusts the heavier component of a mixture preferentially toward the outer wall of the centrifuge, causing the lighter one to concentrate toward the center. Small separations were observed, but the energy requirements were too high for further pursuit of this device. Incentive for the task had come from reports by intelligence sources that South Africa used a similar system in a uranium isotope separation program.

It was during this period that our second son, Thomas, was born in May 1951. The company newspaper offered no further advice in reporting the event, but we received sage family counsel from Newton and Hazel Underwood that we have treasured and shared extensively. We have always referred to these two guiding pieces of wisdom as the "Underwood Principles." (1) Parents must so often say "no" to a child that whenever possible they should look for opportunities to say "yes" to a reasonable request. Nevertheless, the more frequent "no" should be very certain, almost irrefutable, to be reversed only in the event of a serious error. Even then, the reversal should be accompanied by an understandable explanation. The important feature is consistency. If children learn what to expect, they will respond accordingly, for they naturally wish to please parents. If reliance on parents is established while young, it can continue on a diminishing scale, even through the teenage period. (2) Parents should never allow a child to play with something unsuitable as a toy. Matters of safety are too obvious for explanation, but, in addition, children should learn to respect the property of others, especially that of adults. (I was reminded of my mother's problem with her sewing machine attachments.) For example, a young child cannot distinguish between an outdated magazine appropriate for tearing apart and one of value. Respect must be instilled for magazines and books as objects for pleasant reading. We later developed another principle of our own: At times the needs or interests of any family member may require priority over the those of others. A child's need for parental attendance at a school program might be more important than the need for end-of-day relaxation. Later, use of the automobile for an important date might take precedence. Often, a mother's or father's need for quiet rest or for doing important work or the entire family's need to accomplish a common task might claim priority over a child's demand for individual attention.

These simple principles served us well. We missed the Underwoods when Newton decided to pursue biological studies associated with the new Nuclear Engineering Department established by Clifford Beck. He was to develop a program for studies of radiation effects on living organisms, based on his considerable background in biology through previous work at Vanderbilt University.

In the summer of 1952 we drove west to meet Elaine's parents in Wyoming for a vacation. After a brief stop with my family in Nebraska, where the boys would enjoy an extended visit with grandparents, the mission was to visit two major ranches that bred and marketed registered Hereford cattle. These were the WHR near Cheyenne and that of Colorado's Governor Thornton at Gunnison. Mr. Causey's interest was in bloodlines relevant to his sideline of breeding registered Herefords on their fine North Carolina farm; ours was primarily in scenery, but my farm background with cattle was not forgotten. We had a pleasant drive through the beautiful Rockies with a stop at the picturesque town of Fairplay, located at some 10,000 feet elevation. At the Thornton ranch the governor happened to be observing his farm from an aircraft while we were talking with his herdsman. Informed by radio of our presence, Governor Thornton invited us to lunch so he could discuss cattle with Mr. Causey, even then a widely respected Hereford breeder.

When trout was served rather than beef as we expected, the governor's aide insisted on deboning Elaine's fish. Elaine, a world-class expert in food preparation, controlled her chagrin as she watched the aide so badly mangle the poor fish that Governor Thornton ordered it destroyed and requested another trout for Elaine, who thus waited for service. This time the governor himself did the deboning honors with overbearing male courtesy, but not with the finesse that Elaine, left to her own devices, might have demonstrated. The governor's cattle, however, were truly impressive, and I was pleased on return to Nebraska to note that my brother was improving livestock on the Trauger farm.

My association with the Oak Ridge National Laboratory (ORNL) was developing sporadically in 1952. I had previously consulted with Charley Winters and Jim Cox, who were designing a filtering system for air used to cool the graphite reactor as it exited to the stack and to the atmosphere. Although the reactor fuel elements had functioned reliably, failures could result in a wide distribution of radioactive particles. Standards had changed since the war; the risk of even small releases was unacceptable. Although barrier could stop radioactive particles, flow resistance was too great for this application. Through barrier-related studies, I was familiar with the chemical-warfare filters developed for military use, and that technology was adapted for the graphite reactor and other ORNL applications. These provided good filtering and have been widely used for many nuclear reactor facilities. I also aided St. George Tucker Arnold of the ORNL Biology Division in exploring use of barrier-type materials to filter tiny biological organisms. Unfortunately, the metals then used in barrier construction killed the organisms, precluding further study.

The K-25 Laboratory was asked to develop an isotope separation process for lithium working in concert with a parallel project at Y-12. We mutually agreed that molecular distillation of the metal would be appropriate. We found little applicable prior experience so this presented an interesting task that fell within the scope of isotope separation programs at K-25.

Working with lithium metal is a challenge. It reacts readily with oxygen and also with nitrogen at the elevated temperatures required for distillation. In reacting with water, lithium forms lithium oxide, releasing heat and hydrogen. A favorite college-chemistry laboratory demonstration of the time was to drop a small amount of lithium into water to produce a spectacular explosive reaction. Thus, it was necessary to encourage discipline among ourselves in safety rules. I had a favorite demonstration to focus technicians and others on the reactivity of lithium and the potential dangers of our work. A pea-sized piece of lithium metal placed on an insulating-type brick when barely touched with the tip of a torch flame produced a brilliant white flash. That and the ensuing depression in the brick made a lasting impression on personnel.

We conducted this work with lithium under pressures of the Cold War; our work had to be fast, rigorous, and accommodating of the unusual properties of lithium. As the lightest metal, one-half the weight of water, and a high liquid-surface-tension coefficient, lithium will reach unusual elevations in corners or at other irregularities of a vessel. To demonstrate that the chosen techniques were valid, we distilled small quantities of lithium onto a cooled surface in a vacuum chamber and obtained a close approach to the theoretical isotopic separation factor of 1.08 for a single stage of separation. The challenge was to build multi-stage, cascade-type equipment to demonstrate a practical separation process. For lithium, this required operation of evaporating and condensing surfaces, respectively, of 600 and 400 degrees Celsius (approximately 1,100 and 750 degrees Fahrenheit.) The still had to operate in a near-perfect vacuum to prevent reaction with air and thus produce only pure lithium liquid and vapor. Airlocks were designed to introduce lithium and to remove samples for mass spectrometer analysis. We first built a small circular still having four stages formed in a spiral, to demonstrate the process. After some modifications in design, these small units performed well, producing stage separation factors of about 1.04. The process was thus demonstrated, but a pilot-size unit scalable to a cascaded plant was also required.

John Keyes joined the group part-time to assist with data analysis, and we added four technicians to operate the equipment as the distillation units were improved and increased in number and size. These men, Jim Foreman, Van Blankenship, Les Hutcheson, and Chester Culvahouse were among the best technicians I have known. They worked at a fast pace, contributed good ideas for equipment design, and willingly accepted rotating shift schedules when long operating periods were required. All were respectful of the high temperature equipment and chemical reactivity of lithium, but Chester seemed less comfortable in this work than the other three. This discomfort may have stemmed from several accidents he had experienced in personal activities, none of which seemed attributable to his actions or to disqualify him from our work. The most bizarre of these accidents occurred on his drive to work one morning. A loud explosion suddenly shook his Plymouth, hurling the engine from the car into the ditch. The cause was never determined. Regardless, Chester performed well, although he was the one most likely to awaken me at 3:00 A.M. when overly concerned about a problem.

With the successful small-still experience, we proceeded to construct a large unit of eight stages intended as prototypical of production equipment. This prototype, two feet wide and sixteen feet long, was designed with heavy top and bottom plates to withstand atmospheric pressure when evacuated. The entire assembly, perhaps weighing a ton or more, consisted of a heated bottom-tray evaporator and a cooler cover to condense the lithium vapor. The vacuum seal around the perimeter was curved to accommodate thermal expansion differences and to avoid lithium holdup and unwanted transfer in this flawlessly welded stainless steel unit. Staging was effected by attaching small U-shaped strips on the underside of the cover to lead the condensate by surface tension and gravity into troughs at the sides of the tray. The troughs transported condensate enriched in lithium 6 to the succeeding forward stage, and the lithium 7 component flowed around baffles toward the opposite end of the tray. Thus, eight-stage enrichment was accomplished.

The still had a six-inch diameter pumping line to maintain the intense vacuum produced by the large diffusion pump and mechanical backup.

At several stages sampling stations were installed, including one with a viewing port located near the center. (The brilliance of molten lithium when viewed through a high vacuum far exceeds the brightest surfaces of silver or mercury.) A connected filling pot served to introduce block-form lithium, by melting and then transferring it into the still. The solid lithium metal was received in airtight cans, much like those of canned vegetables. The process required opening a can quickly and slipping the block of lithium through an argon-blanketed chute into the pot.

The still was heated with up to sixty kilowatts of electrical energy through exposed heating coils mounted on ceramic blocks. The condenser radiated heat to the small building outside the lab that housed the still. This arrangement made for a comfortable environment inside the uninsulated metal structure on cold winter days, but in summer it was unbearably hot, often over 120 degrees Fahrenheit. Key temperature measurements were relayed to the air-conditioned laboratory to minimize operator time and exposure to weather conditions, but other readings required occasional visits to the still. Since our lab room was at one end of an early air-conditioning system, it was always particularly cool in summer. The thermal shock on moving from the laboratory room to the metal building was sometimes distressing.

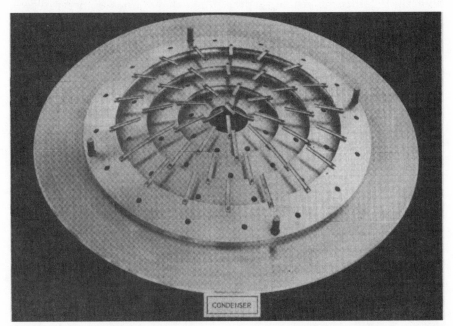

Spiral still for separating lithium isotope by molecular distillation. K-25 photograph.

Some of the dangers involved with the equipment were unknown at the time. Asbestos insulation was used extensively, an acceptable technique at the time. In addition, when operating at low rates of evaporation the condenser was covered with a woven asbestos blanket to reduce heat loss to maintain the condensate as a liquid. The blanket was thrown over the still much as one spreads a bed sheet. Each maneuver produced clouds of airborne asbestos fibers. Given what we now know about exposure to asbestos, that technique would never pass muster today.

At the end of each run with a small still, we observed the lithium configuration in the distillation tray. To preserve surface contours and to prevent oxidation during observation, each still was first flooded with argon gas and, after cooling, with mineral oil. Small stills were easily cooled and opened for direct observation, but the large tray was kept sealed, fearing that rapid oxidation could result in a fire too large to control. Observations and measurements were made through the sampling ports using long handles on dental mirrors, special tools, and focused lighting.

Small stills and their fill containers were cleaned remotely by flooding the vessels with water from a small fire hose. The water reacted with the lithium, diluting the caustic lithium hydroxide, and cooled the tray to limit the exothermic reaction, preventing an ignition temperature for hydrogen in air. This procedure worked well for small units, but how to clean and cool a large one?

Immersion in a pool would exclude oxygen and provide water to both react with the lithium and cool the tray. A colleague suggested Watts Bar Lake on the Clinch River near the K-25 plant site. Located there was a large crane recently idled after unloading barges of heavy components for the gaseous diffusion plant additions. Could this lake be used to clean the lithium stills? We computed the maximum concentrations of chemicals that could enter the stream flow to find them below damage to aquatic life or risk for people using the water even for drinking. It took only a few weeks to obtain permission for this mode of cleaning from the Tennessee Valley Authority, the Army Corps of Engineers, and the Atomic Energy Commission. Today, such approval no doubt would require months or longer, if it could be obtained at all.

For the cleaning operation, the still's openings were fitted with temporary seals, including rubber stoppers, for transport to the crane site. With the unit well secured by slings, the closures were removed, the still was swung far over the lake and rapidly lowered deep into the water. Shortly, hydrogen bubbles appeared at the water surface, which became ever more turbulent as the reaction proceeded. The workmen moved toward the shore for a better look and were displeased when ordered to stay a safe distance away. After an hour or so the bubbling subsided, and in another hour the still was withdrawn and found to be well cleaned.

On a later occasion we cleaned the filling pot, and the results proved more exciting. The pot was a foot in diameter and about two feet high with a thick cover attached to the pot wall by a single-pass weld, quite adequate for vacuum equipment made of stainless steel. The cover accommodated a pumping line and filling pipe. At the lake site the pot was held by a sling fastened around the fill-pipe flange; as with the stills the pot was lowered deep into the water nearly 90 feet from shore. The bubbling proceeded quietly for a few minutes. Suddenly, we were startled by a heavy, muffled explosion and saw a vertical column of water drench the crane boom some twenty-five or thirty feet above. When all bubbling ceased, the sling was found to have retained only the cover, fill-pipe, and flange. The pot-to-cover weld had been severed and the pot propelled to the lake floor. Apparently, enough air had entered during transport by the crane to react explosively with hydrogen when the reaction reached the ignition temperature. In subsequent cleaning operations, the workmen showed no further interest in watching the bubbling at close range.

Overall, the stills had operated well, and we considered the technology ready for the design and construction of a plant to obtain lithium 6; the large still had consistently achieved stage separation factors of 1.06. Some refinement of the welded closures and, possibly, further test operation of the experimental units would have been required for confirming plant design features, but these seemed resolvable during the design phase. However, the mercury ion exchange columns of a competing process were working well, were easier to build, and cheaper to operate because high temperatures were not required. Thus, a proper decision was made to discontinue work on lithium molecular distillation. (Our effort was not wasted, for a few years later our final still and equipment were used effectively to produce isotopic enrichment of another metallic element.)

———•◦•———

Our family lives had been busy with the activities of two small boys and the furnishing of our abode. During evening hours we enjoyed listening to records on newly acquired high-fidelity sound equipment. We also had fun playing records we had made ourselves during the war.

Yvonne, one of Elaine's New York apartment mates, had privileges to use recording equipment at the studio where she worked. We had joined friends there one evening to record music and songs, as well as stories of our Vermont trips. The results were hardly worth preserving for intrinsic value, but the recordings were fun to do at the time and later provided pleasant reliving of events of our New York era.

We designed a radio cabinet to house the new equipment and found a talented local furniture maker, M. S. Handshaw, to fabricate and finish it. He had designed his own small workshop to fit on a slight hill above his house at the end of a dirt driveway. Mr. Handshaw was noted for his sign on the Oak Ridge Highway, "Furniture Built Here" and its post-script, "I Do Not File Saws." We were pleased with the cabinet and designed matching end tables and a coffee table for him to make. We came to learn that Mr. Handshaw would build only from designs he liked. This slender, wiry man lived quite comfortably on his army pension, so this work was not a necessity for him. If a submitted design did not meet his approval, the job was not refused but quietly placed on a shelf and never touched. My map-case design was so treated; months later I retrieved the drawings from the shelf and destroyed them, having come to agree with the old man's judgment.

Grandparents in "A" house living room: (left to right) Byron, Roscoe and Isabel Causey, Tom, and Charley and Ethel Trauger, 1952.

Mr. Handshaw's eccentricity was further illustrated when a woman asked him to make a table with smooth surfaces, then scar them with small tools she provided to make it appear antique. He was incensed, suggesting that she take the request to his competitor down the road: "Tell him to make the very best piece he can; it will be just what you want." The last item he made for us was a breakfront cabinet for china or books. The lower cabinet doors were to have burl veneer panels, and beveled glass panes would adorn those at the top. Elaine had a lengthy discussion with Handshaw over her insistence on beveled glass instead of an etched design he favored. Finally, he refused to talk with her. With Elaine present, he would say to me, "You tell her. . . ." Nevertheless, he liked the design sufficiently and made a handsome piece of furniture with beveled glass in the doors.

At this time we also took on a very different venture—raising cattle. Elaine's father had sold his wholesale meat business and turned fully to his former part-time avocation: raising, buying, and selling registered Hereford cattle. When Dad Causey found a small herd for sale, we purchased some of the cows to graze on the North Carolina farm of Hargrove Bowles, a wholesale grocery business executive. On occasion, Bowles's young son accompanied his father and me as we walked in the pasture to observe the cattle. This was Erskine Bowles, who grew to be a successful businessman, and, more recently, chief of staff for President Clinton. For us this venture was more social and financial than farm experience, since Mr. Causey handled most of the field management.

With the possibility that we someday might become responsible for Dad Causey's large herd of fine cattle, we decided to gain experience by establishing a small part-time cattle business ourselves. Charley Walker, who headed the laboratory services at K-25, had a farm near Crossville, in eastern Middle Tennessee, with plenty of good grass and only one cow, for milking. We purchased five old registered Hereford cows from a farm near Maryville, Tennessee, to place on the Walker farm. With a young bull purchased from Dad Causey's Rolling Hills Hereford Farm, we found ourselves in the cow-calf business. We provided the animals, Mr. Walker the feed and care, and we would share the calves as in the Bowles operation. Both contracts were designed to enable the farm owner to enter the cattle business with minimum capital investment and provide a cash return to us as calves were sold. Mr. Walker eventually was to buy the heifers to build his own herd and displaced our interest.

In our third year of this enterprise we learned the importance of thoroughly investigating animal history before purchasing. Unknown to us, our purchased cattle had been exposed to concentrated fluoride effluents from the Aluminum Company of America plant at Alcoa, near Maryville. One by one, the old cows lost their teeth. Since they could no longer eat pasture grass and required finely ground and costly meal, it was necessary to sell them quickly on the beef market at a considerable loss. We retained ownership of the heifers for a while because the Walkers were not yet ready to build their herd. When our two cattle ventures had nearly become stable, we faced a drought. Middle Tennessee experienced the driest weather on record, and the Walker farm grasses withered; we had to move the cattle.

I consulted with Patrick Murphy, then the first Agricultural Agent of Knox County, Tennessee. He agreed to explore farm possibilities in the area and a few days later proposed placing the cattle on his farm near Knoxville. We moved them there and initiated a marvelous friendship with the Murphy family. Those cattle prospered until well after Mr. Murphy retired. Eventually, however, their care became burdensome to him and we sold them to his neighbor. Overall, we were fortunate to break even financially on our Tennessee cattle venture.

If I had not already known that farming was not for me, that would have become clear when a severe drought also struck the Greensboro, North Carolina, area. We then moved those cattle to river bottom pastures owned by friends in South Carolina, where they were well fed and provided us a modest financial return.

On closing the cattle business, I was relieved, gratefully, of the burdens of maintaining registration records, account books, and tax forms. Those forms once precipitated an encounter with the IRS. Our books were first set up on an accrual basis to conform with that used by Mr. Causey. That system was too time-consuming for one whose principal interest, at that time, lay in separating lithium isotopes, and I, therefore, proposed switching to a cash basis. The IRS required submission of a detailed plan because changing an accounting system was a ploy some used to cheat on taxes. The local IRS representatives were sympathetic to my need but said they could not accept our simple plan for the change. After a long session of questioning, and perhaps just to get me out of the office, they kindly offered to find a substitute plan for my consideration. The plan they eventually proposed no doubt fit some other businesses but would not work for us. When this was pointed out, they found another one that also did not fit. Their third proposal was essentially the one I had first submitted. I was lavish in my compliments of their insights while quickly signing the document.

Vacation time was for visiting grandparents in Nebraska or North Carolina, leaving little extended recreation time for our little family unit. Many years would pass before we found or took time for fully relaxing vacations, free from work and family responsibilities. During whatever time might have been available for recreation of that sort, I took graduate courses and Elaine was very busy caring for the family and house, teaching nutrition classes for adults, and doing volunteer work with church groups, Girl Scouts, and Campfire Girls. As partial compensation I spent several weekends building a toy train large enough to accommodate four small children and consisting of a locomotive, tender, box car, and caboose. Each car carried the painted logo BTTRR, for Byron and Tom Trauger Rail Road. Tom was then small enough to sit inside the box car with its roof board hanging to the side. The cars made of hard rock maple were durable, and that train has served many children at play, even to the present. I am thrilled that our granddaughters and neighbors' children have enjoyed riding and propelling it.

Our "A" house had steadily filled not only with the furniture made by Handshaw but with a large chest-type freezer purchased by an even trade of a bull calf. Because that sixteen-cubic-foot freezer was too large to pass through any doors of the little house, we had to remove two adjoining windows of the boys' room to permit easing the freezer from the truck bed, through the window opening, into the house. We also needed a washing machine, but with the oversized furnace the house provided no suitable space for such conveniences. A kitchen cabinet was therefore removed and the washer installed beside the refrigerator. Because we could not afford to lose the storage space, we hung the removed cabinet from the ceiling over my desk in our bedroom. Incidentally, both moves were made without a permit; Ed Granek had moved to another city. The boys' room, eight-by-eleven feet, then contained a double bed, a chest-on-chest, sewing machine, Ironrite clothes ironer, baby bed, and the freezer.

When near the breaking point from crowding, we were offered a larger, "Type 32" house, built near the top of Delaware Avenue after the war. We easily filled the nearly new, three-bedroom house with furniture, and, to avoid continued crowding, the freezer was placed on the formal entry porch, away from the street. This house lot bordered a "green belt" and provided a spectacular view of the Cumberland Mountains from every room. It was better than Elaine's hopes had been for the "A" house location.

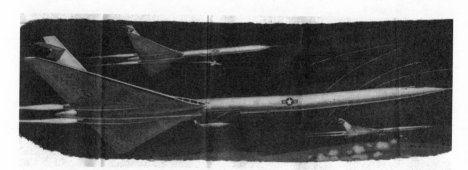

This artist's conception of a nuclear-powered aircraft appeared in Parade Magazine *on September 18, 1955.*

EIGHT

Nuclear Power for Aircraft and Electricity

————————————

C old War phobia of the late 1940s led to serious consideration of the now unthinkable: developing an airplane powered by an operating nuclear reactor. It was one thing in 1940 to consider nuclear powered aircraft. But it was a quite different prospect in 1950, when the realities of radioactivity from fission products had been fully realized in weapons tests and through the Hanford site operation with large reactors and fuel reprocessing operations. However, a broad program of aircraft design and development had been established with nine aircraft companies working under the coordination of the Fairchild Engine and Airplane Corporation as prime contractor to the U.S. government. At that time, about 650 persons were employed on the project by Fairchild and others, many in the Oak Ridge area. Jack Hunter, who had worked with me in the Nash Building, was prominent in that effort. The program was to consider a broad range of nuclear power applications to aircraft propulsion.

The Aircraft Nuclear Propulsion project (ANP) was created as part of the national defense program. A variety of scenarios were no doubt available to justify the project, but they were secret and difficult to envision. Capability for an aircraft to stay aloft for a long time without refueling would clearly be an advantage for military purposes. But the obvious and overwhelming problems of added weight for shielding and exposure to radiation for the crew had apparently not been fully evaluated. Of course, allowable nuclear radiation exposures were not as stringent as they are today. Even so, the consequences of an accident on landing and any failure over friendly territory seemed to be a totally unacceptable risk.

By 1954 the engine project had been transferred to the U.S. Atomic Energy Commission (AEC) and was divided between the General Electric Company and a combination of the Oak Ridge National Laboratory (ORNL) and the Pratt & Whitney Aircraft Engine Company (P&W). The aircraft design effort remained under Air Force control and was conducted by the Marietta Aircraft Company of Georgia. General Electric was pursuing a direct-cycle power plant in which a nuclear reactor core replaced the burner portion of a jet engine directly. Thus, nuclear fuel plates forming the reactor core were located between the compressor and the turbine. This offered an efficient power plant, but the large radiation shield for a nuclear engine presented a major design problem. The concept carried an additional disadvantage: if a fuel element failed, radioactive materials would be released to the atmosphere wherever the plane was located.

In the other approach ORNL and P&W had joint authority. P&W was concerned primarily with the engine, and ORNL the reactor; I will refer to ORNL as developer of the reactor, even though both organizations participated in the total effort. The ORNL approach was an indirect system having a liquid-fueled reactor located in the fuselage. Piping would transport hot, nonradioactive fluids to radiators located in engine(s). This isolated the reactor and its shielding from the propulsion but required more equipment, some of which would operate at high temperatures, up to 1,500 degrees Fahrenheit (815 degrees Celsius).

The ORNL role was to design, build, and test a prototype nuclear heat source called the Aircraft Reactor Test (ART). In this prototype a liquid fuel consisting of fluorides of zirconium, sodium, and isotopically enriched uranium would be encased in a spherical assembly that included a solid neutron moderator of beryllium metal. This geometry was chosen to minimize the shield size and weight. A centrifugal pump propelled the fuel through a central fissioning chamber inside the hollowed beryllium, then over a multitude of small heat-exchanger tube bundles wrapped inside the outer surface of the primary vessel. A helical configuration was required for the heat exchanger to conform with the spherical shape of the reactor assembly and to provide flexing to accommodate metal expansion for temperature changes during operation. Thin metal shells directed the fuel flow through the many passages. Pumping a mixture of sodium and potassium (called NaK for their chemical symbols) through the heat exchanger tubes extracted heat energy from the fuel and carried it through piping to the engine's finned radiators. Art Fraas was the principal designer.

Other auxiliary systems of heat exchange and piping were required to cool the beryllium, which would otherwise overheat from gamma radiation, and to service the pumps. Following a series of tests, the commercial alloy Inconel, marketed by the International Nickel Company, was pronounced suitable for containing the fuel, a mixture of sodium, zirconium, and uranium fluorides that generated fission energy at temperatures up to 1,500 degrees Fahrenheit (815 Celsius).

The first step by ORNL was to build and operate the Aircraft Reactor Experiment (ARE), a small pilot reactor to validate the general concept. It was to include basic features of the nuclear engine power system, including the fuel, structural materials, and some control features. Its role would be to gain experience with this entirely new concept and to test the system for stability and materials compatibility. For their purposes, however, designers of the ARE were not troubled with the crucial matter of compact design, nor were they concerned with potential for corrosion after extended operation, from accumulation of fission products in the liquid fuel.

Having established some credibility with high temperature systems, our little group was asked to assist with the ARE. I was invited to the ANP offices of ORNL for a briefing on this secret task. We were to assemble prototypes of the pumps and heat exchangers for the ARE and to test their performance as a system, using electrical heaters to achieve the required temperatures. The project was urgent; starting with the date of my visit, Saint Patrick's Day 1954, we were requested to have the test stand completed with the components ready to operate three months later.

We were in a favorable position for this priority task. Because the gaseous diffusion plant-expansion work was scaling down, many services were available for design, fabrication, and construction.

Components provided by ORNL required welding for the construction of the pump and the serpentine heat exchanger that we would test. Assistance by Bill Manly and Pete Patriarca of ORNL Metals and Ceramics and Bob Affle of the Instrumentation and Controls Divisions was vital, for the technology and equipment we had to master differed from the challenges in isotope separation.

This task also required training a new operating crew. The four technicians completing the lithium-still work could not properly do both. More importantly, this was an opportunity for designated reactor operators to gain needed experience. The new team consisted of Chuck Cunningham, Bob Affle, J. J. Harriston, Roscoe Reid, and two or three others. We stress-tested the system by putting it through thermal cycles: once the operating temperature had been achieved, doors on the heat exchanger were opened, and the use of fans accelerated cooling; then electrical heat was reapplied to regain the operating temperature as fast as possible. This did not simulate the maximum conditions that the reactor could impose but provided a reasonably good measure of expected performance and integrity. The pump functioned well and thus supported the decision to continue at full speed with reactor construction.

During that summer Art Miller, assistant project director for the ORNL-ANP, and others suggested that I join the project directly. Art lived two doors from us on Delaware Avenue and came over one Saturday when I was mowing the lawn to persuade me to join ORNL and assume a large task in testing the ANP fuel system. This offered interesting work and professional opportunities, including utilization of the Materials Test Reactor (MTR) in Idaho. That would be a major challenge, but should I join a project as seemingly foolish as putting a nuclear reactor in an airplane? The Air Force's insistence on developing this aircraft in defense of our country seemed compelling. But being quite aware of the hazards of radioactive fission products, it looked unacceptably dangerous for a nuclear reactor powered airplane to fly from a home airfield to reach enemy territory. It was even worse to consider the potential accidents in landing on the return. This reality was quite different from the idle speculation about nuclear-powered airplanes and automobiles in the heady days of 1940.

Early the next week I went to see Dixon Callihan, who was conducting nuclear criticality experiments for the aircraft project. We talked in a remote underground room that isolated very sensitive monitoring equipment from the shielded cells that housed and protected technicians from his criticality experiments. Dixon observed that the liquid-fuel reactor, even if never used in an airplane, was a potentially attractive system for electric power production. I liked that idea, and, after pondering the opportunity for a few days, agreed in September 1954 to join the project and ORNL.

My task was to build experimental equipment called a loop to model the ANP reactor and to test it at reactor operating temperatures in an intense radiation environment. Demonstrating that the alloy and fuel would retain their stability after much of the uranium had fissioned was necessary because the products of fission could also cause corrosion. The MTR in Idaho was at the time the only reactor in the world having a neutron flux high enough to accommodate testing the fuel system of this nuclear engine's core. As mentioned in chapter 7, the MTR had been designed by ORNL to be built in Oak Ridge but had been constructed in an Idaho desert, chosen for its remoteness. Thus, our experiments would involve much long-distance communication and travel, first to assure design compatibility with the MTR and its operating rules and, later, to assure proper installation and operation of equipment.

The molten-salt class of reactors is often described as a pot, pump, and heat exchanger; in the ART these were to be packed tightly into the spherical configuration to minimize shielding size and weight. At the MTR it was necessary to fit equivalent components into a beam port, a horizontal, six-inch-diameter penetration of the shield. This extended to the wall of the reactor vessel containing the fissioning fuel and its cooling water. Although the beam tube was twelve feet long, the active part of the test equipment was limited to four feet, with auxiliary equipment and shielding at the rear. The task was to design a miniature pot, pump, and heat-exchanger assembly to test compatibility of the ART primary containment alloy with the fuel. The assembly was required to fit within the confines of a water jacket that would slide into the MTR-3 beam tube. A. A. Abbatiello, who had cooperated with me in designing the barrier test devices of 1943, was available to help meet these challenges, and a team of engineers, whose roles will be described later, were to build and operate the loop.

For want of a better name, we called our device the MTR-ANP Loop, just loop for short. Because the system could fail after the loop was heated and filled if the fuel dropped to 1,000 degrees Fahrenheit and, thus, freeze, reliable heaters and temperature measurements were necessary to establish and maintain the desired test conditions. A principle of double containment was followed rigorously; no single failure could cause release of radioactive materials. Compressed air cooled the heat exchanger and controlled the loop operating temperature before discharge to the MTR cooling duct; that air then passed through filters and into the atmosphere. Failure of the heat exchanger was unacceptable because that would allow release of fission products to contaminate MTR ductwork and release radioactive gases. This unlikely precipitous failure of a straight, thick-wall tube was prevented by an over-sleeve of a tightly fitting concentric wall that provided double containment. Integrity of both the inner and outer wall was continuously monitored by a gas flow through small passages machined into the inner surface of the outer wall. Leakage of either wall would be discovered immediately in this early example of a double-wall heat exchanger. The hydraulic motor that drove the pump was mounted behind a bulkhead with an oil-lubricated shaft seal for containment. Abbatiello properly chose to make the assembly drawing full scale, even though it was twelve feet long. It was difficult to use but served well in meeting precise dimensional requirements.

In mid-October of 1954, with the design concept on paper, Harry Gray of P&W and I traveled to Idaho to discuss plans with staff members of the Phillips Petroleum Company, which operated the MTR for the AEC. We met with Brad Lewis, a senior manager responsible for experiments, Howard Watanabe, staff member assigned to our project, and Fred MacMillan, chief of reactor operations. Near the end of lengthy discussions that had emphasized the complexities of this loop, Lewis asked, "When can we expect you to be here?" I replied, "In July." Brad leaned back in his chair, hesitated for emphasis and asked, "In July of what year?" We answered, "In 1955 of course," at which point he tilted forward to say he would look for us in 1956 and expect us in 1957. Experiments conducted by a variety of agencies coming to Idaho during the MTR's nearly three years of operation—including those of the Naval Reactors Program's Nautilus submarine test-loop project under Admiral Hyman Rickover—had consistently experienced delays.

While flying back to Oak Ridge, we discussed the tasks and cautions in detail, sobered by the MTR staff predictions, but returned with renewed determination to meet the ORNL/P&W and Air Force commitments. I also recall Harry Gray's attention to the engines of the DC-4 aircraft in which we flew between Salt Lake City and Chicago. Harry had been a designer of those engines and of the famous J-57 engine, which saw much early military and airline jet service. Every sound of those engines told him something of their condition, and he was inclined to discuss the possibilities for trouble. Traveling with Harry was pleasant, but on subsequent trips I insisted on avoiding seats close to the engines.

The team to design, build, test, and operate the loop was quickly assembled. John Conlin was already designing a small turbine-type pump and the hydraulic motor and system to drive it. Lew Carpenter led the work on the heat exchanger and water jacket. Jack DeVan and Albert Taboada provided metallurgical planning and assistance. Bill Ferguson, Vic DeCarlo, and Paul Gnadt worked with various aspects of the design, assembly, and testing. Ken Stair and Joel Bailey of the University of Tennessee assisted part-time with analysis of mechanical and heat transfer features. Bob Affle was assisted by Howard Burger and Bob Hyland in designing the complex instrumentation required both to serve the loop and to interface with the MTR. Reliable and fast-acting instrumentation was vital because the high power-density fuel and the high operating temperatures left little margin for error in loop control. As an ultimate safety measure, the design required that this instrumentation function with loop safety system to withdraw the loop assembly from the high neutron flux or to shut down the reactor if necessary.

David Haines, Chris Bolta, and Marty Cooper were assigned from P&W to assist with construction and operation. Dale Magnuson joined the team for several months, bringing us together for the third significant period of cooperation. Here again, craftsmen were to be of vital importance, both for their skilled work and in making useful suggestions for improvements to the complex and crowded assembly. Bill Coward and Ed Pollard were invaluable craftsmen assigned from the Y-12 Plant Maintenance Division, as was Bill Mason, their foreman. Technicians Jim Kingsley, Bill Montgomery, Clarence Stevenson, and Charley Wallace followed every detail of the construction, assisted by others we will meet later as loop operators.

We needed to know how much heat would be produced in the forward end of the loop assembly by gamma rays streaming out of the MTR core. This heat would add to the energy from neutron fissioning and require accommodation in operation and for data analysis. An experiment was required since this would be the first use of the HB-3 beam hole. The gamma heating rate was approximately that predicted by the ORNL reactor designers, but the experiment produced a

surprise. The measurement graph of gamma heating was a linear function through about one-half reactor power; then it deviated noticeably,
from one straight line to another straight line. We shared consterna-
tion with the reactor operators: what was wrong? We then noted that
the change occurred when measurement of reactor cooling-water flow
was switched from one flow meter to another. The second flow meter
was either improperly designed or miscalibrated. The deviation was
within the technical specifications for safe MTR operation, but,
clearly, the reactor had not been working at the exact and prescribed
high-power levels for nearly four years.

A team of four—Ray, Floyd, Clarence, and Steve—were recruited
early to assist with loop assembly and with refinement of operating
procedures, as preparation for work in Idaho. Chuck Cunningham,
who had been an operator of the pump and heat-exchanger stand at
K-25 and of the ARE, also joined the team. As construction of equipment to interface with the loop was started in Idaho, Howard Klaus
moved there to assist as our experienced person on-site for the duration of this and other ORNL experiments. Assignment in Idaho was
viewed by many as a hardship tour, much as Oak Ridge had been
during the war, but Howard was pleased to enjoy Idaho's excellent
trout fishing on weekends.

Three loops were built in overlapping sequence: one to be tested in
Oak Ridge, with an extended section of resistance-heated pipe substi-
tuting for the fission-heated coil; the second to operate at the reactor;
and the third to follow as a backup to assure early success in the
project. Shortly before the loop was to be shipped to Idaho, the electrical heaters failed on the bench-test loop. We deemed it unwise to
ship a loop to Idaho until the cause of failure could be determined;
however, project management insisted on proceeding. Their premises
were (1) the problem might be unique to our bench test, and (2) if at
all possible the urgent schedule imposed by the AEC and military
advisers must be met. The loop and auxiliary equipment were shipped
on a C-47 aircraft, owned by the AEC, and piloted by Hap Wilson.
Several of the engineers and I also rode in a few seats on one side of
the fuselage beside the loop and other boxes of equipment.

The trip took twelve hours with a stop in Kansas City for refueling
and lunch. While crossing Kansas on that summer day, the aircraft
repeatedly rose and fell as we flew over hot, shining, harvested wheat
fields and cool, dark-green plots of corn. Everyone but Hap became
airsick, including the copilot stretched out on the floor. I did not lose

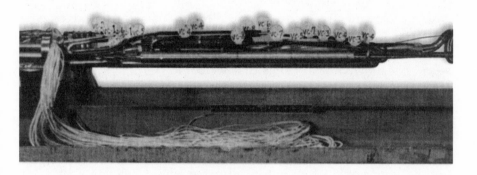

ANP test loop in assembly steps. ORNL photograph.

my lunch, as did many, but could retain it only by sitting quietly while watching through the window to anticipate the next vertical excursion.

Hap Wilson was a superb pilot with much experience, including having flown in the military and for South American Airlines. He was distressed on one flight to find we were nearly half a mile off course at a visual checkpoint over western Nebraska. His tolerance for error contrasted sharply with that of the crew of an Air Force C-54 on a later flight to Idaho. Because I enjoyed amateur navigation and those flights provided me the opportunity, I found them nearly one hundred miles from their projected course. Although we arrived safely with plenty of fuel, the late arrival required paying overtime for local workers to unload the aircraft.

After arriving in Idaho, we began our tests. As we feared, the electrical heaters failed in loop no. 2; the problem observed in Oak Ridge was generic. The heater terminals failed because they were necessarily exposed to temperatures above their normal uses. Because we could not change the terminal locations, we designed and built new heaters with improved terminals that functioned well for all successive loops. Although the schedule was not maintained, owing to the premature move to Idaho and the subsequent redesign work, we showed Brad Lewis that we could be there in July, just as we had predicted, and there was satisfaction in that. Also, some of the crew had a brief opportunity to visit the Yellowstone and Grand Teton National Parks on summer weekends. But for Bill Ferguson, Bill Scott, Bill Montgomery, and most of the engineers and technicians mentioned previously, that was small compensation for the long hours of strenuous effort made in meeting a difficult startup schedule.

Loop no. 3 was completed and shipped to Idaho in September. It was installed without major difficulty. The principal problem for personnel was the extended time away from home and the long hours of work. Also, housing was in Idaho Falls, fifty-five miles from the MTR. I recently reread one of my letters home that reported six hours of bed sleep in the previous sixty-seven hours. In those more youthful days, I could relax instantly and awaken refreshed after fifteen minutes sleep. The naps were on the one couch in an office, when available, or on a desktop. A luxury on the 12:00–8:00 A.M. shift was taking naps on the small cot in the ladies' rest room; I entered there with trepidation, even though no women then worked at the reactor during those hours.

Our team could be heard rejoicing when loop operation started under nuclear power on October 8, 1955. That was exactly one year from our meeting with Brad Lewis during which he said he did not expect to see us until 1956 or 1957. We were congratulated as the first team to operate a major loop in the MTR. It functioned smoothly for 462 hours, then operated erratically and caused the sensitive control system to "scram" the reactor (shut it down). The loop was removed for examination in hot cells, where the rotating seal was found to have leaked excessively, causing a buildup of nearly black carbonaceous debris along the shaft. Those pumps and seals had operated with no measurable leakage for up to five thousand hours in tests without radiation.

Although located away from the intense radiation source, the lubricating oil and rubber mounts for the pump's drive motor had been severely damaged. The fuel showed no deterioration except for the fissioning of uranium, and we found no corrosion of the Inconel piping or pump parts. Since there was little existing experience with organic materials exposed to neutrons or intense gamma radiation, these effects were informative to nuclear aircraft designers and the fledgling nuclear industry. The test had been successful, but a longer period of operation was needed. We therefore fitted later loops with such improvements as could be made.

In December Elaine accompanied me to Cleveland, Ohio, where I consulted with the Air Force staff at the Plumbrook Reactor where they were testing other nuclear aircraft components for radiation damage. From Cleveland we went to Hartford, Connecticut, where I met with the P&W staff. They were pleased to have a firsthand account of the irradiation test loop operation, and their ANP experimental work was interesting to me because it complemented that at ORNL. I also enjoyed a visit to their jet engine manufacturing plant. We spent the weekend in New York City visiting friends, enjoying Christmas pageantry, store decorations, and attending services at Riverside Church.

The test of loop no. 4 occurred in February 1956. This one functioned more smoothly than no. 3 and produced equally reassuring results for the performance of the fuel and materials of the reactor. Loop no. 5, scheduled for April, could not be started because a zirconium oxide plug had formed in the fill-line to the pump sump. The cause of that formation was never discovered, but it was considered unique to the small lines of the loop and not relevant to the aircraft reactor. Loop no. 6 was run in June and July, completing this test series. Following irradiation, each loop water jacket was disassembled in a large hot cell at the MTR and the operating components shipped in heavy shielding casks to Oak Ridge for detailed examination. Oscar Sisman, Ernie Long, Stewart Dismuke and technicians of the ORNL Solid State Division were responsible for the final disassembly and evaluation. Examination of the piping, pump, and fuel showed that neither the irradiation nor the fission products had caused corrosion or other deterioration of the primary pot, pump, and heat-exchanger loop. These tests assured the project that this type of nuclear reactor could produce energy at the temperatures and power levels required for the proposed operational lifetime of such an aircraft.

———•◦•———

At home, during the years of the lithium and aircraft programs, the boys were a delight. Only twenty-one months apart in age, they were inseparable and enjoyed many young friends on Delaware Avenue. We purchased a pet, choosing a registered Shetland sheep dog because of its resemblance to my old dog, Possie. This one was named Toby, "To" for Tom and "by" for Byron. I had suggested Byto, but was voted down in family council. Each of us developed a special relationship with Toby. I taught him to do tricks, as might have been expected. Elaine had him "sing" along with her while held on her lap; they especially liked "Dixie." The boys romped with him in the house and on the lawn. We also enjoyed occasional family outings in those days. A favorite was to spend a weekend at the Mountain View Hotel in Gatlinburg, enjoying its good food, swimming pool, and proximity to the Great Smoky Mountains for hiking and picnicking.

During this period Elaine and I were much involved in church work. I progressed through many committee roles to chairman of the Board of Stewards, and she was equally active in the women's program. When the church decided to build a proper church structure and leave the theater, Elaine was chosen as a member of the building committee. As the only woman, she was particularly instrumental in assuring the beauty and practicality of the stately structure, which was located on the corner of Tulane Avenue and the Oak Ridge Turnpike. In addition she taught classes in food and nutrition at the YWCA and in the Oak Ridge Schools's Adult Education Program.

Elaine also assisted the Oak Ridge Associated Universities by entertaining mature foreign students enrolled in courses on nuclear energy technology. On one evening I was to go by the dormitory provided for them and pick up one of the foreign students to bring home for dinner. His name escaped me. I could only remember that it started with the letter "S." The desk clerk reviewed the student roster and found two, Adrian Schuffelen of the Netherlands and Alfred Schurch of Switzerland, describing both as pleasant and lonely. So with Elaine's permission, I invited both to dinner. We spent not only that evening but several weekend days with these gentlemen, both of whom were agriculturists interested in the use of radioisotopes to trace plant functions. We particularly recall one evening discussing languages and the game of Scrabble. Schurch said that his family liked to allow words of any language. Schuffelen scoffed, "That would be too easy," but Schurch replied that only one part of speech was allowed, adverbs for example. Their facility with languages impressed us greatly.

My next assignment, to manage the ART operation, brought mixed feelings. On the one hand, although my familiarity with the loops and with the MTR provided applicable experience, my interests lay more with experimentation than operation. On the other hand, development of a compact reactor of 60,000 kilowatts power was a challenge. The small ARE, for which we had tested the pump and heat exchanger at K-25, had run well but had few of the sophisticated features of this large unit. To gain related experience from others, I returned to Idaho to consult with the Westinghouse and AEC people who were operating the prototype power unit for the Nautilus, the first nuclear submarine. Even though that reactor was of an entirely different type, its core was also compact and of a high power density. I was favorably impressed with their success.

We considered devising a computer control system for the ART, both to meet the short time-response requirements of this very high power-density reactor and in anticipation of using it in an aircraft. After study, we concluded that computers then available were not reliable enough for the aircraft or the test. As education for reactor operation, I became more familiar with the multitude of component tests in progress under Bob McPherson and Bill McDonald. These included the spiral heat exchangers, primary radiators for the engine, secondary radiators to cool other components, and pumps. Special tests were required for valves that would hold the hot molten fuel in the reactor and reliably release it to a cooled, criticality-safe and separately shielded tank as an emergency measure for the ART. Instrument systems, pumps, and pump seals were also under test, as was the beryllium-moderator material for the effects of exposure to irradiation. The moderator tests were conducted at the MTR, both by Irradiation Engineering and by Jerry Keilholtz, Oscar Sisman, and Jim Trice of the ORNL Solid State Division. Manufacture and preassembly of components for the reactor test were underway, and we made plans for extensive training of the ART operators.

By the early summer of 1957 the development of missiles and the apparent limitations of a nuclear powered aircraft led to cancellation of the ORNL engine project. The direct cycle program continued for a short time, and the Air Force pursued the airframe design effort for several more years. The aircraft would have been the size class of the present C-5A, the enormous Air Force transport plane. I have little doubt that with further development, either of the two nuclear engine types could have delivered enough power to sustain flight, but maneuverability would have been limited by weight and engine temperature. It was a disappointment not to operate and control the challenging ART, but there was great relief from concern over making such a large effort for an aircraft of dubious practicality and safety.

In retrospect this ill-fated project had produced much useful information for other nuclear programs and for U.S. industry. Improvements were made in materials, instrumentation, control-systems, and heat-exchange technologies. Also, ANP facilities were useful for later nuclear projects, for example, the Tower Shield Reactor Facility (TSF) at ORNL. Four towers, each 450 feet tall, were constructed and rigged with cables from which a spherical-shaped reactor could be suspended to test separate experimental radiation shields for the reactor and aircraft crew compartments. This suspension, two hundred feet above the ground, freed the measuring instruments from the effect of neutrons interacting with the earth. Such elaborate means for testing were important in order to optimize the shielding and to minimize the aircraft weight. The TSF reactor uses a curved plate-type fuel and a control system of novel design. An early use of the facility was to test a J-57 jet engine under reactor radiation-exposure conditions that would occur in the aircraft. The noise of that engine running day and night could be heard at our home ten miles away. Irradiation effects, which were identified for possible correction, were similar to those experienced with the radiation-damaged oil and rubber components of our loops.

With termination of the nuclear airplane project, the Tower Reactor was securely housed at ground level inside concrete shielding, with provision to test materials and shielding configurations for many reactor types. It operated there for forty years, contributing steadily to shielding designs and configurations for many present and perhaps future power reactors. Even though no plans are currently being pursued with the intention of developing an aircraft reactor, the ORNL reactor project became a major contributor to nuclear technology. The TSF, now shut down, remains on a standby status.

If nuclear fission power is ever to be used in deep space exploration, operation of this or a similar reactor would be required to help create and validate the design of more sophisticated shields. It also has potential for use in biological and medical studies and for neutron irradiation treatment of patients. The TSF towers have been used repeatedly to test the integrity of heavy shipping casks used in transportation of radioactive materials. The tests assure vessel survival during severe accidents. The heavy casks are lifted to a height of thirty feet and dropped onto a short, six-inch diameter steel post welded into a thick steel plate backed by a ten-foot cube of concrete. Casks used today survive intact and leak tight even when striking the post on the most vulnerable area of their closures and then exposed to gasoline fires, as might occur in transport accidents.

The Aircraft Nuclear Propulsion Project had been exciting, absorbing, and demanding. Thus, I spent little time reflecting on the philosophy of propelling an aircraft with nuclear power. The USSR eventually put several small reactors into earth orbit, but those were lifted into space with chemical rockets, put into operation by remote control, and crashed into the ocean when the mission was completed. To apply nuclear power to aircraft takeoff and landing is more problematic. I had a favorite, somewhat lighthearted, comment on the subject that sometimes offended project enthusiasts: "All we must do is: get the nuclear aircraft over enemy territory; it can fly there forever, for they dare not shoot it down."

The molten-salt reactor (MSR), derived in part from the ART experience, was to produce both electric power and new fuel, uranium 233, by neutron capture in thorium, an element found abundantly in nature. I was enthusiastic about this concept and hoped that it might validate my rationale for joining the nuclear aircraft project. This was a time when, based on steady growth throughout a half-century, projections for future demand of electrical energy indicated dramatic increase. Thus, it appeared that breeder reactors such as those that convert the more plentiful U-238 to plutonium as the MSR would be

Nuclear reactor and aircraft-crew compartment, suspended for testing at the Tower Shield Reactor Facility (TSF). ORNL photograph.

necessary if nuclear energy was to meet the potential demand. In the molten-salt reactor the plentiful element thorium is converted to U-238, a fissionable element. The original molten fuel was fluorides of lithium, beryllium, and enriched uranium, with thorium added later as the project developed. Neutron moderation was by use of graphite in a cylindrical reactor configuration.

The fuel salt also required a new container material. After some experimentation, a suitable alloy was developed of nickel, molybdenum, copper, and iron, with other minor alloying components. It was named INOR-8 for the International Nickel Company, a major participant, and Oak Ridge. The digit indicated the number of major recipes tried in finding this alloy. However, since Hank Inouye was a principal developer, we also thought of the alloy as IN(ouye)OR-8. It has subsequently been marketed commercially by the International

Nickel Company as Hastelloy N and is widely used for high temperature applications.

Laboratory bench tests had shown this fuel and alloy to be stable and compatible, but the MSR system also required testing under irradiation. We first adapted the water jacket of the MTR loop to accommodate several small fuel containers of INOR-8, with suitable cooling for static fuel-irradiation tests. Later, two loops of the MTR design were built of INOR-8 and were operated for longer periods than those required for the ANP. This again involved much travel and telephone communication with the Idaho reactor people, and Howard Klaus remained there to manage the details. Those tests indicated that the fuel and the new alloy were compatible and stable under extended irradiation exposure. Other work was done to certify heat exchangers, pumps, valves, and instrumentation appropriate for the concept of a commercial reactor design. A small reactor, the Molten Salt Reactor Experiment (MSRE), was built and operated successfully in the facility built for the ART.

The MSR concept appeared to have a useful place in the complex of nuclear plant concepts under consideration through the 1960s. Although less compact than the ART, it was small enough to be built

Closeup of suspended reactor at the TSF. ORNL photograph.

in a factory and transported to the site, thus avoiding the cost and complications of the field assembly required for the larger light-water-cooled reactors. The MSR could also be operated in conjunction with its own fuel reprocessing system, with capabilities ranging from essentially manufacturing and maintaining its own fuel supply to providing sufficient excess fuel for additional reactors in an expanding nuclear economy. However, unlike other fission reactors—the solid fuel elements of which could be certified in test reactors and the basic cooling and steam-generating systems of which resembled established power plant practice, needing only prototype confirmation—the MSR's radioactive liquid-fuel system demanded demonstration as a complete unit.

During that period, the size of all electric power-generating plants was increased greatly to achieve economies of scale. This advantage is not easily obtainable for the MSR, in which the entire primary circuit is radioactive. To increase the design size of an MSR would likely require costly tests and demonstrations above and beyond anything required of a conventional nuclear plant, challenging justification for investment in a plant embodying such new features. These factors, together with the rapid expansion of a seemingly viable light-water reactor industry, led to loss of political favor and financial support for the MSR.

———————

When in 1953 Elaine's father had sold his wholesale meat business to concentrate solely on the breeding and raising of Hereford cattle, the expanded business required more land than that of the Causey home place. Eventually, he was renting pastures and placing cattle on farms as far away as Southwest Virginia and Eastern North Carolina. We grew increasingly concerned over the amount of travel required of him in his venture as he grew older. When a tract adjoining his farm became available in 1956, we purchased it jointly with Elaine's sister, Eloise, to help reduce Mr. Causey's dependence on distant acreage. It was then cleared of timber and seeded to grass for his use. Until our own cattle at the South Carolina farm were sold, we were "farmers" with land and cattle, but never with both at the same place.

When Elaine and I married, we received a three-hundred-acre tract of land known as the William Underwood farm as a wedding present from Elaine's family. Although the tax assessors of Moore County,

North Carolina, found us immediately, we did not so easily find the land. Sometime during our first years in Oak Ridge, we resolved to find the property, knowing only that it was near Sanford, in Lee County. The first search confidently undertaken with general directions was fruitless, as we drove through a wooded, lightly populated area where no one we could find had heard of the Underwood family. Two more trips with family guidance were required to locate the old building site on a rough, unpaved, sand-clay road. We were fortunate that Russell Gordon, a very old African American who lived in the only house on that unmarked road, was able to identify the tract and its approximate boundaries. The road remains unimproved today, although it now has signs at each end identifying it as "Underwood Road."

Our property included about sixty acres of cut-over pine, with the remainder in nearly valueless hardwoods. Early on we decided to have the property surveyed because we could not find the boundary markers placed nearly a century before. The first surveyor we hired worked for several days and gave up because he could not find a starting point. A more experienced surveyor then undertook the task but had to survey an adjoining property to establish one corner. He also discovered that a neighbor was farming some of our land on an encroachment made several decades earlier.

When we had some hardwood trees cut in a reforesting program, I was curious about their age. Growth rings of stumps and the distribution of trees indicated that they were naturally seeded following the Civil War. We know little about the family at that time and have no written history of the farm, but the area must have been prosperous in the early 1800s. The Old Plank Road, first paved with split logs from which its name derived, is nearby. The House in the Horseshoe, a museum of the eighteenth century, is nearby across the Deep River.

One can postulate that the Underwood family had grown cotton, the prevalent early crop of that region, and judging by the floor area of the collapsed house may have been somewhat affluent. Thus, the tree-ring record can be read to suggest a family retrenchment occurring when their slaves were freed in 1865 and cultivation became restricted to only the sixty acres found in southern pine trees. Although furrows from the last plowing could be discerned, the farm had apparently been unoccupied and untended for about fifty years; the old house was a pile of decaying wood with hearth and chimney stones partially tumbled at each end. Trees near other abandoned home sites on Underwood Road also seemed about five decades old.

The thin, sandy soil may have become so depleted by a century and a half of cotton farming that the people migrated westward about 1900; we know, for instance, that the Underwoods moved to Bonlee in the Piedmont, an area of richer soil.

In exploring the tract, we were surprised and distressed to discover a hand-dug well that had been covered with pine slabs during the last and perhaps only timber cutting, probably in 1938. These rotting timbers were well-concealed with forest debris and offered a serious danger to any person or animal venturing there. Our first improvement was to contract with Mr. Gordon to transport the chimney rocks by his mule and wagon and fill the well. We have now planted loblolly pines periodically in approximately thirty-acre tracts using good forestry practices. Hardwoods have been retained as needed to protect the waterways and to sustain wildlife.

I have greatly enjoyed the experience of learning some forestry, of observing the growth of beautiful trees, and of developing a profitable timbering operation having desirable environmental features. In recent years we have benefited from harvesting mature trees, having spent most prior timber income on taxes, land preparation, planting, and forestry consultation. Regardless of the direction of expenditure of the income, we have always followed each harvest with immediate reforestation.

In Oak Ridge Byron was such an excited learner in first grade at Elm Grove School that he spent evenings teaching Tom to read. We watched but neither encouraged nor discouraged the activity. In the following year the school became aware of Tom's skill when he read a story on the blackboard to the kindergarten class in the absence of the teacher. This resulted in many placement tests for both boys and endless conferences for us. Tom was finally placed in first grade in the afternoon after his regular morning kindergarten. The following year, with Tom in grade two, the boys were only one year apart in school and established even closer bonds with more friends and activities in common, as they progressed through elementary and high school years.

All real property in Oak Ridge remained in AEC ownership until late in 1956, when a firm commitment was made to sell houses and other properties with a first-purchase or refusal option available to occupants. We chose to purchase and immediately expand the three-bedroom house by adding two bedrooms, two baths, a new kitchen, and a family room. In part, this expansion was to accommodate anticipated family growth. Elaine was pregnant, this time experiencing a very uncomfortable term. We lived in the house during construction, since the addition was at one end and required minimal changes in what we began to call "the Old House." The workmen worried about Elaine as she climbed over boards to follow the progress.

We stayed anxious during the difficult pregnancy, but we were quite unprepared when little Harold was born in the Oak Ridge Hospital on February 26, 1957, as a "blue baby" destined to live only a few hours. We were also overwhelmed and grateful for the outpouring of concern from friends, neighbors, and colleagues. Byron was distressed over the loss of a new brother, but Tom, at age five, seemed somehow unconcerned. As the boys and I were returning home from making burial arrangements, Tom quietly asked, "When will we bring the baby home?" In response to my statement that the baby had died and would never come home, he replied, "But when people die on television, we see them there again." After explaining that television shows were play-acting and that in real life people who die can never live again, Tom burst into uncontrollable sobbing. In fact we all cried so profusely that I parked the car to regain composure.

Returning to the work scene, ORNL wondered and I worried about what should follow the ill-fated ANP Project. We assembled a team to visit Washington to show AEC the capabilities developed in nuclear programs that included designs for the graphite (low-intensity) test-reactor of the MTR, as well as the bulk-shielding and Geneva Conference "swimming pool" reactors. Somewhat coincidentally, the U.S. Congress became aware of a British gas-cooled nuclear reactor project of commercial size and design. Two graphite-moderated reactors had been built at Caulder Hall in the north of England to produce plutonium for weapons but also electricity for their national grid. These were called "Magnox" reactors, after the name of the magnesium alloy that contained the uranium fuel. In a Magnox reactor nuclear heat was transferred from the finned containers of uranium, called fuel elements, to steam generators by pressurized carbon dioxide gas. The steam-driven turbine generator of each reactor produced 50 megawatts of electricity. Very small amounts of electricity had previously been produced by the homogeneous reactor in Oak Ridge, and somewhat more in a sodium-cooled reactor by the Argonne National Laboratory at its auxiliary site in Idaho. However, the United

Kingdom project was the first in the world to produce commercial quantities of nuclear-generated electricity. A logical congressional question was, "Why does the United States not have a power reactor?" They then mandated that the U.S. AEC should build a reactor capable of producing electricity, as the British had done.

Gas cooling for nuclear reactors was not a new idea. Pressurized helium gas had been considered for the graphite reactor in Oak Ridge and for the plutonium production reactors at Hanford. Helium was attractive for its inertness, excellent heat transfer properties, and transparency to neutrons. However, as noted previously, Eugene Wigner and others decided that the technology for using helium was too difficult to perfect during wartime and chose air-cooling for the Oak Ridge reactor and water-cooling for the Hanford reactors. A small helium-cooled reactor concept, proposed by Farrington Daniels, had also been studied at ORNL soon after the war. Now, under AEC direction, ORNL conducted a review of the relative merits of gas and water-cooling for reactors. The team—headed by Bob Charpie with Mike Bender, Grady Whitman, and others—found the two competitive and also concluded that the technology for helium use to have advanced since World War II.

As a result, the AEC proposed to model a project like that of the British Advanced Gas-Cooled Reactor (AGR) then being designed for construction at Windscale, England. Uranium oxide contained in stainless steel cans had been chosen for fuel elements to permit higher temperatures and hence higher thermal efficiency than for the Magnox design. This, too, was moderated by graphite. Following ORNL advice, the U.S. reactor was to be cooled with helium instead of carbon dioxide. This project, the Experimental Gas-Cooled Reactor (EGCR), was managed and contracted by the Oak Ridge Operations Office of the AEC in cooperation with the Tennessee Valley Authority and was to be located adjacent to ORNL. The Laboratory was requested to carry out several specific tasks. Among these was the design of the fuel elements and their qualification by irradiation testing. My group, by then known as the Irradiation Testing Section of the Reactor Division, was to conduct the fuel-element tests. We then could use the Oak Ridge Research Reactor (ORR) that had been authorized in 1953 for completion in 1958.

An important step in validating the EGCR fuel elements was to test the compatibility of the uranium oxide ceramic pellets with their 304L stainless steel containers (called fuel pins) during irradiation.

Devices called capsules were chosen to provide suitable test conditions. Eight capsule test facilities were installed at the pool-side face of the ORR that is adjacent to reactor fuel elements, to obtain a suitable neutron intensity. Fuel pins were immersed in the sodium-potassium alloy, NaK, that was pressurized to seven hundred pounds per square inch, the EGCR coolant operating pressure. The desired high-temperature operation was achieved by the nuclear heat of the fuel and the thermal insulation of an annular gas-gap between the NaK containment vessel and an outer vessel cooled by the pool water. This design also provided the secondary containment necessary for safety. Temperature was controlled by varying the concentrations of argon and helium in the gap; these gases respectively have relatively low and high thermal conductivity. The program was soon expanded to include a large loop test with helium gas circulating at the EGCR pressure and temperatures. This loop in the ORR was of a very different configuration from the MTR units but would be equally complex and could repeatedly accommodate more definitive tests of fuel pins having different designs.

By 1958 much work had been done by others in the development of commercial water-cooled reactors based on technology from the Naval Reactors Program led by Admiral Hyman Rickover to power submarines and aircraft carriers. The Light-Water Cooled Reactor (LWR) concept was approaching commercial deployment. With encouragement from Bob Charpie of ORNL, Bob Pahler and Dick Kirkpatrick of the AEC chose to pursue the EGCR concept with a vigor that might put it on a technological par with the LWR. The EGCR fuel elements could operate at higher temperatures than was available through LWR technology and thus obtain advantages in thermal efficiency and fuel utilization.

Many variables had to be considered in the design and testing of the EGCR fuel pins. (Seven fuel pins, six symmetrically supported around one, comprised an EGCR fuel element.) Although much fuel-pin experience was available from LWR testing and operation, those pins were smaller in diameter, had different container materials, and operated at lower temperatures. Uranium dioxide fuel has low thermal conductivity, and the EGCR pellets would therefore operate at high center temperatures, from 3,000 to 4,000 degrees Fahrenheit. Thus, they distort by differential thermal expansion, can crack, and can interact adversely with the container. Pellet shape, method of manufacture, operating limits, and stability all had to be explored, and other fuel-pin designs such as packed fuel powder were to be evaluated.

This experimentation and tight schedule required an additional fourteen capsules that were to be installed in the large, new AEC Experimental Test Reactor (ETR) facility in Idaho. Later, several capsule tests of a slightly different design were planned for the General Electric Company's privately owned reactor in their Vallecitos facility near Pleasanton, California. They also proposed to construct a large helium-cooled loop that could test scaled models of several EGCR fuel elements and their supporting structures. In addition smaller capsules were designed for operation by others in the low intensity test reactor (LITR) at ORNL to study variations in fuel composition. It was an ambitious program that challenged the team of John Conlin, Frank McQuilkin, Chuck Cunningham, Victor DeCarlo, Al Longest, Ron Senn, and many other members of the Irradiation Testing Section.

The Atoms for Peace program initiated by President Eisenhower included a nuclear ship, the NS *Savannah,* commissioned with a water-cooled reactor power plant. Envisioning that the ship would be successful, the Maritime Reactors Branch of AEC chose to develop a more advanced fuel element for replacement reactor cores. We were asked to test the fuel (provided by commercial suppliers) in a pressurized water loop to be installed in the ORR. The Maritime Branch also directed that we adapt the design of a test loop under construction by the Naval Reactors Program at the ETR. Ira Travis Dudley, recruited from the molten-salt reactor program, and I studied the drawings and design features of the ETR loop to conclude that it was not suitable for the Maritime Reactors work. We were dissatisfied with its design and were later vindicated when it was never made to operate satisfactorily. Responding to persuasion, the Maritime Reactors staff agreed to let us design a test loop that would meet their requirements and fit into the ORR operation.

We contacted Naval Reactors to learn more from their broad experience. They were most cooperative in sharing their experience with problems encountered, even though we had not chosen their latest design. Through visits to their contractors at the Westinghouse Bettis facility, near Pittsburgh, the KNOLLS Atomic Power Laboratory in New York State, and the Canadian Atomic Energy Limited facility at Chalk River, Ontario, we observed the best available technology. Most of the contacts were quite normal, but one day at KNOLLS, each time entering a room, I found people responding to my knock with rapid and unusual movement. After several such encounters, I commented on the phenomenon: "Why such commotion?" The answer: "Admiral Rickover is here today; we start moving, just in case, whenever the doorknob turns."

Armed with the best technology, Dudley and Don Tidwell proceeded to design and construct a highly reliable loop that, by utilizing two fuel element positions in the ORR core, could simultaneously irradiate six maritime reactor fuel pins. That loop functioned well for six years without missing a scheduled operation to qualify the fuel for use in the ship. Unfortunately, the NS *Savannah*'s second core was not needed because this beautiful, streamlined vessel, designed for both passengers and freight, proved uneconomical and was retired after making many ports of call. The loop was still so clean and leakproof at the end of its fuel-test operation that the shielded enclosure could be entered in street clothing without significant radiation exposure. The equipment, except for the in-reactor portion, is still in place at the ORR, which also has been shut down since 1986. However, the shielded room is prominently marked as hazardous, not primarily because of the danger of radiation exposure but because we had insulated the piping, valves, and tanks with asbestos.

Parallel to the EGCR work, Murray Rosenthal and others were studying fuel designs for a High Temperature Gas-Cooled Reactor, one not restricted by the temperature limits of stainless-steel-clad fuel pins. They favored a vessel filled with two-inch-diameter graphite spheres containing enriched uranium and cooled with helium. Two designs were considered. In one, tiny spherical particles of uranium dioxide or uranium carbide were coated with pyrolytic carbon to retain fission products and keep the helium cooling system relatively free from radioactive contamination. In the other, the fuel particles were mixed intimately with the graphite, allowing volatile fission products to escape to the coolant. The level of radioactive contamination in that helium system would be high (as with the molten-salt reactor). We were to conduct irradiation tests for both fuel types.

Fuel embedded directly in the graphite was chosen for the first test because it was easier to prepare. We adapted our MTR capsule equipment design to accommodate a suitable specimen and irradiated it for a three-week reactor cycle at temperatures anticipated for graphite fuel, up to 1,400 degrees Fahrenheit. Upon examination in the hot cell, the specimen was found badly distorted. Swelling of the fuel had occurred from the increased volume of fission products and from pressure exerted by trapped gaseous or volatile components. Other forms of less porous and stronger graphites were tried in successive tests, but none could be relied upon for use in a reactor.

It was apparent that this simple fuel element would not be satisfactory; it would be necessary to coat and thus seal small particles of the fuel. For the next test, spherical fuel particles a bit larger than a ball-point-pen tip were coated with two layers of pyrolytic carbon. The first coating was made soft and porous to accommodate increased fuel volume during fission with the outer shell, made stronger and impervious to penetration by the fuel and volatile fission products. Graphite structures embedded with hundreds of these spherical fuel particles proved stable when irradiated.

We had learned of a similar high-temperature gas-cooled reactor study in the Federal Republic of Germany, and the AEC invited representatives of that program to visit ORNL. The Germans were designing a helium-cooled reactor to use spherical graphite fuel elements much like those conceived by ORNL. Since there were no test reactors in Germany, they were especially interested in our fuel irradiation experiments and were shocked when shown photographs of the distorted specimens. Those results meant that their proposed fueled spheres, having uranium carbide dispersed in the graphite, could not be used. However, coated-particle fuel was already under development in England and at the Battelle Memorial Institute, General Atomics, the Minnesota Milling & Mining Corporation, and ORNL. Much of this work was for reactors envisioned for use in outer space.

Coated-particle fuel was chosen for the German Arbeitsgemeinschaft Versuchs Reaktor (AVR). After much study and negotiation, the Germans chose Union Carbide to manufacture the fuel in a privately owned plant at Lawrenceburg, Tennessee. Irradiation Engineering was to test and validate the fuel for the AVR using capsules to be installed in an ORR core position. Eight fuel spheres were encased in each "eight-ball" capsule test. Irradiation Engineering was then busy with fuel testing for four reactor concepts: two cooled with helium, one cooled with water, and one fueled by molten-salt.

———•◦•———

This was a particularly busy time for us with my heavy work load in Oak Ridge, Idaho Falls, and at Pleasanton, California, and with two young boys at home. Elaine was again pregnant, in pursuit of our goal of raising four children. This term was uneventful, except for concern for the unknown cause of Harold's death. On February 23, 1959, Caroline Sue was born as an apparently well-developed and healthy baby. We were exuberant, then fearful when the doctors observed symptoms of the previous problem. The medics this time were prepared to effect immediate blood transfusions, then the only known cure for this "blue baby" problem and were hopeful of success. The nurses insisted that I hold and carry her about the hospital during the brief time when she was not subject to treatment; I am forever grateful for that privilege. Caroline was a beautiful baby, with much flaming red hair, a vision that is deeply embedded in memory, but she died the next day despite all efforts to save her.

Baby Harold's problem had been diagnosed as an unknown blood factor mismatch. Even though both Elaine and I are O+, the apparent subtle factor of this problem had not been discovered and, at least to our knowledge, is not known today. Researchers at Children's Hospital in Boston, Duke University, Johns Hopkins, and the National Institutes of Health all received many blood samples from us and some from each of our sons. The studies continued over many months as they repeatedly requested blood and family history. Elaine, even more devastated than I by the loss of Caroline, was to have serious post-partum difficulty. She suffered continued hemorrhaging that extended through several doctor's appointments, but further treatment was delayed in optimistic hope of natural recovery. The delay nearly cost her life. I had worked late on April 8, when the first EGCR capsules were being installed during a scheduled shutdown of the ORR. The installation was to be carefully executed to minimize reactor down time. At 2:00 the following morning, as the work was proceeding well, I arrived home to discover that Elaine had been hemorrhaging profusely. Still the optimist, she had not telephoned for me to come home and could not go for help. On assessing the situation, I called Dr. Fortney rather than retiring as Elaine insisted, then on his advice ordered an ambulance and telephoned a neighbor to stay with the boys through the night. The ambulance, the doctor's car, and mine proceeded through Jackson Square at sixty miles an hour on the way to the hospital.

Elaine was close to going into shock from loss of blood and none could be found to match her type. Finally, one pint appeared satis-factory, although questionable; there was no more available, and the need was so critical that we agreed to a blood transfusion. The only mistake I can note among the several medical decisions of that night was to leave me alone with Elaine as the transfusion progressed. Midway, she reacted strangely; I called the nurse, who quickly stopped the blood flow.

Following a medical conference, a D&C procedure was recommended to stop the hemorrhaging as they continued to search other hospitals for additional blood. The procedure went well, but the bleeding continued when the packing was removed, and Elaine was returned to the operating room for a hysterectomy. It was 5:00 A.M. By then additional blood had been found, and Elaine was strong enough to tolerate additional transfusions, although the matching at first remained imperfect.

During this surgery, I made further arrangements for the boys to go to school and consulted with the capsule installations and with the many other activities at ORNL, Idaho Falls, and Vallecitos; each was advised to carry on since I might be unavailable for some time. At mid-morning, the surgeon, Dr. Robert Bigelow, with Dr. Julian Reagan and Dr. Guy Fortney, emerged to tell me that Elaine had survived the surgery but was not expected to live. She had been taken to an ante-room outside the intensive care area so that I could be with her for the estimated twenty minutes of life remaining.

Subcutaneous bleeding was evident, and its subsidence or progression indicated life or death. I watched intently and prayed. Nurses came frequently but could do nothing but console. As an experienced observer, I saw signs of stabilization and even slight reductions in the red blotches. I slowly became encouraged by her increasingly steady breathing. After about an hour she was wheeled to intensive care, and I was sent to get food and rest. The doctors, however, requested that I not leave the hospital for a few more hours. I telephoned Elaine's parents to tell them of the crisis, describing the prognosis as very discouraging but not hopeless.

I spent some time that evening with Byron and Tom, explaining the seriousness of their mother's illness but reassuring them as best I could that she would be well. They were to stay with Evelyn McQuilkin, a neighbor, close friend, and the wife of Frank, who was still working frantically at the ORR to get the capsules into operation. Elaine's cousin, Benson Causey, also arrived late that evening and was a great help, as comforter and confidant. I spent the night in the hospital. By morning Elaine's condition had stabilized, but the doctors concluded that, as they had feared, the transfusion mismatch had caused a renal shutdown. Temporary use of an artificial kidney would be required if she was to survive. Baptist Hospital in Knoxville had recently obtained an artificial kidney for dialysis, and Elaine was to be their third patient. On the one hand, it was not reassuring that the first two had died; on the other hand, they had suffered from kidney disease.

We would move to Baptist Hospital at the proper time, allowing Elaine to gain some strength before transport. She was in a special room near a nurses' station with no visitors except clergy. One was our dear friend, Bill Pollard, who, in addition to his position with the Oak Ridge Institute for Nuclear Studies (ORINS), had studied and become an Episcopal priest. At one point, however, Elaine was less than comforted when she interpreted one of his prayers as the administration of last rites. Our own pastor, Mervin Seymour, was also there and was supportive and encouraging. There was no further problem with matching blood, but, of course, she no longer had much, if any, of her own.

The ambulance ride to Baptist Hospital on the third day of hospitalization went well; Elaine was soon exhausted and slept most of the way. One of the doctors, a renal specialist, absently commented at her bedside that she had no chance to survive. I was furious, fearing that she had heard his remarks and, perhaps rashly, lectured her doctors about such bedside manners. Fortunately, she had not been awake, but her optimism was losing out to depression, and a positive outlook seemed essential for recovery.

The poisons normally removed by the kidneys reached a critical point after three more days, requiring use of the artificial kidney machine managed by Dr. Harwell Dabbs. I stayed with Elaine for the twelve-hour procedure. The medics constantly fussed with that early model that resembled a clothes washing machine in size, appearance, and sounds it emitted. From time to time, after reading instruments and blood-sample analyses, they added a bit of one salt or another to the secondary circuit to fine-tune the process. Their obvious inexperience was not reassuring. We also worried about the lack of blood flow to her left arm below a cannula, because her internal bleeding did not permit a proper insertion. The arm turned blue, and there was serious concern that she might lose it. The clotting time for her blood had been essentially infinite when the inter-hospital move was made, and she continued to be hemophiliac.

In preventing loss of blood from the arm, flow below the penetrating needle was severely restricted.

Upon Elaine's return to her room, Dr. Freeman Rawson, in charge of Elaine's treatment, requested that I not leave her side for more than twenty minutes at any time, a restriction that continued through nearly all of the next three weeks. This was necessary even though, at the doctor's recommendation, we had private nurses day and night in addition to those of the hospital. Dr. Rawson wanted me there to communicate with her and to observe every symptom of problems or progress. She was so weak that others could not understand her or properly evaluate subtle changes in her condition and attitude. Elaine had no physical strength; both a nurse and I were required even to turn her over. I was to receive all of the hospital laboratory reports, read the charts, and participate in the daily medical planning meetings of doctors and nurses. This was most unusual in those days, when such information was seldom shared with the patient or family. We struggled through two more procedures with the kidney machine, required at about three-day intervals, before indications of kidney recovery became evident.

Midway through the Baptist Hospital stay, Elaine's mother and Aunt Annice, who were caring for the boys at our house, were on their way for one of their visits and noticed Dr. Fortney as he passed them on the road, heading toward the hospital. He was simply hurrying to meet commitments with many patients, but they feared that Elaine was in trouble. He did not recognize them but soon became aware of a car following him at a greater speed than he expected from two older ladies. Aunt Annice was driving and the doctor, in describing the pursuit, commented that the driver could just see over the steering wheel, but followed his every turn. When they met at the hospital, the ladies were relieved that Elaine was stable, the doctor that his pursuers were identified. Even then, Mother and Aunt Annice were not permitted to visit Elaine, except to make their presence known; she was too weak to attempt conversing except softly with me.

The severity of Elaine's condition was illustrated two years later when she had occasion to see Dr. Rawson. He had spent three full days with her during dialysis and had seen her every day of the five-week stay at Baptist Hospital. He could not recognize this vivacious patient. We are deeply grateful for the medical skills and care received at both hospitals and particularly note Dr. Roberts who, without choosing improbable words of encouragement, conveyed needed hope to Elaine.

So many friends and relatives sent flowers and helped with the boys that it is impossible to attempt proper recognition; however, our appreciation cannot be overstated. Not the least was ORNL's generosity in granting time away from work. Even though I used up all my allowable personal leave and vacation time, I was granted additional days with pay.

At the end of the fifth week, Elaine could be moved to the Oak Ridge Hospital for an additional five weeks of recovery, there to experience painful withdrawal from massive doses of cortisone received as part of the treatment. Although she was slow in gaining strength, the boys and I were able to return to a more normal routine at our house, where Elaine's mother and aunt were still helping.

It was good to return to work and find that the many tasks had been handled well. The one serious trouble spot was at Vallecitos, where the large test loop and capsules were being installed. Bill Ferguson had spent many weeks ably following the work there. He had advised me by telephone during Elaine's last week at Baptist Hospital that problems had developed for which he might need assistance. As the situation evolved, I decided that I would have to go to the Vallecitos facility as soon as possible, during the last week of Elaine's hospital recuperation in Oak Ridge. The decision was difficult, but the Vallecitos situation seemed urgent, and Mrs. Causey's presence at the house made it possible.

I flew to San Francisco and drove to Vallecitos. That evening and the next day, Bill and I observed the problems he had noted, and we found even more. Late that afternoon we met with Bob Coyle, the Vallecitos customer representative. I had become acquainted with Bob four years earlier, when we had worked on similar projects at the MTR in Idaho. Bill and I recounted the defects observed in the Vallecitos installation. Most were specifically pertinent to our loop, but we also had found a mixing of control wiring between our equipment and another experimenter's system. Other serious concerns were primarily with the reactor facility. Bob's face became increasingly ashen as we discussed the list of defects, also described in a written memorandum. As a competent engineer, he had understood the seriousness of every word. We called it a day and left for dinner. I then retired early to recover, not only from jet lag but from near exhaustion.

Nevertheless, the phone rang at 2:30 A.M. It was Bob asking that we come to the reactor site immediately for a meeting with his management. I replied, "Bob, are you out of your mind? Call me again at seven and we will be there at eight." After several exchanges, he pleaded, "I have a company vice president from San Jose and other officials standing over me. Please, as a personal favor, come now!" I then agreed, "Okay, Bob, we'll be there in a few minutes." That proved to be a tense negotiating session that lasted far into the day. We insisted not only that all of the defects be corrected but that an intensive search be made for others that might have been missed. A plan was developed for a systematic review of the equipment and for repair and revision as needed. I hurried home on overnight "red eye" flights to see Elaine and the boys.

The seriousness of the problems involved was demonstrated later by General Electric's firing the plant manager and committing to make the changes required to bring the reactor operation and test facilities up to the demanding standards of nuclear energy technology. This was difficult, but the company also had entered the commercial nuclear power business and previously had planned to build a nuclear-powered airplane engine. Compliance with safety standards was becoming recognized as vital in this new and pioneering industry.

Although Elaine was now at home, many more weeks passed before she was strong enough first to walk again, then to leave our bedroom surroundings, and finally to venture downstairs to the kitchen. The confinement had been a great frustration, for she loved to cook—and still does to my good fortune. That period was also busy for me, what with cooking and other housework to do, spending additional time with the boys, and completing ORNL work delayed by the extended absence. Elaine's left arm, injured during connection to the artificial kidney, slowly recovered its normal color but had little strength and no sense of touch. Reentry to the kitchen was dangerous; she once sensed contact with a hot surface by the odor of burning flesh. In due course she regained feeling in the arm, but with that came chronic pain and extreme sensitivity to cold. That year, and for several more, she wore a long glove on the left hand and arm, even in warm weather. I am pleased to say that events and stresses in the remainder of that year pale when compared to those of the first half, both at home and at work. We were, and are, grateful.

Installation and operation of the ORR helium cooled loop under John Conlin and John Franzrab had progressed well with none of the problems encountered at Vallecitos. The capsules in the ORR had become "work horses" for irradiation testing, and the installation of fourteen capsule positions in the ETR was nearing completion. The small capsule experiments in the LITR were producing needed information about the behavior of uranium dioxide fuel as a function of temperature. The pressurized water loop, described earlier for the NS *Savannah* fuel tests, was also complete and operating well. The team was functioning with precision and camaraderie.

In the spring of 1960 I was requested to visit several nuclear installations in Britain to obtain more of their gas-cooled reactor technology. Oscar Sisman and Ed Storto were to accompany me to function as a team. Ed was designing large test loops to be installed in the EGCR, and Oscar was in charge of evaluation of the hot-cell experiments by Irradiation Engineering. It was an extensive trip, reaching from the Dragon Project reactor at the U.K. Windscale nuclear facility near Bournemouth in the south of England to their Dounray facility near Thurso on the north coast of Scotland. I proposed to Elaine that she go as a celebration of her recovery. She protested, fearing that she might not be strong enough. I pointed out that she could rest each working day and that our sons would stay at home with an experienced woman baby-sitter. Elaine agreed, and the trip was planned, largely through the U.S. Embassy in London.

Because the AEC would reimburse us only for my fare, Elaine and I took a turboprop aircraft, which cost substantially less than the faster jet. Ed and Oscar, who flew by jet, waited at the airport for us to arrive, thus losing their time advantage. Also, the longer flight time had afforded us more sleep. We took a cab to the embassy to find that travel plans had changed. We had been scheduled with a day to relax in London, then fly to Wick near Thurso. However, as it turned out all flights were booked; we were to take the night train to Inverness, spend a day and night there, and board a local train to Thurso. At the rail station the compartment tickets read, "Trauger and Storto" and "Trauger and Sisman." Some jesting ensued, but I firmly exercised team leader authority to revise the assignments.

During the train ride we watched the countryside slip by with its remnants of wartime "victory gardens" and then farms and pastures with sheep, sheep, and more sheep. Even city views were pleasant. At dinner we commented on our enjoyment of this countryside and our plan to rent a car and see Loch Ness on the next day, Sunday—only peripherally aware of a gentleman sitting nearby.

At breakfast our neighbor at dinner introduced himself as Robert Urquhurt and invited us to join him for a drive around Loch Ness. In a heavy Scottish accent he observed, "Since we had liked the English countryside, we must see the greater beauty of Scotland" and explained that we could not rent a car. This was the first day of Scotland's fishing season, and rental cars would all be booked. This also explained the unavailability of flights to the north. We accepted; he was to pick us up after lunch at the Station Hotel Restaurant.

Shortly after dessert was served, a waiter came to inform that our host was waiting. We thanked the waiter, casually stating that we would be out soon. The maitre d' then appeared in a flurry announcing, "Sir Robert Urquhurt is waiting for you," in a tone indicating that we should not tarry. We soon discovered that Sir Robert was the Commissioner of Crofters for Scotland, the equivalent of Secretary of Agriculture in the United States. He had come to Inverness to take care of some business without his wife, who was in a London hospital recovering from a fish-bone penetration of her throat. Sir Robert graciously led us to his Rover, commenting that it was the poor man's Rolls-Royce.

We drove across the moor on single-lane roads, some closely lined with hedges and fitted with distantly spaced "turnouts," which provided room for vehicles to pass. The drivers, including our own, played "chicken" to see who could bluff the other into waiting at these meeting points. This contrasted with the charm of a quiet walk on soft, thick peat beds at one brief stop. We followed the western shoreline of beautiful Loch Ness with Elaine speculating about the monster, a myth that Sir Robert viewed with disdain. We briefly visited the ruins of Urquhurt Castle but were told that our host's family did not trace to that of the castle. He then took us for "full English tea" at a quaint cottage restaurant where our hostess was a former handmaiden to the Queen Mother.

Elaine was apprehensive when asked to pour but did very well; at least there was no indication of our having breached British etiquette.

The last stop, after passing the Cullodon battle ground, was to Cullodon House, where in 1746 Prince Charley slept fitfully during his last night in control and where Cromwell, who replaced him, slept well on the next. The owners were away but the house was open, and Sir Robert, as a close friend, seemed comfortable in giving us a tour that included even the famous bedroom. (The matron, member of a prominent Scottish family, was a pilot who had ferried airplanes from the United States to Britain during World War II.)

Our tour ended at Sir Robert's farm, Meldrum of Inshes, where he proceeded to show us the house. He particularly wanted Elaine to see a newly remodeled bathroom patterned after some they had seen in the United States. Sir Robert then excused himself to gather eggs from the hennery while we observed the sturdy barns and fences of the farmstead. Later, while relaxing with a fine wine in the deep chairs of a comfortable den, we invited him to be our guest for dinner at the hotel. He first declined, stating that he had not brought proper clothing for dining. I commented that present attire was what I had to wear (one of two suits for the trip). He glanced my way and accepted; his was a beautiful woolen tweed. During dinner, Sir Robert explained his desire to entertain Americans: his children had been sent to the United States for safety during the war. He had been prominent in military intelligence, and his family would have been in jeopardy in the event of a German invasion. Their children had received excellent care for which he was most grateful.

On Monday we took the early train to Thurso across the flat northern plain of Scotland. As I stood in a walkway beside the compartments quietly observing a passing village, I was suddenly startled. The adjacent window had shattered with a sharp and resounding crash to spread shards of glass across the aisle. Other passengers alerted by the sound, gathered quickly to ask, "What happened?" I had noted some boys playing within view of the tracks a few seconds before, but it would have taken a Mickey Mantle to throw a stone that far, and no rock could be found on the floor. Someone then noted a scarred penetration of the compartment wall's ceiling beam, where a bullet might have passed through or be lodged. My already-racing heart accelerated. We informed the conductor, who stopped the train at the next local station to report the incident. In due course my pulse returned to normal, and we received no further word of the event.

We were met by a Dounray staff car and driver at a whistle stop short of Thurso. The team was taken to the Dounray facility and Elaine to our hotel. At lunch Dr. Atkinson, in charge of experiments, and Dr. Hurst, plant manager, described the overall program for nuclear reactor development in Britain. In many respects it was remarkably parallel to that of the U.S. AEC but was focused only on gas-cooled reactors and the sodium-cooled fast breeder concept. The American program at that time—in addition to the liquid metal fast-breeder, MSR, GCR—had others of organic fluid-cooled designs, a sodium-cooled thermal reactor, and several variations of water-cooled reactors, including one cooled by heavy water. The organizational structure of the U.K. Atomic Energy Authority (UKAEA) was also similar, except that their "commercial" reactors were to be built and operated by the authority or the government utility, the Central Electric Generating Board (CEGB). On the second day, we visited reactors and other facilities to become familiar with their designs.

The Dounray Test Reactor, a counterpart to the MTR in Idaho, was just starting up. Their experimental program was similar to ours for the ORR with a bit more emphasis on studying possible failure mechanisms in fuel elements. However, we did not receive enough information for it to be useful, and I commented on the need for more information if the exchange was to be fully effective. We had provided Atkinson with much detail during his earlier visit to ORNL. That evening we were entertained at dinner by the Hursts and others. It was an excellent meal with stimulating conversation that extended well past ten o'clock. I was then presented with a thick notebook providing much of the requested information. I was accorded the privilege of reading any or all of the notebook, as long as I returned it the next morning. Fortunately, I had brought along a then modern, but heavy, voice recording machine for preparing a trip report. After the others retired, I sat alone in a huge chair before the dying fire of the hearth in the hotel lounge to dictate many key pages of the book to plastic records. At about 3:00 A.M. that May 16, as I put a shilling in our room's heater control for warmth and brushed my teeth, I noticed that dawn was already breaking across the world above fifty-eight degrees north latitude.

That day we visited machine shops and facilities for equipment assembly that were sufficiently different from ours to be interesting and informative. At lunch we described advances in our work since the time of Atkinson's visit and discussed potential opportunities for further exchanges of information. We also visited the Dounray Fast Breeder Reactor, under construction, and were told of plans for experimental test loop facilities for their Advanced Gas-Cooled Reactor as counterparts to the EGCR loops that Ed Storto was designing. The visit ended with a brief walk along the shore, from which the Orkney Islands were visible on that especially clear day. Mrs. Atkinson and Mrs. Hurst had entertained Elaine at luncheons and had taken her to craft shops and to see other points of interest. We were grateful that she was able to participate in comfort. With perfect weather for the visit we were frequently admonished not to mention that pleasantry in England, for the Dounray staff were pursuing additional compensation for living in the harsh climate of northern Scotland.

That evening a thick fog quickly rolled in from the sea to close the Wick airport and thwart our flight to Glasgow. A personable young female chauffeur from the UKAEA was commandeered to drive us to Inverness, where we could catch the late night train to Glasgow. Our driver attended some errands on the way, including a stop at her home for a warmer sweater as protection from the increasingly chilly evening. The stark exterior of the house and weathered surroundings attested to the severity of that climate on the shore of the North Sea. We reached the train on time and arrived in Glasgow for breakfast.

Another UKAEA driver then brought us to the Scawfell Hotel at Seascale, England. The road skirted the beautiful Lake District, but I was so tired that sleep frequently interrupted the view. Late that afternoon Geoff Grenough took us for a drive to Wastwater Lake, then to see Scawfell Pike, and then to dinner at a very old and charming English pub. The next morning the team went to the Windscale works, now called Sellafield, to learn about the Calder Hall reactors, the project that had precipitated U.S. congressional interest in gas-cooled reactors. The British nuclear facilities were dual-purpose, to produce weapons-grade plutonium and to generate electricity.

Elaine had been left at the tiny seaside village with instructions to meet the four o'clock south-bound, whistle-stop train with all of our luggage. We were to board the train at Windscale and after picking up Elaine and the baggage, proceed to Manchester. Elaine lightheartedly complained that she had not come along to be our porter but handled the task well, with help from the manager of that small hotel. Our hotel in Manchester was of the old English style: guests left shoes outside the room for polishing, the toilet and bath facilities were down the hall, and a maid was in attendance on each floor.

The next morning the team went to Warrington and on to the research facility of Risley. Elaine had the day to explore the then smoky, dirty, manufacturing city of Manchester. Risley was and is still an engineering center for designing and testing large equipment. We were interested in the design of the British AGR and the tests to support its design. The experimental irradiation loop designs were also available for study. This was directly useful information because both the reactor and nuclear fuel test facilities represented experience applicable to the EGCR. We then visited nearby Culcheth, where the use of beryllium for fuel cladding was being explored. The U.S. AEC exchange agreement with the U.K. provided for us to test their beryllium-clad fuel elements in the ORR pool-side capsules, and we developed technical plans for that work. Next, we were free to spend the weekend in London.

The Royal Hotel near Hyde Park was a choice location for us as weekend tourists. We first enjoyed seeing London from the famous double-decked buses and learned to look the "wrong way" when crossing the street. Elaine and I then found the convenience of the underground. My primary touring goal, however, was to see the works of Sir Christopher Wren. His beautiful churches and public buildings had been a fascination in my reading about London in such works as *Stoddard's Lectures.*

I walked through the nave of Wren's Saint Paul's Cathedral, in awe of its grandeur and beauty. At the front the American Chapel was of particular interest for its intricate wood carvings of North American birds and plants and because it was a tribute to the Americans who served during the Battle of Britain. While studying the stately, slightly spiraled, carved wooden columns that support the altar canopy, I was vaguely aware of an elderly couple speaking softly near the wall carvings. As I stared in admiration at the altar's columns and canopy, the man moved to face me, and in a voice crepitant from age looked up to say, "It is wonderful to see one's own work admired." Startled, I looked down at this hawk-nosed, wrinkled old man, to think, "Is this Sir Christopher Wren reincarnated?" He seemed the image of portraits made in Wren's later years.

Henry Young then introduced himself, apologized profusely for intruding, and explained that he was one of the carvers of that beautiful woodwork. He had been a machinist who carved wood as a hobby before retiring. When the Cathedral advertised for carvers to redecorate the chapel, he applied, was chosen, and became leader of the team that produced the panels and columns. The work had been funded by contributions from children and others to repair the only bomb damage sustained by the cathedral in World War II. The wood-carving had been done in the North, and this was Henry Young's first opportunity to see it in place. His noting of details enhanced my appreciation of the patterns and identified features that otherwise might have been missed.

Mr. Young also showed where he secreted his initials, despite strict orders that carvers were not to so identify themselves. He and his wife had come to London to place recent carvings for sale in Selfridges department store. I immediately thought of engaging him to carve something of beauty for us. Mr. Young was interested, but we could not decide what he should carve. We exchanged names and addresses at the close of a delightful thirty-minute visit.

On Sunday morning, Elaine wanted to attend the City Temple Methodist Church. She was fond of writings by its pastor, Leslie D. Weatherhead. We arrived early at the Holborn Viaduct address and took seats to view the beautiful stained-glass windows at either side of the nave. A parishioner soon joined us and struck up a friendly conversation. We commented on the newness and beauty of the nave and then were shocked by her tears that followed. She explained that her husband had been in attendance with her when the church was bombed in 1944; he and many of their friends were killed and the church interior destroyed. Her account of that night of terror brought World War II to us as a deeply personal experience. Reconstruction had just been completed, hence the new appearance of the walls and furnishings.

It was equally surprising that we were to hear Leslie Weatherhead's farewell sermon to the congregation on the occasion of his retirement. The sermon title was "Where Are You Going to Live?" The double meaning was first an echo of the question posed to the Weatherheads upon leaving the parsonage. The deeper message was a scholarly discourse on the concept of the afterlife. We met the great theologian on the steps of the church to enjoy several minutes of pleasant conversation after most parishioners had departed. We were deeply moved by the experiences of that weekend.

The train to Bournemouth for our visit to Winfreth and the Dragon Project—so named because of the reactor's extraordinarily high temperatures of operation—passed through the New Forest Region; the term "New" may date to the twelfth century. We saw it partially forested, partly in grass and populated by wild horses. According to history this region provided the great oak trees felled to build the ships that defeated the Spanish Armada in 1588. We stayed in Durley Hall, a hotel purchased by the UKAEA for visitors to Winfreth. The amenities of this establishment are illustrated by how toast was served in the dining room. It was brought before other food and placed in a heavy silver rack that sat on a marble window sill. Cold toast was a certainty, especially during winter.

The visit to Winfreth was particularly useful. Their work had included extensive studies of coated particle fuels for later use in the small high-temperature reactor being constructed there. They had experimented with fuel-particle coatings of silicon carbide, other ceramic materials, and pyrolytic carbon and included combinations of coating layers. Their helium-cooled Dragon Project reactor was supported financially by a consortium of fourteen European countries, most of whom also provided technical support by assigning engineers and scientists to work there. The director was Compton Rennie of England, whose skilled management leadership made the international team function efficiently. A key to success was the management decision to award purchases for equipment and supplies to member countries under strict rules related to their financial contributions to the project and with firm specifications for products produced. We also discussed technical areas for possible cooperation with ORNL. Leslie Shepherd was to be their principal contact for us, and Roy Huddle would serve as a specialist in coated particle fuels. ORNL, as a minor partner, would assign staff members to the project without other financial obligation.

From Bournemouth, we traveled north to Harwell by car through the courtesy of Allen Saunders. He drove a bit out of the way to include a brief visit to Stonehenge. That famous site was not then a tourist center, so we opened and closed what appeared to be a farmer's gate, walked about with no other visitors, and marveled at the ancient people's ability to transport and erect the massive stones. There were no signs or other descriptions, but we could envision the alignment of key stones with the probable sunrise location at the winter solstice. I took special pleasure in this quiet event, pleased to see this monument much as it had stood through many centuries.

Harwell at that time was a nuclear research facility much like the ORNL but has since become more commercial. There we met the director, Walter Marshall, later knighted for early contributions in nuclear energy development. He also served several years as director of the U.K. Central Electric Generating Board. Others with whom we exchanged nuclear fuel-testing details were Peter Murray, director of their Metallurgy Division, and John Sayers and Oliver Plail, whose roles were similar to ours. We visited the DIDO and PLUTO, reactors that were similar to the ORR in function though not in design. There we met Bob Jackson, head of engineering, and some of the reactor operators. In general, the Harwell facilities were simpler and probably less costly than ours. They were serviceable but we thought did not offer safety margins as wide as those we required.

Having completed their visit to Britain, Ed and Oscar returned home. I was to spend Friday at the Saclay nuclear research center near Paris to discuss plans for irradiating French experimental fuel elements in the ORR capsule facility. Elaine and I would then enjoy a weekend in Paris before returning home. However, while at Risley I received a telephone call from the director of the Petten Research Center in the Netherlands, asking me to visit them. That required travel on the weekend and less free time in Paris. I had accepted, pending ORNL approval and if the Dutch would provide some special entertainment for Elaine; they agreed, as did Bob Charpie by telephone from ORNL and Elaine. Our time in Paris, though brief, was pleasant, providing a bit of sightseeing and a show at the Lido. Having seen many photographs of the Eiffel Tower, I thought visiting it was not a priority, but I probably took as many photographs as any other tourist.

In Amsterdam we stayed at the Rieu de Lieu hotel, visited the Riiksmuseum to see the Rembrandt collection, and had dinner at the Lido restaurant. There we were served an excellent rijisttafel of Indonesian food. Elaine, the foods expert, was fascinated by one seemingly exotic morsel in a stir-fry type dish. Our language skills were inadequate to understand the waiter's answer to our question, so the chef appeared with an uncooked rooster comb on a napkin. Sometimes it is best not to ask questions, but the food was superb.

On Sunday we traveled by train to Wageningen, the home of Adrian Schuffelen, the Dutch gentleman we had entertained in Oak Ridge. We were honored that one of his daughters bicycled thirty-six kilometers from the hospital where she was employed to visit with us as friends of her father. We spent a delightful afternoon with Adrian in their home, which bore its construction date of 1609 inscribed in stone beside the door. The walled garden reflected their interests in plants and natural beauty.

Later, Adrian led us to the overlook above the Rhine River where during World War II he had been an observer for the Allied armies and reported on German troop concentrations and movements in the wide Rhine valley far below. That had been an especially dangerous espionage mission in the midst of German occupation troops. The Rhine impressed us with its winding traverse of the plain, but we shuddered to think of the harrowing and heroic role our friend had performed while hiding in the crevices of the bluff. Equally daring was his transmittal of the information through the Dutch and French underground network. Occupying Nazi troops had displaced the Schuffelen family from their home, occupied it, and severely damaged the interior. Valuable pieces of art and furniture were stolen, presumably shipped to Germany and never recovered. Adrian Schuffelen said that he had regained equanimity to communicate with Germans in their language, but if the conversation turned to the war of fifteen years earlier, he could not speak unless both parties resorted to English.

At Petten I discussed with J. J. Schnepvangers their experiments' designs and procedures that could be used in their reactor. Because their reactor was nearly an exact copy of the ORR, our experience was most relevant. For example, I observed that the access they had provided for pool-side experiments differed from the ORR access and would be difficult to use; their reactor was later modified to correct the problem. The Petten Laboratory was supported jointly by the Dutch government and the Organization for European Cooperation and Development. Like the Dragon Project it was to be staffed by people from many countries.

Meanwhile, Mrs. Schnepvangers entertained Elaine at the fabulous Alkmaar Cheese Market (with Elaine driving our rental car) and with an afternoon tea at her home. The Snepvangers and most of their friends traveled only by bicycle and public transport. Indeed, on leaving Petten, we drove to Amsterdam amidst afternoon rush-hour bicycle traffic, arriving in time for dinner and to pack for the flights home.

One autumn Sunday afternoon, while relaxing under our dogwood trees with their leaves and berries in full color, my thoughts turned to Henry Young and Saint Paul's Cathedral. With an idea in mind I rushed to the house for a sketch pad to design bookends that he might carve. They were to feature a dogwood flower, its leaves and stem to be carved into the back of each piece. Henry Young responded to my order by sending his own sketch, but it was of the English dog rose. Included was a statement that he had already started the carving, but without obligation for purchase if the price was not satisfactory. The cost was reasonable, but I very much wanted a dogwood design because the American Chapel carvings were of North American plants and birds.

Our letters had crossed over the Atlantic. When he discovered his mistake, he agreed to carve the dogwood pattern. I was greatly pleased with the set and also purchased the dog rose bookends as a gift for close friends. We later bought several other carvings; our favorite is a cheese tray adorned by a cleverly designed mouse. I corresponded with Henry Young for several years. He was an excellent writer of interesting letters, though he had no more than a high school public education. When those letters stopped arriving, we could only presume that he was seriously ill or had died; both he and Mrs. Young had suffered several illnesses. During later trips to England, I tried to locate the Youngs or some of their sons through the deans of Saint Paul's and Durham Cathedrals. The latter had known the Youngs well over many years, but the search was to no avail.

Elaine's arm and hand continued to trouble her with pain that at times was excruciating. A nerve block was tried, and several medications proved unavailing. We then scheduled a visit in February of 1961 to the Mayo Clinic in Rochester, Minnesota. The tests and examinations were comprehensive but yielded no solution beyond the suggestion of exploratory surgery, which Elaine rejected. Although we were favorably impressed by the clinic, it provided no help.

Despite the handicap, Elaine continued to assist friends with their designs of new or remodeled kitchens and had large dinner parties for friends, church groups, and foreign visitors in our home. Her dinners were never catered. The largest gathering at our tables up until that time accommodated forty-nine guests. Other guests included Harold Bosley, the commencement speaker at my college graduation and then senior pastor at Riverside Church, and Ralph Sockman of Christ Church in New York. We enjoyed reminiscing and learning of new programs at those two churches that we had enjoyed two decades earlier. Each pastor had been a speaker for special services at the Oak Ridge First Methodist Church.

Elaine taught courses in foods, nutrition, and the economics of home management both in YWCA and Adult Education Programs. She also entertained many guests who had come to ORNL for work and experience. One was Kiyoshi Inoue of the Power Reactor and Nuclear Fuel Development Corporation of Japan. He had come to work with Irradiation Engineering at the ORR to gain familiarity with nuclear engineering technology. Because Kiyoshi's wife, Mitsuko, and their two small sons, ages two and four, spoke no English, Elaine spent much time helping them with transportation, shopping, learning the language, and understanding American culture.

In the spring of 1961 Robert Charpie left the Laboratory and Oak Ridge for other work. He had been a member of the first Oak Ridge Board of Education, established after incorporation of the city in 1957. Charpie suggested that I run for another position on the board to be vacated by attorney Frank Wilson, who had been appointed a United States district judge for the Eastern District of Tennessee and would move to Chattanooga. With a strong interest in the schools I decided to run, having just finished one year as chairman of the Parents Advisory Council, which, before city incorporation, had served in lieu of a school board and assisted Fred Peitch, the AEC official responsible for schools. My plan to disband that body, which had grown redundant, succeeded; the city already had an elected board, and the PTA Council had become quite effective. With that background I was elected to the Board of Education for the remaining two years of Wilson's term and was reelected for a full four-year term.

This was a busy time for the Oak Ridge Schools; we completed a racial integration program, built two new schools, and made major additions to the high school. Integration proceeded smoothly, except for the usual concerns over inevitable reassignments of students to different schools. I am particularly proud of having found, through the assistance of Professor Goslin of Vanderbilt University, Edward Whigham to become the first superintendent of schools hired by the board. He led the schools through the first three years of this period, helping to establish the commitment to excellence in education that had been the mission of Forest Blankenship, organizer of the schools in 1943. Whigham was recruited away to the Dade County, Florida, school system and was followed by Jack Davidson, who also served well. Oak Ridge schools continue to be recognized as among the best in Tennessee and the nation.

Adriann Schuffelen and his daughter with Don Trauger at the Rhine River Valley overlook, the Netherlands, 1960.

In the summer of 1962 a surprising letter came from Vance Rogers, president of Nebraska Wesleyan University, informing me that I was to be honored with the Alumni Achievement Award at a fall convocation. The award was sponsored by the NWU Alumni Association, the Methodist Church, and the University. We traveled to Lincoln for the event of November 2, and while there I also spoke to the Alpha Gamma Beta Physics and Engineering Club that I had addressed on nuclear energy twenty years earlier as a student. The subject again was nuclear energy, but I focused on the realities of working with test reactors such as the MTR. The trip also provided for a short visit with my family.

On returning to Oak Ridge, a new problem was reported from Vallecitos. The contracted loop had gone into operation but had malfunctioned, and a stringer of fuel pins was damaged, invalidating the test. Two immediate questions arose. Should the tests be continued? That answer was yes; the loop was to confirm the EGCR fuel element design, and the problems with the General Electric equipment were correctable by proper operation. The second concerned who would pay for the extra work that resulted from the mishap.

I reflected on the earlier contract negotiations between Hagan Bond, attorney for the Union Carbide Nuclear Division and, thus, attorney for ORNL, and a battery of General Electric attorneys involved in preparing the contract between ORNL-Union Carbide and General Electric. I had known Hagan only through his daughter, who had been in a church youth group that Elaine and I had sponsored. He was the stereotype of a country lawyer and, indeed, was from the hills of Tennessee. Hagan was slow of speech and seemingly slow of thought. When those original negotiations began and I thought about Hagan confronted by three polished, big-city attorneys, I was concerned. But Hagan Bond was not; he met every challenge with a sage and considered position. After extended negotiation, the completed contract had placed responsibility for failures of the type now confronting us on the subcontractor, which wound up saving the government nearly a million dollars that was required to correct the mishap.

The Vallecitos loop then operated well with a new set of fuel pins for the required test period. The test loop demonstrated that the basic assembly of fuel pins in a multi-pin element of the EGCR was stable. In addition to achieving success with this task, our finding of defects and assistance in their correction had contributed significantly in strengthening this major industrial enterprise for its role in the commercial nuclear field.

In the ten years from 1954 to 1964 our Irradiation Engineering group of from ten to twenty engineers and six to ten technicians built and operated many test facilities and conducted major experiments. First were the three loops for the ANP project and, later, two for the molten-salt commercial-reactor concept, with several capsule tests for each. For the EGCR-related programs fifty-three capsules had been operated successfully in the ETR or ORR and thirty-seven in the LITR. Tests in the ORR Helium Loop numbered fourteen. In addition tests were completed for the French (with two still underway in 1964, the terminus of this summary), and three had been completed for the British. For graphite fueled reactors the tally was ten in the LITR, seven in the ORR, and six in the MTR. Much of the LITR work was conducted by Oscar Sisman and Giles Morgan, both of the ORNL Solid State Division. The Maritime Pressurized loop in the ORR had operated for nearly five years, and a dozen experiments had been conducted, with two still underway in 1964. It had run as consistently as any of the Naval reactor loops and much better than most, in part for having benefited by their early experience. This was a highly dedicated and competent group of people who achieved remarkable success in a pioneering field.

During this time my secretaries were Shirley Hendrix and then Peggy Ward. They kept good reports and met the group's needs, whether at our Y-12 offices and shops or with regard to reactor experiments at X-10, Idaho, and Vallecitos. Although the group status had been raised to section level, it never numbered more than thirty people; yet more than fifty persons participated directly at different times. In addition many engineers, scientists, and technicians from other divisions provided dedicated support in physics, metallurgy, chemistry, experiment design, reactor operation, hot-cell examination, and instrumentation. The excellent shop and assembly work by craftsmen was essential to the safety and success of this program. The total budget for Irradiation Engineering at times exceeded that of any division of the Laboratory except the Reactor Division, where it was administratively located.

In the spring of 1964 Bill Manly, who had succeeded H. G. MacPherson as director of the Gas-Cooled Reactor Program, left for another part of the Union Carbide Corporation. I then became director of the advanced GCR work. I found leaving Irradiation Engineering difficult, but the section was to be well managed, first by Harold McCurdy, who had applicable experience, and later by John Conlin, who had participated from the beginning. Peggy Ward came with me, and the section gave her a well-deserved tribute. This work continues today under Ken Thoms, though it is less extensive because nuclear research programs are greatly diminished. My new assignment was for a very different role but one with adventure and new associates in many places, both domestic and foreign. It was to involve reactor designs that could fulfill my dream of widely used, public-benefiting, safe nuclear energy.

PART THREE

1957~2100

Energy Sources, Resources, and Projections

The German AVR Reactor Building. Courtesy of AVR GmbH.

NINE

Gas-Cooled
Nuclear Reactors

he United States Congress had authorized the U.S. Gas-Cooled
Reactor Program in 1957, and by 1964 it had grown into a
complex of projects spread from California to Pennsylvania.
Results of the program's research and development were shared on an
international basis by national laboratories through arrangements
with governments in Europe, North America, and Asia. ORNL participation included laboratories and commercial facilities in England,
France, Germany, Japan, and several countries of the European
Dragon Reactor Project located in England. Staff members from the
various institutions were expected to cooperate within terms of
exchange agreements, and individuals were often assigned to work in
organizations other than their own.

As the lead national laboratory in the U.S. High-Temperature
Gas-Cooled Reactor (HTGR) Program, ORNL was assigned important research tasks. I was pleased to participate more broadly in an
effort that also included testing of fuel and materials irradiation.
Some ORNL work was in support of the General Atomics Company

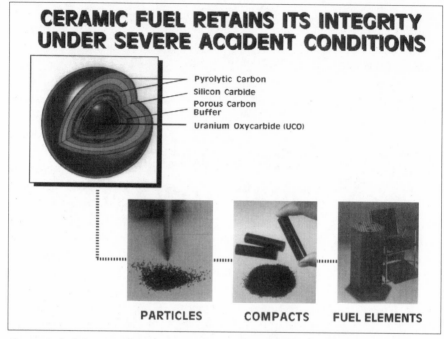

Ceramic fuel for GT-MHR. Figure courtesy of General Atomics.

in its design and construction of the Peach Bottom Demonstration Reactor in Pennsylvania. General Atomics was also designing a first commercial-size HTGR of 330 megawatts of electrical power or MW(e). In that reactor's design, coated fuel particles are embedded in small, cylindrical, graphite pellets that fit into holes drilled into large, hexagonal blocks of graphite stacked to form the reactor core. The helium coolant, pressurized to 750 pounds per square inch, passes through alternate holes in the blocks and transfers the fission energy to steam-generating heat exchangers, and then, repressurized by gas circulators, the helium returns to the core. This assembly is located inside a concrete vessel designed to provide containment and radiation shielding. Steam turbine generators produce electricity.

The ORNL role for the U.S. commercial HTGR effort was to review the design of selected components and conduct supporting research as needed and appropriate to the Atomic Energy Commission (AEC) program. ORNL was also responsible for development and testing of advanced coated-particle fuels, for graphite development, for other materials studies, and for assistance in evaluating new reactor design features. The HTGR fuel-irradiation test work continued to be my responsibility because much of it was funded through this program. ORNL would also continue assistance to the AEC's Experimental Gas-Cooled Reactor, then under construction in Oak Ridge, with Grady Whitman as director for that work.

The HTGR has many merits in safety and efficiency. Inherent properties of the materials and their configuration cause the nuclear fission reaction to extinguish if temperatures increase beyond allowable limits. In addition, the coated particle fuel can accommodate temperatures far above those required for efficient generation of steam and can achieve temperatures high enough to operate a gas turbine to drive an electric generator. Thermal response to accident conditions is slow because so much graphite must be heated before the fuel is in jeopardy, thus reducing the need for fast response by safety systems. This "thermal ballast," so to speak, also gives the reactor operator time to evaluate an abnormal situation and plan an appropriate response.

Large HTGRs gain additional safety from their massive, steel-lined, prestressed-concrete pressure vessels having thick walls, bound for strength by a multitude of cables strategically placed so that the failure of one or several will not jeopardize integrity or safety. The gas circulators, steam generators, and emergency cooling systems are housed inside the vessel. Thus, safety is ensured in large part by inherent design features of the fuel, coolant, and primary containment vessel. As a further precaution, a secondary, more conventional containment building encloses the entire system.

Light-water reactors, in contrast, operate in vessels and piping systems at high pressures contained in single-membrane steel vessels and use Zircalloy (metal) fuel containers that do not have a wide temperature margin before failure. They are made quite safe by multiple emergency cooling systems and extensive procedures for operation and lifetime surveillance. Their efficiency is limited by the temperatures and pressures allowable for the vessel and fuel containers, thus restricting their use to steam-electric systems of modest efficiency. In contrast, the energy efficiencies achieved by the HTGR are comparable to those for fossil fuel plants, and the high temperature capability can accommodate many process-heat applications, an important objective in early Japanese and German programs. I was pleased to work more directly with the flexible and inherently safe HTGR, which seemed to fit with my earlier expectations for nuclear energy.

A Gas-Cooled Fast-Reactor (GCFR) concept had been introduced later than the HTGR, with General Atomics responsible for its design. This was to be a breeder reactor (producing more new fuel than it burns by converting U-238 to Plutonium). This reactor would also use helium to cool the fuel, but metal cans were to contain uranium oxide in fuel elements similar to those of the experimental gas-cooled reactor. The incentive for this concept derived from the early concerns about availability of uranium ore, because the GCFR offered better capability for breeding than the sodium-cooled breeders designed elsewhere by companies in the United States as well as in Europe and Japan. This work at ORNL was initiated by Mike Bender but was later assigned to me. It was never a very large effort, and the lead was eventually moved to the Argonne National Laboratory, where it was joined to the Sodium-Cooled Fast Breeder Reactor program.

Our GCFR role was primarily to test fuel-element concepts in the Irradiation Engineering Section facilities of the Oak Ridge Research Reactor, and to evaluate devices proposed for safety systems. That evaluation was made difficult because the reactor would operate at a high power density and, under accident conditions, had only the helium to absorb heat directly since there is no graphite moderator. Failure of the primary heat removal system would cause the fuel container temperature to rise rapidly, producing serious damage. The GCFR necessarily would require emergency-shutdown and cooling actions comparable to those for the LWR and would need even more rapid operation of emergency systems. A significant measure of safety could be achieved by housing the reactor in a prestressed-concrete pressure vessel, thus minimizing the probability of losing coolant. Because safety issues began drawing increased emphasis, uranium ores were becoming more plentiful, and the HTGR seemed more promising, this reactor lost favor and its program was terminated.

National laboratories are particularly suitable for pursuing large programs like development of the HTGR technology. Their multiple resources and talents can be assembled in teams to address a project in a balanced effort of the required disciplines. The management plan created for the Aircraft Nuclear Propulsion Project had been adapted for the GCR Program by my predecessors, Robert A. Charpie, H. G. MacPherson, and William D. Manly. In this system, program management is overlaid on a conventional line organization. Permanent line divisions are organized along conventional disciplines such as chemistry, physics, metallurgy, chemical engineering, mechanical engineering, and others, including health physics and craft support. The role of line management is to ensure that employees' needs are met, that equipment and facilities are adequate, and that the quality of work is state-of-the-art.

Program management provides strategic planning and direction for the work and maintains a focus toward the objective by distributing and controlling funds. ORNL programmatic work could be placed with an appropriate division, divided among divisions, or even contracted outside the laboratory. This maintained a competitive environment to meet requirements of the sponsor effectively.

The small program staff thus functioned in somewhat the same way customer service groups operate in corporate structures. The Atomic Energy Commission was effectively the sole customer for the ORNL Gas-Cooled Reactor Program. Thus, we were to accomplish tasks specified by Bob Pahler and his assistant, Dick Kirkpatrick, of the AEC Headquarters GCR Program Office at Germantown, Maryland.

My secretary, Peggy Ward, and I were the total ORNL designated staff for the HTGR program. However, group leaders from the several participating divisions met weekly with me to plan and coordinate the effort. Bill Harms and, later, John Coobs managed HTGR fuel and materials development; other participants were Jim Scott, Troy Washburn, Frank Homan, Ron Beatty, and John Prados, a consultant from the University of Tennessee. Walt Eatherly and Ray Kennedy conducted research on graphite. Tom Kress, Dick Lorenz, Bob Wischner, Oscar Sisman, and others studied fission product retention and potential release from various coated particle-fuel designs, as well as the challenge of transporting such impurities in helium systems. Lester Oakes and, later, Syd Ball assisted with instrumentation and control system designs.

Several of these task leaders also effectively assisted with program management and conducted specific tasks within their divisions. Steve Kaplan soon joined the GCR team with responsibility for safety in reactor design and operation. As the program grew in size and complexity, John Coobs was made assistant director to coordinate all materials work. We cooperated with Don Ferguson and Pete Lotts, leaders of the Thorium Utilization Program, since the HTGR design of that time was expected ultimately to operate on the thorium-uranium 233 fuel cycle.

In addition we cooperated closely with General Atomics in high temperature gas-cooled reactor research through many contacts there. Norval Carey, their Washington, D.C., representative, was especially helpful by assisting in our interaction with the company and by interpreting government policies and legislation. That cooperation developed into a valued friendship with Norv. We also cooperated extensively with Walter Goddel in the design and testing of coated-particle fuel, with Tom Johnson and Walter Simon in reactor design, and with many others. Presidents of General Atomics during this period, in order, were John Landis, Richard Dean, and, currently, Linden Blue. This work at ORNL was initiated by Mike Bender but later was assigned to me.

In September 1964 I was assigned to a team that would review gas-cooled reactor work in Europe. Members were Bob Pahler, Bill Larkin, and Art Larson of AEC; Mike Bender, Grady Whitman, and I, ORNL; William R. Cooper, Tennessee Valley Authority; and Corwin Rickard, General Atomics. The visit started at the Winfreth site in Southern England, where the Dragon demonstration HTGR, the first in the world to operate, had just achieved criticality. Al Goldman of ORNL, who was following John Coobs as assignee there for two years, also participated. Our discussions were primarily with Dragon's Les Shepherd and Roy Huddle, whose staff had responsibility for fuel and materials development. George Lockett and John Dean provided insightful briefings concerning the engineering problems encountered in developing a new reactor system.

In addition to the work on reactor and fuel designs, the Dragon staff discussed extensive studies to select materials and develop surface treatments for use in the dry helium of this very high temperature reactor. These problems relate closely to those for equipment used in space, where water and oxygen are similarly absent. Metal oxides can be protective and help to prevent galling of bearings and sliding surfaces.

We were impressed by the extensive progress of the Dragon Project during the four years since my earlier visit. This was particularly significant considering its complex international management structure. We also admired the collegial attitudes of the project staff members; their humor and innovation were exemplified by the slogan "Put a Dragon on Your Grid." A cartoon figure of a fire-spitting dragon balanced on the wires of an electric-utility transmission line graced stationery and some technical report covers to emphasize the reactor's high temperature capability. The objective was to produce a power plant for electric utility companies that offered thermal efficiency and exceptional safety features.

From Bournemouth we traveled north to the Windscale Works near the Lake District where the United Kingdom (U.K.) Advanced Gas-Cooled Reactor (AGR), on which the EGCR was modeled, had been operating for more than a year. In discussions with John Moore, Ken Saddington, and others we learned much from their experience that could facilitate EGCR completion and operation. For example, the AGR helium piping loops, which provided separate coolant systems for irradiation tests of very advanced fuels that might fail, had already produced valuable data. Our more extended operation of the ORR helium loop was of particular interest to them, and we exchanged useful insights on the design and function of helium-irradiation test loops and about reactor operation.

The next stop was at Risley near Warrington, where discussions with Gordon Brown, Goeff Greenough, and Hans Kronberger centered around the overall nuclear reactor program of the U.K. That evening, we were entertained at a formal dinner at which we were favored with trifle, a traditional English dessert. When someone asked how it was prepared, the hosts had no answer. I then described the dessert, its preparation, and variations, as recalled from one of Elaine's parties that featured English desserts. Hans Kronberger, our senior host, took delight in chiding his staff for their ignorance, though he too had been silent when the question was asked.

The offices of the Nuclear Power Group at Knutsford, south of Manchester, made some of us envious. The management and research staffs were housed in a superb English manor house that retained its beautifully paneled walls and much of the original furniture. It was a striking contrast to our spartan spaces in World War II buildings. S. A. Galib, R. D. Vaughan, and Derek Smith discussed the merits of various reactor types, including steam-cooled, Magnox, helium-cooled, water-cooled—both pressurized and boiling—and sodium-cooled systems. This "think-tank" exhibited a great depth of knowledge about each of these technologies.

The Nuclear Power Group was part of a manufacturing company. In that capacity they proposed that Bill Cooper persuade TVA to seek a fixed-price contract from the British group to build a Magnox-type reactor. To what extent that discussion was serious was not clear to me, and TVA did not follow up on the suggestion. In retrospect, even though that Magnox reactor probably would not have met U.S. safety standards, the working arrangement could have been beneficial for TVA. Cooper's inquiry might have led TVA to consider obtaining more outside assistance. Many difficulties in the TVA nuclear program apparently stemmed from the agency's assuming a very broad role in a new field without adequate and relevant experience.

We also visited English Electric at Whetstone for discussions with Everett Long and others. There I had the pleasure of seeing the monstrous but well-designed machines by which U.K. reactors were refueled on-line (with the reactor in operation), a practice not followed in the United States.

On returning to London, our planned meeting with the Central Electric Generating Board's technical staff was nearly thwarted by a remarkable mistake. The Board at that time was the national electric generating utility for nearly all of England. Nevertheless, we had a problem finding its offices, for the cabby misunderstood the address given in our American English and deposited us in the Southwark section, on the wrong side of the Thames, and we dismissed him before discovering the error. This is my only knowledge of a mistake by a London cabby. I once was fascinated by a young man on a motor scooter who drove a short distance, stopped to check an address, consult a book and repeat the maneuver at another stop. Curiosity led me to ask what he was about. He was preparing for the examination required of all London taxi drivers, a test demanding incredibly detailed knowledge of the city. He was memorizing the home addresses of several hundred prominent people, to be prepared to take each one home on sight. Our own dilemma was compounded when, during that morning rush hour, we could not find another taxi. Fortunately, a telephone call explained our delay, and we enjoyed a pleasant walk across London Bridge to the new offices of the utility near the Cathedral of Saint Paul.

In meetings with K. P. Gibbs, W. B. Kemmish, and others we learned of the bids just received for building the Dungeness B nuclear power station, the first of several commercial AGRs they would order. In turn we informed them of plans for the first commercial size HTGR in the United States, then proposed for the Rochester Gas & Electric grid and later to become the Fort St. Vrain Reactor in Colorado. That day we also visited the United Power Company of Heston. This consortium of English General Electric, Fairey Engineering, and Combustion Engineering, Ltd., was building the Magnox plants at Hunterston, Scotland, and Trawsfynydd, Wales.

The AEC team members, Rickard, and I then flew across the English Channel to Cologne, Germany, and drove to Jülich. Discussions there related to the Arbeitsgemeinschaft Versuchs Reaktor (AVR) and plans for the manufacture and testing of its fuel in the United States. The meetings were with Professor Rudolf Schulten, principal originator of the AVR concept; Günther Ivens, the reactor operator; D. Cautius, the project director; and others. Also included were representatives of Euratom, an umbrella nuclear organization that linked the work of several European nations. The AVR reactor building was under construction, and components were in manufacture. I was particularly impressed with the complexity of the design. For example, the elaborate on-line refueling system that was to recycle thousands of the fuel spheres through the reactor during operation could also identify and remove spent elements and add new fuel spheres as needed. The research laboratory at Jülich was comprehensive and bore resemblance to ORNL.

I then traveled alone to Petten in the Netherlands to see how the installation of experiments in their reactor was progressing and to meet with Gene Hise, an ORNL temporary assignee there. It was interesting and reassuring to see the resemblance of their equipment to that in our ORR, but the differences were also instructive. They showed me their copy of a report I had written entitled, "Some Irradiation Test Facilities at the Oak Ridge National Laboratory." It was well worn, even "dog eared," and they referred to it as "their Bible."

In appreciation for my assistance, the Dutch arranged a guided tour on Saturday to the Delta Works, one of the world's largest engineering projects, the latest of many to reclaim and protect their land from the sea. My host, a retired engineer, had been a principal designer of this elaborate system. First envisioned some years earlier, the project gained support when a disastrous flood driven by a North Sea storm occurred on February 1, 1953.

The Delta Works's primary functions are to provide protection for low-lying areas, create canals for shipping, and provide lakes that store fresh water from rivers. Large sluice gates that facilitate the escape of ice from rivers, including one of the Rhine River Delta streams, were most impressive, as were the roads, often built atop the dikes, and bridges. The massive dams, gates, dikes, and locks to accommodate shipping were also impressive structures. Looking out over one panorama, I reflected on the great expenditure of energy involved in the movement of earth and in building these concrete monolithic structures, and I was reminded of the centuries-long relationship between the Dutch, the sea, and their lowlands. Those installations are indeed a long way from the simple dike portrayed in the proverbial story of the Dutch boy who plugged a leak in the dike with his finger to prevent erosion and potential flooding of the countryside. The inscription on a major barrier-dam support "Een volk dat leeft, bouwt aan zijn toekomst" (a vigorous nation builds for its future) seems an appropriate motto for the Netherlands, which has accomplished much from a small endowment of land and natural resources.

Following the tour we dined at a quaint hotel, where I was to spend the night. My host described it as typical of nineteenth century Europe and, it was, no doubt, built in that time. The dining room was dusky with draped windows, walls of mahogany paneling, heavily upholstered chairs, and massive tables and, along with the lounge, was reminiscent of Rembrandt paintings. The guest rooms were located around a large second-floor lounge. My room, however, was tiny with no closet; the only space available for my suitcase was first on, then under, the bed. That night, stricken with a traveler's malady, I was further confounded on discovering that the electricity had been turned off. The toilet was far away, beyond the large lounge scattered with chairs over which I repeatedly stumbled in the total darkness. Never again have I traveled without a small flashlight.

At the last stop in Europe I was to pay a brief visit to the Centre d'Etude de l'Energie Nucleaire of Mol, Belgium. On arrival in the nearby city of Gheel I was puzzled by the many signs I saw on curbs at various corners around the city, each of which displayed only a prominent exclamation point. While dining at the hotel, intriguing figures painted high on the walls told a story that shed light on the purpose of the signs. The story was of Dymphna, a spirited Irish princess who had gone to Gheel in the year 600 A.D. to escape repression of her social philosophy and activities and persecution by her family. While in Belgium, she is said to have performed several miracles that helped the mentally ill. After a time, word of her whereabouts and activities reached Ireland, and her father arranged to have Dymphna returned and executed. She was later canonized as the Patron Saint of the Insane. In the thirteenth century an advanced system of care for the mentally impaired was devised at Gheel. Afflicted individuals who could be accommodated by families were placed in homes by a sponsoring organization. A physician was assigned for each area to assist both patients and host families. The practice continues to this day, and the street before each participating home is marked by the sign featuring an exclamation point to warn motorists that a pedestrian living there may not exercise due caution. The inspiration for devising this remarkable system of care is attributed to Dymphna.

My visit to the Mol center featured its nuclear test reactor, BR-2, where AVR fuel spheres were to be irradiated in addition to those in the principal program at the ORR. The reactor also had other large test loops and was becoming a major nuclear research facility for Europe. The visit included the Mol nuclear-fuel reprocessing facility where Earl Shank was assigned from the ORNL Chemical Technology Division. Mol was developing advanced methods of reactor fuel processing and of the manufacturing of ceramic fuels, called mixed oxides, from recovered uranium and plutonium (a mixture of UO_2 and PuO_2). Overall, the visit to Europe had provided valuable information about very advanced work on high-temperature reactor technology and many features of other nuclear system designs.

Driving to a Brussels hotel that night for an early flight the next morning, I arrived in the city on a very dark night, missed a sign to the "Centrum," and soon was lost. There was no one to be found for help on seemingly deserted streets. In desperation I hailed a passing taxi. Upon describing my plight to the driver in a combination of German and English, he suddenly said, "Follow me." We took off at high speed through the narrow winding streets. I was determined not to lose sight of my benefactor, but the adventure made for great lumps in my throat as our speed often reached 110 kilometers per hour on streets where speed limit signs read 30 or 40. Was the taxi driver playing a game to have me arrested by a gendarme? It was a great relief when the hotel appeared on the right.

Because of the late arrival, the dining room was closed and after searching the streets, I found a small bar for a sandwich and a beer. Glancing about the dingy bar, my thoughts abruptly turned to Mom, the teetotaler, who would have frowned on my being there and particularly on the beer, which seemed necessary to down the dry and unappetizing sandwich.

After a short night and quick flight, I arrived at London's Heathrow Airport, only to face more distress. The clerk stated that my reservation to New York had been canceled and that obtaining a seat on the flight would be impossible. My response was to remain at the window until the ticket was restored. When the clerk moved on to another location to assist in boarding duties, my lot improved with the replacement. Finally, when approaching the gate, I was pleased to meet the first clerk, who could not believe that I was boarding.

I was pleasantly surprised when Elaine met me at the Knoxville airport that evening but soon sensed that something was wrong. Elaine guided us to a less congested place to tell me that Mom had died unexpectedly on the previous day. Her death had occurred at about the time I had thought of her while in the Brussels bar. We flew to Nebraska early the next morning, a somber trip in contrast to the visit we had planned for a month later. More than a year had passed since we had taken my parents to the Black Hills, a trip they had enjoyed as they reminisced about the earlier journeys there. At least I had that pleasant memory to ease the pain of the sudden loss.

By the end of 1964 Byron and Tom were doing well both in studies and extracurricular activities at Jefferson Junior High School. Elaine and I had discussed a trip to Europe with the boys and concluded that at ages fourteen and nearly sixteen they would know enough geography and history to appreciate European cultures. Both had studied Spanish for several years, including while they were in the Elm Grove Elementary School. (I had initiated the Spanish program at Elm Grove while president of its PTA. We started with a program open to students who wished to participate, using unpaid volunteer teachers. Because Oak Ridge had a wealth of people skilled in languages, finding qualified instructors was not a problem. Later, an elementary school language program became part of the official curriculum.) With this background, I presented brochures of a vacation trip to Spain as my Christmas gift for the family to be delivered in June, at the close of the school year. The timing, but not the trip, was a surprise to Elaine as well. Enthusiastic planning started early in 1965.

After our travel schedule was fixed, I was invited to an International Gas-Cooled Reactor meeting in Brussels on dates midway into the trip. For me to be in Europe and not participate, as the new ORNL HTGR Program Director, would have seemed inappropriate. The family reluctantly agreed that I would fly from Madrid to Brussels for the meeting, with Elaine and the boys staying in Spain. At almost the moment those plans were made, Jim Lane returned from Europe with a new problem. While in Brussels he had learned of my registration for the Brussels meeting and took the liberty of naming me as a participant in a similar meeting in Vienna during the following week. The meeting was of interest to me but unwelcome as an intrusion during vacation. We again consulted travel agents and brochures, discussed family interests, and created a new plan that delayed our vacation until after both meetings.

In June we flew to Brussels for the GCR meeting where we met many European colleagues, several of whom Elaine had entertained at our dinner table. During meeting hours she and the boys visited historic sites such as the Waterloo battle grounds. From Brussels we flew to Zurich for the weekend. The boys were fascinated with the feather beds, comforters, and other amenities of our small Swiss hotel. Our friends, the Alfred Schurch family of nearby Kusnacht, entertained us at dinner, and the next day Alf gave us a grand tour around Lake Greifensee and beautiful Zurich on the Linth river.

Dale Magnuson was on assignment from ORNL to the International Atomic Energy Agency in Vienna but was not involved in my meeting. We met only for dinner, but Elaine and the boys spent much time with his family in touring the Hofburg and other historic sites. In contrast I mostly saw the interior of the Agency, its conference rooms and lecture halls; but one evening we enjoyed a performance of *Don Giovanni* at the rebuilt Vienna State Opera House. Watching Tom's rapt response to the music and acting made the cost and inconvenience of his music lessons seem highly rewarding. The conference provided my first opportunity for significant interaction with scientists and engineers from the Soviet Union. Their openness and candor in discussions were both surprising and pleasing. The Soviet technical progress in nuclear power plant development was impressive, but their flagrant disregard for radiation exposure and other hazards disturbed me.

Our family vacation took control of itinerary at the next stop, Venice. We stayed at a small hotel near the Rialto Bridge and enjoyed leisurely strolls and boating through the city of canals. We were late for the afternoon flight of pigeons into Piazza San Marco, although it was besieged with the birds at our arrival. We dodged pigeons and their missiles while touring the magnificent square, as we admired the colossal bronze horses above the entrance to the Basilica San Marco and the view from the San Marco tower.

The next morning Elaine and I slept late and when we awoke found that Tom was missing from our suite. His absence in an unfamiliar place was disturbing, but knowing Tom's penchant for exploration we were not greatly alarmed. He had wanted to see the pigeons come in and had walked to the other side of the city for their morning arrival. Reunited, we were loath to leave Venice, but after a visit to the glass factories on Murano Island, we flew to Nice via Milan.

We spent two days in Nice during which we inadvertently allowed the bath to flood, not having observed the absence of an overflow drain. When we summoned the maid for help, she responded "O la la!," or something to that effect, and promptly began a vigorous mopping exercise. Having brought that emergency under control, we headed for the beach. The pebbled shore was disappointing, particularly because the stones and waves were interspersed with globs of thick, tar-like oil. Apparently an unidentified warship had cleaned its fuel tanks at night while nearby.

Tom, Byron, and Elaine with Alfred Schurch at the Swiss Federal Institute of Agricultural Research in Zurich, 1965.

Nevertheless, the beach pollution had not deterred the many sun bathers, clad in bikinis, who carefully avoided even the water's edge. We also visited the quaint and well-preserved medieval city of St. Paul to have lunch at the famous restaurant Colombo de Or. Choosing from its appetizer menu of seventeen items provided us with ample sustenance delightfully flavored, but even without entrees our travel budget was strained—and remember, this was 1965, before the financial exchange rates made Europe more expensive to Americans.

The French were building a large nuclear laboratory at Cadarache, north of Nice, and I particularly wanted to see their test reactor, a major feature of the site. After renting a Renault, we drove to the Canyon of the Verdun and on to Aix-en-Provence, the lovely and interesting city near Cadarache. Our hotel, the Paul Cezanne, displayed copies of many paintings by the famous artist. Elaine and the boys explored the city, while I spent a day with French technicians. Their new Pegase reactor was well designed, and by its placement in a large pool of water could accommodate several nuclear fuel test loops similar to those of my earlier work. (Some years later, one Pegase loop accommodated fuel tests of direct interest to ORNL and the U.S. Gas-Cooled Reactor Program.) This laboratory, a new and proud addition to the French civilian nuclear capabilities, was nestled in the foothills of the French Alps. I was treated to lunch in the great hall of a castle (owned by the French Commissariat a l'Energie Atomique) located on the south bank of the Durance River. From Aix we motored to Marseille for a flight to Madrid.

By then we had perfected an airport routine: Elaine changed money, Byron and Tom secured the luggage, and I rented the automobile. We were soon maneuvering through the rushing traffic of Madrid to find our hotel. One wrong turn in the maddening swirl took us on an unexpected tour of the central postal facilities, but we arrived safely. The hotel proved pleasant until the water supply was shut off, because of a severe drought, just as we each suffered a traveler's distress. Water was available only during evening and early morning hours; we were ill at mid-day. Fortunately, we soon recovered to enjoy the Prado and sights of Madrid and surroundings. The bullfight was a must on Tom's agenda; we particularly enjoyed watching the bull win the fifth fight until the matador was seriously injured. The pageantry was superb, but the experience quelled any further interest in the sport, if the term "sport" is even appropriate.

Our principal goal in Spain was to visit the Andalusian region, especially the Alhambra; Toledo on the way and Córdoba on the return were delightful stops. The month of June can be hot in southern Spain, where we experienced temperatures and dryness reminiscent to me of the great U.S. Midwest drought. As a more or less grown-up farm boy, I was interested in the wheat harvest then under way. The technology and energy systems for harvesting ranged from cutting the stems with a scythe and attached cradle, to fully modern grain combines. We observed more than one hundred years of farm technology and energy evolution in traveling only a few dozen kilometers in an agricultural region.

We once stopped in blazing heat where men were wielding scythes and saw that they were tying the bundles with wheat stems. I stepped into the field, dressed as a fairly well groomed tourist to tie a few bundles by the technique, similar to theirs, learned from my father. The workers watched with interest and seeming surprise. After I placed several tight bundles into orderly grain shocks, the workers' apprehension over my activities changed to enthusiastic applause. Then, when I turned with camera in hand to photograph the site, I noticed two horsemen on a ridge in the camera's view but several hundred yards away. They quickly reined their horses to attention, to make for a better shot. The fast camera shutter easily would have arrested their motion on film, but such was the courtesy we often experienced from the Spanish people.

The Alhambra was even more beautiful than expected, but we had been disappointed not to get advance reservations to stay in the National Parador, San Francisco. Through Elaine's insistence in talking with the management, she and I obtained a beautiful room overlooking the gardens, but the boys had to be tucked into tiny attic spaces. We were thrilled to visit portions of the Alhambra previously occupied by Washington Irving and Samuel Clemens, to remember their writings, and to learn more of the Roman and Muslim periods.

Our last overnight stop before returning to Madrid was at the Torremolinos Parador near Málaga, a modern motel-like facility on the beach. Tom, as a swimming enthusiast, wanted to test the Mediterranean Sea, which I found unbearably cold. We soon observed that the water was severely polluted and literally raced to an outdoor shower in the hotel courtyard. Arriving at full speed, we grasped the valve, and to our dismay the pipe cracked near the ground sending a fine spray over our feet. I steadied the pipe and sent

Tom for Byron, who with his better command of Spanish could best summon help. Alas, the pipe quickly broke completely, sending a fountain of water to arch high above the grounds as it spread gently in the breeze. We then backed away to join the rapidly assembling crowd that was enjoying the spectacle. After the crisis was under control, we and the manager examined the pipe to observe that it had become badly rusted, clearly relieving us of any liability. This trek through southern Europe remains in memory as one associated with the water of plumbing problems.

At home during the years 1965 and 1966 I was busy with HTGR program management and an Oak Ridge school-building project. Elaine was giving lectures on foods and nutrition, and the boys' activities included achieving Eagle Scout awards and recognition in piano competition. With Byron on the basketball team and Tom in the band, we spent much time with school activities in addition to my school board activities, which averaged about forty hours per month. A highlight of the year 1965 was the boys' success in receiving first place awards in a science-project competition. Those awards qualified each to enter the 1966 state contest sponsored by the Tennessee Academy of Science. Byron had measured the growth of pine trees as a function of soil type and terrain. Tom chose to study hyper-filtration (often called reverse osmosis) that removes salt and other minerals from brackish water. His project was to observe the efficiency of the process as a function of water pressure, using membranes provided by the General Atomics Company. Under the guidance of Kurt Kraus, his advisor and an ORNL expert on water filtration, Tom's determinations provided new information on membrane performance. He won the first place award in the state competition, and his written paper was published in the Academy's journal.

The growing HTGR program was busy with construction underway for the Peach Bottom demonstration reactor and design of the commercial 330 MW(e) Fort St. Vrain unit in progress, both contracted by General Atomics and with R&D work at several sites. ORNL supported each project with consultation, testing of fuels, and review of component designs.

I was particularly interested in a program for making concrete more suitable for pre-stressed concrete pressure vessels. The British had pioneered use of such vessels for their AGR at Oldbury, and General Atomics had chosen that container design for Fort St. Vrain. The ORNL program in concrete development, led by Jim Corum, Pat Callihan, Grant Stradley, and later David Naus, focused on understanding the stress patterns in those vessels. It also included development of improved formulations for concrete in order to increase strength and accommodate service at higher temperatures. Concrete formulations of this kind, of course, are more precisely formulated and more highly reinforced with steel rebar than the concrete of sidewalks and building construction. I was intrigued to discover that the formulations resembled the most lasting of Roman concrete materials.

Construction of the experimental gas-cooled reactor was completed early in 1966. The fuel, qualified by ORNL irradiation tests, had been manufactured commercially and was ready for loading into the reactor. On the Monday morning when fuel was to be brought from storage at K-25, guards had been deployed to assume responsibility for a fueled reactor, and the plant operators were ready. Milton Shaw, AEC's director of Reactor Technology Development, then ordered a one-day delay followed by several more postponements and finally cancellation of the project.

My belief was and still is that the EGCR was an ill-conceived design. On the one hand, it offered little, if any, advantage over light-water reactors then well along in deployment and was not as efficient or versatile as the HTGR. On the other hand, it seemed unwise to cancel a project with most of the expenditure made and relatively little of the potential value gained. Even limited operation of the EGCR would have revealed a design problem with the steam generators, which were identical with those of the Peach Bottom plant. That defect would have been discovered early in the EGCR operation, and Peach Bottom could have been modified during construction to avoid the later, long, and costly delay in repairing a radioactively contaminated plant.

In the spring of 1966 Roger Carlsmith and I participated in two large engineering conferences in London. One was on fast-breeder reactors and the other on the HTGR. Those meetings included formal papers and informal discussions during which scientists and engineers exchanged ideas and experiences. Communication between people working in a technical field serves to strengthen the overall endeavor and to minimize its cost. We were not surprised to receive many penetrating questions concerning the EGCR cancellation. European governments are more consistent than ours in completing projects that have been authorized and are underway; we had some difficulty convincing attendees that no substantial defect had been found in the design or construction. That concern was important to the British, because many EGCR features were similar to, or copied from, their AGR designs. French interest in the cancellation was even higher, because they were carefully reevaluating future nuclear program directions. That review eventually led to cooperation with the Westinghouse Corporation in constructing pressurized water reactors and subsequent abandonment of their GCR program.

The Fast Breeder Reactor Conference was first. (The term "fast" refers to the velocity of neutrons. These reactors do not employ a moderator to slow the neutrons, thus enhancing plutonium production for future use as fuel and justifying the term *breeder*.) Very interesting papers were presented by the Russians on their design for the Beloyarsk BN-600 sodium-cooled fast breeder. It was and continues as the largest reactor of its type designed to produce both electric power and heat for desalination of sea water by distillation. Britain and France, respectively, announced details of the Prototype Fast Reactor at Dounray and the Phenix at Marcoule. Both were to be power-producing models of commercial breeder reactors. These reactors use sodium as the coolant for UO_2-PuO_2 fuel contained in stainless-steel cladding. Many conference participants seemed confident that fast reactors would become cost-competitive. This advantage has not yet been achieved and is considered unlikely until other energy sources are more limited and costly. For the breeder reactor to reach its full potential in supplying electrical energy, economic, political, and cultural circumstances will have to change, allowing for reduction in the use of fossil fuels, a fuller acceptance of nuclear power as safe, clean, and effective, and a commitment to the reduction of atmospheric carbon dioxide.

Between the fast-reactor conference and the HTGR meetings the following week, we had a free weekend to enjoy London. Unfortunately, our travel office had not found lodging for us for Friday and Saturday nights, and we could find nothing ourselves in London. Upon arrival on the previous Tuesday, I had to promise in writing before registering in a dusty, third-rate hotel to leave on

Friday morning. Cassius Clay (now Muhammad Ali), who had been on our flight to London, was to defend his world boxing championship there on Saturday night. Clay, flying first class, occasionally walked through our tourist cabin, his impressive physique impossible to ignore. Compounding the hotel problem was the World Soccer championship game, to be played on Sunday. As the week progressed, London's park benches and public loos, respectively, looked increasingly hard and malodorous. On Friday morning I appealed to members of the Dragon Project who solved our problem.

At the suggestion of George Lockett, Roger and I traveled by train about forty miles west from London to Goring-On-Thames to stay at the Miller of Mansfield Hotel. Legends vary concerning the hotel's name and origin. The one I like best describes King Henry III having enjoyed a meal of superbly prepared venison at a Mansfield miller's hostelry; then, upon learning from attendants that the deer had been poached from his Sherwood Forest, the King was torn between rewarding his host as a chef and beheading him as a poacher. The solution: move the miller far from Sherwood Forest and establish him at Goring nearer to Windsor Castle in a new tavern.

Although portions of the present hotel building with its stone walls and thatched roof date back at least to the sixteenth century, its setting near the banks of the Thames beside a Norman Church and an old stone bridge seemed to give the historic legend credibility. I greatly enjoyed walking along the river's towpaths as a peaceful respite from London and the conferences. On a brief stroll through the town I met by chance John Sayers with whom I previously had discussed the design of irradiation experiments at both Harwell and ORNL. He invited Roger and me to his home that evening. His lovely wife presented a delicious dinner in the garden setting of a charming home.

At the HTGR conference speakers from the Dragon Reactor Project touted their success. Skeptics were also present, who emphasized the difficulties in developing a new fuel-supply system and an accompanying nuclear fuel reprocessing technology, then thought to be required. Although the conference was lively, it lacked substance, except for Dragon Project technical papers and those describing the British Dungeness B reactor as a prototype for their projected large nuclear energy program. In separate meetings with the British Central Electric Generating Board staff, the discussion was similarly optimistic concerning their advanced gas-cooled reactor program. They expressed little concern over the EGCR cancellation; perhaps they understood American inconsistencies better than the French. Contacts there included K. P. Gibbs, manager of Design and H. M. Carruthers, Advanced Reactor Evaluation.

These meetings were held on the top floor of their tower that overlooked much of Saint Paul's Cathedral. The conference room windows offered an excellent view of portions of the cathedral roof and towers, although, by law, the height of the office-building had been limited to be lower than the top of the church dome. The cathedral's beauty was stunning, having been recently cleaned of soot, which over centuries had accumulated from the burning of soft coal in inefficient stoves, in places to a thickness of more than four inches. London also had been cleared of an aggravating element of the persistent fogs for which it was well known through the introduction of gas heating, made possible in part by the discovery of oil and gas in the North Sea.

Following the London conferences, I read a paper for a Harwell Laboratory meeting on irradiation-testing technology and conferred with staff members of the British Concrete and Cement Association at Slough. Their excellent modeling of the huge concrete cooling towers for power plants was impressive, both for the precise scaling of concrete and steel elements and the development of "micro concrete." They used extremely fine sand for aggregate and formulated special cements. The scale model of a 320-foot-high tower was over twelve feet tall and used concrete for the walls only a fifth of an inch thick. The models were so carefully constructed and instrumented that they were suitable for testing wind-load effects for full-scale towers.

Continuing to Paris, I was greeted by M. Levallee of Électricité de France, retired chief engineer of their nuclear design center at Tours, who added that stop to my agenda. There we reviewed designs for French advanced reactors, not then generally known in the United States. Before continuing to Chinon, my host proposed having coffee at the station. The waiter, recognizing me as American, prepared a cup of American coffee, but my host, observing this, ordered the coffee replaced with "good coffee." Because I prefer a rather weak brew, the French coffee was a jolt to drink and difficult to swallow with feigned appreciation in deference to my host—no sleepiness that afternoon. At the Chinon and St. Laurent nuclear sites, respectively, we observed early Gas-Cooled Reactors EDF-2 and 3 in operation and the newer and larger EDF-4, where construction was nearing completion.

The French are often maligned as being unwilling to share their technology, but I have not found that to be true. Although government officials often express reluctance, research staffs are proud of their accomplishments and quite willing to share information with visitors who can offer a critique of the work. I also visited Jülich, Germany, where the AVR was nearly ready to operate with the American-made fuel that we had qualified. (This would be the first extended reactor operation with fuel that ORNL Irradiation Engineering had tested. The molten-salt reactors had not operated for long periods, and the EGCR had been canceled.) The excellence of German engineering and manufacturing had been well deployed in constructing the Arbeitsgemeinschaft Versuchs Reaktor, a "pebble-bed" reactor of unique design.

A pebble-bed reactor core, as the name implies, consists of spheres about two inches in diameter embedded with nuclear fuel and literally poured into a steel vessel lined with blocks of graphite to insulate the walls from the hot fuel. Pure graphite spheres are interspersed with those that are fueled and in the periphery to obtain maximum utilization of neutrons. Graphite blocks in an outer ring also serve to intercept neutrons that otherwise would escape from the bed and cause them to diffuse back to the core and fission more enriched uranium. This device in nuclear terminology is called a reflector and is also employed in the prismatic fueled HTGR designs. In the German AVR, helium flows upward through the bed to encounter the steam generator used to boil the water and further heat the steam. Steam, so produced, is piped to a steam turbine in a separate hall, external to the reactor containment building, where it powers a turbine-driven electricity generator.

Soon after I arrived home in late May, Byron received word of an American Field Service (AFS) assignment to spend the 1966 summer in Japan, where he would stay with the Takeuchi family in Nagoya. In a hurried exchange of letters between Elaine and Byron's "Japanese Mother," plans had been made for his stay, including clothing sizes for "proper Japanese attire," to be worn in the home. Upon his arrival Mrs. Takeuchi discovered that the kimono she had made for Byron was several sizes too small. She had not believed that a sixteen-year-old boy could be so large and accordingly scaled down the dimensions; she quickly made a second attractive garment. Since Byron had short notice for the assignment and had spent a previously scheduled week at Boys State, he had gained no Japanese

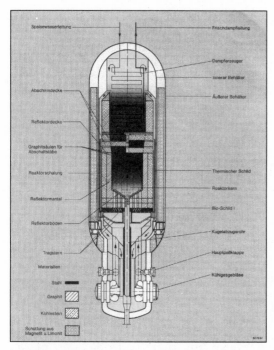

Sectional view of the German AVR Pebble-bed Reactor. Courtesy of AVR GmbH.

language facility; neither Mr. nor Mrs. Takeuchi spoke English. Although his two "brothers" in the family spoke English, they were away most of the summer at Nanzan University. Clearly, Byron had to learn spoken Japanese quickly if he was to eat well and enjoy the assignment. He succeeded.

Two features of his summer impressed us greatly. The AFS students were met by U.S. Ambassador Edwin O. Reischauer, who spoke to the group and met each one individually. We became admirers of Reischauer. Later, when we learned of his wise counsel concerning the Vietnam War—advice not heeded by several U.S. administrations—our understanding of that sad adventure differed markedly from that held by most Americans for a long time. Had the U.S. government followed the advice of this eminent scholar of the Far East, that very costly war might have been avoided.

Mr. Takeuchi, whom Byron addressed as "Otosan" (Father), had been a member of the Japanese Air Force during World War II. The Takeuchi family knew of my wartime work, but neither they nor their acquaintances showed any animosity toward Byron. On the twenty-first anniversary of the devastating Hiroshima nuclear blast, the family visited a park where that event was being recognized. Byron wrote, "I saw many tears and much anguish, but felt no hostility toward me as an American." That understanding and friendship also has been extended to us repeatedly by Japanese friends and colleagues who also have known of my early nuclear role. Our friendship with the Takeuchi family continues to this day, bridging oceans, continents, and cultures.

The year 1966 also brought a period of intense cooperation with the Fort St. Vrain reactor designers. The ORNL role was to help apply research results and other GCR technology experience to the design of that HTGR reactor. For example, A. A. Abbatiello again demonstrated his genius in mechanical design by solving a problem with the drive mechanism for the reactor control elements. I found it interesting and exciting to debate on friendly terms the design features of a new nuclear reactor.

As the months progressed, I argued with an increasing sense of caution for freezing the design and proceeding to validation and testing of components. Representatives of General Atomics frequently pressed for introduction of innovations through research and, as the official designers, they sometimes instituted them. On one occasion I interrupted such a discussion, suggesting that we trade sides of the table because I, representing a national laboratory, should have been arguing for more research and they, as private industry, should have been insisting on tests to demonstrate that the hardware components had no flaws or defects.

To a lesser extent ORNL was involved with other small and experimental gas-cooled reactor units at Los Alamos, Idaho Falls, and Peach Bottom. We also cooperated with the German AVR project, although in a way necessarily more formal to comply with international agreements. The AEC established a Technical Advisory Committee charged with coordinating all US-GCR work. It was headed by Bob Pahler, with members Al Goodjohn, of General Atomics, and me. I was to serve on behalf of national laboratories. Others were brought in on an ad hoc basis, but the triad was held responsible. This task required much time and travel as we participated in briefings and reviews and made recommendations concerning work at all of the GCR sites. Our suggestions generally were followed; however, our recommendation that the Turret Reactor operation at Los Alamos be terminated was ignored. During a telephone conversation with Dick Kirkpatrick of AEC Headquarters, I asked if Turret was still operating. He replied, "Just a minute—I'll look out the window." He returned with, "It is operating; the flag is at full staff." His meaning was anchored in a political reality: unless and until a certain New Mexico senator died, that reactor project would continue.

———•◦•———

Nineteen sixty-six was also a year of family crisis, as Elaine's mother became increasingly distraught and her doctor in Greensboro seemed unable to help. With concern for possible harm to herself and others, we sought psychiatric treatment for her at the University of North Carolina Medical Center. On the day scheduled for commitment to the mental health hospital, I was designated to accompany her in an ambulance. Understandably resented by the mother-in-law I loved, I will never forget that trip. The hospital staff welcomed us, and we left trusting that she would receive effective treatment. As weeks passed, our confidence eroded as treatment became more aggressive with minimal, if any, progress. Elaine and I requested consultation with our private physician, Dr. Guy Fortney. On hearing lay descriptions of the symptoms, he suggested that a thyroid disorder might be the cause of her illness. We telephoned his suggestion to the UNC medical center with a request for tests. Mother Causey responded well to thyroid treatment; soon she was well enough to be released, to enjoy life—and to love me as before.

An expanding Oak Ridge student population and deterioration of some of the later schools built in wartime required the construction of two additions to the high school and replacement of a junior high school and the Linden Elementary School buildings. During one trip to North Carolina I persuaded the Duke School of Architecture staff to assist as my unpaid, personal consultants. Their principal recommendation was that for the large projects envisioned we should hire a major firm selected from competitive proposals. They also encouraged teaming the large firm with a local architect, who would follow the work and represent the board. We hired internationally recognized architects, but I failed to convince colleagues on the board also to employ a local designer/manager.

As the board member with engineering experience, it fell my lot to review the architects' work. I was frustrated to discover design errors made by the well-respected Houston firm, particularly because they specialized in school buildings. For example, in the first design submitted for the new Jefferson Junior High School (now a middle school) the plan required students and faculty of first-floor classrooms to enter the building, climb to the second floor, and then descend to the first level again. This and many less obvious errors were discovered through long evenings spent in review of drawings. With effort, we completed four good school buildings. My last act as a member of the Board of Education was to hand Byron his high school diploma in June of 1967; I had chosen not to run for reelection.

Byron aspired to spend some time in a Spanish-speaking country. Through George Watson, an ORNL employee, we had found a Mexican family agreeable to a private exchange. Byron was to spend the summer after graduation with the Salas family in Mexico City, and Eduardo Salas would live with us during the Mexican school holidays in the autumn. In June of 1967 it was with mixed emotions that Elaine and I watched the takeoff of an aircraft carrying Byron and Tom to Cincinnati. From there, Byron flew to Mexico City, by way of a short visit with my family in Nebraska, and Tom to Chicago, where he spent most of the summer as a National Science Foundation Scholar at Northwestern University.

In late August Elaine, Tom, and I took our vacation as a trip to visit Mexico and accompany Byron home. We naively rented a Dodge Dart from Avis in Amarillo, Texas, to drive to Mexico City (instead of renting in Mexico). It was a fine little car, performing as it should, until we parked it in the Tepozotlán (Gold Cathedral) plaza on a very hot day. On the return to Mexico City, the differential produced a piercing, howling sound that increased to painful intensity. The Avis manager promised to repair the car while we toured Mexico City, but day after day passed with no progress on repair; they could not find the needed parts. Although I had checked in advance to be sure that Avis rented Dodge Darts in Mexico, the Mexican cars were built to the metric system; Mexican parts would not fit our vehicle. At nearly the last day before we were to depart for Acapulco, I pressed the manager for the needed service, and he arranged to have parts shipped from Texas.

It was apparent that a commercial shipment would not reach us in time, particularly with its probable arrival on Saturday, when the customs delay could be prolonged. Avis then arranged with a Mexicana Airlines pilot to bring the parts and take them through customs on his exemption. The Avis manager and I went to the airport to meet the pilot;

I covered the arrival gate and he the pilots' lounge. When I met the pilot at the gate, he denied knowledge of the arrangement and advised that the package must be coming through another pilot. After some scurrying, the manager surmised that the parts had been handed to a customs employee for his personal benefit, and through further search found a customs agent with the parts. The manager requested that I not interfere and negotiated for some time in Spanish. Finally, I understood enough to conclude that they were haggling about the price. "How much?" I inquired. It translated to fifteen dollars. I handed the customs agent the money in dollars, he thanked me, and we had the parts. (The manager gave me a receipt for the amount of the bribe, which I later collected from their office in Texas.) The remaining problem was to have the parts installed on Saturday evening.

At the Avis garage we discovered that two essential grease seals were missing. It seemed probable that those parts would not have been damaged and could be salvaged by careful disassembly of the troublesome differential. We also observed that the differential had never been filled with lubricant and had functioned quietly from Texas to Mexico City with lubrication only from grease applied to prevent rusting of its parts prior to assembly. The mechanics were quite cooperative in allowing me to work with them. At the critical point of salvaging the grease seals, I took over to prevent almost certain damage by these inadequately trained men and, fortunately, was successful. At that point I had to depart for a dinner party at the Salas home at the edge of Chapultepec Park, leaving the mechanics to complete the assembly.

The Salas house with its marble floors and staircase was the perfect setting for an elegant occasion. This was a party for Byron, who was leaving after a delightful and educational summer with this exceptional family. (Nine days earlier, we had joined the Salas family in celebrating Byron's eighteenth birthday and also for dinner at their home.) Even in this party environment, I wondered how the mechanics were faring, but they proudly delivered the Dart to the Salas home just before midnight. The car performed perfectly during our pleasant drive to the Pacific coast.

Elaine and the boys enjoyed Acapulco, but I developed a problem. Curious about this famous resort city, I walked a few blocks from the Boulevard Costena M. Alemán and observed severely dilapidated slums. Thereafter, I had difficulty enjoying the incredible beauty of that narrow coastline of modern luxury hotels, beaches, parks, and amusement centers.

A hurricane blew in from the Pacific Ocean on our second afternoon in Acapulco. At dinner in a restaurant buffeted by severe wind and rain, we listened anxiously to the noisy metal roof, creaking beams, and rattling windows and often worried about building collapse as the floor swayed. In the relative quiet as the eye of the storm passed, we rushed to our small motel. There, we were awakened at 4:00 A.M. by an earthquake that shook the beds and rattled the building. Later that morning we ventured forth in the Dart with great care to avoid washed-out streets and debris from the storm. The basement and grounds of the Hilton Hotel were flooded, the roofs of luxury cars showing just above the water in its garage. We drove along severely damaged beaches and up the hillside to Roca Sola. Near there, an overlook in a small park where we had stood to take photographs on the previous morning was gone; the quake had toppled it into Acapulco Bay.

We returned to Texas through Mexico City to pick up some of Byron's possessions. Mr. Salas asked that I come alone to his office for an important discussion. Mr. Salas was a very successful civil engineer who planned to support a college education for each of his six children. Among other accomplishments, he had been the chief engineer for both design and construction of the celebrated railroad through the great Copper Canyon in Sonora. In that capacity Mr. Salas had been the host for President Eisenhower, who complimented that remarkable Mexican accomplishment by riding in the first train through the canyon.

In his office Mr. Salas proposed that Byron attend four years at the University of Mexico, offering to pay all expenses including food and housing. He had noted the high tuition at Duke University where Byron had been accepted and justifiably praised the fine University of Mexico on the outskirts of his city. It had been obvious that Mr. Salas was fond of Byron, but his primary purpose may have been to secure continued companionship for his son Eduardo. Knowing Byron's desire to attend Duke and the advantages for an American to obtain the first degree at a university north of the border, I declined the offer with strong expressions of appreciation.

With Byron away at college Tom was lonely at first, then became immersed in activities. He was elected president of the senior class, hosted Eduardo, gave a private piano recital, presented his paper on reverse osmosis at the Tennessee Academy of Science meeting, and served as president of the Oak Ridge High School Band. As band president he helped organize a school band trip to Washington, D.C. The band arrived by April 5, 1968, the day I was to participate in a late afternoon meeting of the AEC Advisory Committee on Reactor Safeguards. Elaine accompanied me, and we arrived in time to meet the band at the Smithsonian Institution's Museum of History and Technology.

Leaving the museum after the band bus departed, we looked for a taxi to our hotel that was near the AEC building at 1717 H St. NW. Unable to find a cab, we proceeded to walk on Fourteenth Street, puzzled by unusual traffic patterns of autos and pedestrians. We noted that Garfinkel's Department Store was closed, but attributed that to respect for the tragic death of Martin Luther King. Then, when two police cars carrying occupants in full riot gear rushed by with flashing lights, we realized that Washington was in serious trouble and rushed to the hotel. The AEC meeting was canceled and in accord with city and federal advisories, we remained in the hotel for safety. The seriousness of the crisis became quite evident as armored vehicles were deployed in Lafayette Park to protect the White House. Later, the glow of large fires projected on clouds, and smoke drifted through the streets from a nearby part of the city that was burning. We were relieved by a telephone report that the band was safely housed in Northern Virginia.

Tom later made a trip to St. Louis in final competition for a four-year scholarship in electrical engineering at Washington University. On the very day he received notice of winning the Langsdorf Scholarship, he was also awarded a one-year AFS assignment to a northern-hemisphere country. But he could accept only one of the two awards. He had been greatly attracted to the AFS experience through Byron's reports from Japan. We left the difficult decision entirely to him but were pleased when he chose the scholarship.

I then offered to find Tom a German summer opportunity similar to Byron's in Mexico. The Ivens family of Titz, Germany, near Jülich, was a possibility since I had worked extensively with Günther Ivens and had met his son, Klaus, about Tom's age. Two years earlier, Günther, Klaus, and I had enjoyed a Sunday afternoon riverboat trip on the Rhine. Fortuitously, Ivens was to visit ORNL in two weeks; we invited him to accompany Tom, me, and others to climb Mount LeConte in the Great Smoky Mountains National Park on a weekend of his visit. Both he and Tom were pleased with a plan for Tom to spend the summer of 1969 with the Ivens family and for Klaus to be with us in 1970. The arrangement was completed when Ivens returned home to find Klaus enthusiastic about the plan.

Tom's experimental study of reverse osmosis was basic to his becoming a winner in the Westinghouse Science Talent Search program. As a result he was invited to spend the summer after graduation in a work project at the National Bureau of Standards (now the National Institute for Science and Technology). We accompanied Tom to the Bureau campus outside Washington, D.C. The registration progressed well until someone noticed that Tom had only recently passed his seventeenth birthday; the Bureau could not hire anyone under eighteen. Their solution, after much consultation, was to enroll Tom in the Montgomery County School system, as a student from which he could be assigned for training in the Bureau. He found lodging on the campus of the American University and registered for a Saturday course in spoken German at Georgetown University.

During the summer, I visited Tom a few times when traveling past the Bureau for meetings or work at the nearby AEC Headquarters. He sometimes needed help with the problems an inexperienced person encounters in a large institution. Services such as preparation of metal specimens were available but not easily located by someone a bit intimidated by the formalities and paperwork required. Tom finally achieved some useful results, but they were perhaps not commensurate with the impressive plaque that displayed his name on the door of the laboratory room where he worked.

———•◦•———

An early concern for the viability of gas-cooled reactors had been the availability of helium as the preferred coolant. Helium, a rare noble, or inert, gas formed from radioactive decay of nuclear elements such as radium, is found only in a few locations of the world. The United States is fortunate in having natural gas pockets containing relatively high concentrations of helium from which it can be obtained economically. The U.S. government established a program for helium recovery in the 1920s when extensive dirigible air travel was projected. Helium was extracted from the carrier gas and pumped into a depleted gas well for storage. The extraction rate was determined by the sale of gas by the private owner of the well. Unfortunately, several wells containing recoverable quantities were not included and their helium content was released to the atmosphere wherever the gas was burned. Eventually, unless stored or left in the well of origin, this light gas will escape and be lost forever into the upper atmosphere.

In 1968 Scott Carpenter, one of the original team of space astronauts, became concerned over the sustainability of this natural resource. He with others formed a Helium Society to urge improved practices for extraction, storage, and conservation of helium, and I joined soon after its formation. It was a privilege to discuss energy subjects such as the HTGR and other helium-related subjects with this early pioneer of space exploration. In addition to the nuclear-power uses of my interests, helium is also utilized widely in medical applications, in welding, deep-sea diving, and scientific investigations, as well as for the Goodyear and other blimps seen at major recreational events. A centennial exposition was held in Amarillo, Texas, to celebrate the discovery of helium in 1868 (through analysis of the solar spectrum) and to promote both helium conservation and beneficial uses. As part of the celebration a monument of stainless steel was erected to contain time capsules describing contemporary helium storage, its uses, and related technologies. By invitation, I had prepared a paper on helium-cooled nuclear reactors that was encapsulated for revelation a millennium later.

The process of extracting helium from natural gas at a rate determined by sales of the natural gas from those unique helium-bearing wells and then storing it for long periods seemed foolish to me. In that procedure, the expense for removal and extended storage of the helium is incurred long before costs are recovered; also storage of concentrated helium in a well entails the risk of loss through closure failure. A more rational approach would be to sell the natural gas and helium at the rate of helium demand. This might require purchase of the well and its contents by the government or through a government subsidy, both of which seem contrary to U.S. policy and practice for natural resources, but the law has been revised recently. Helium is no longer extracted for storage in the earth but, instead, is extracted on the basis of a more market-based plan.

In 1968 Elaine had noticed a newspaper solicitation for a college-level Rotary Foundation International Ambassadorial Scholarship; Byron applied and won an all-expense-paid program of study for the school year 1969–70, at San Marcos University in Lima, Peru. Rotary Scholars are to be good representatives for their country while gaining personally both through studies and experience with another culture. It was a great opportunity that fit well with Byron's interests and prior student-exchange experience in Japan and Mexico. However, credits could not be transferred directly from San Marcos to Duke University, and therefore, the assignment could have cost a year of progress toward graduation. Byron then discovered an Indiana University resident program in Lima, Peru, that could ensure credit transferability. But departure for Lima, scheduled for early 1969, was delayed by a revolution that removed the elected Peruvian government and installed a military dictatorship. The upheaval, precipitated in part by a controversy over the price of Peruvian crude oil in which U.S. firms were involved and exacerbated by a fishing-rights issue, left Lima in turmoil. San Marcos University was closed by student revolt, and Byron's prospect for study in Peru seemed shattered.

By April the new Peruvian government had stabilized, the Indiana program was moved to the Catholic University, and Byron departed for Lima. The university did not offer housing, and the twelve North American students had to find lodging in private homes. Byron lived with the Fausto Valdeavellano family—mining engineers, devout Catholics, fashion designers, student leaders, self-styled revolutionaries, and lovers of classical music. This family added much cultural and educational experience beyond Byron's formal university program of study in the Spanish language. The experience was exciting throughout, but observing the street celebrations when Peru won a World Cup soccer match was a reported highlight, relayed to us through voice and sounds on a recording tape. Understandably, Elaine and I chose that year to spend our vacation in Peru. Arriving early in October, we also stayed in the large Valdeavellano house to be with Byron, who by then was studying at the reopened San Marcos University.

That house, on Malecón Balta, featured skylights over its three-story central hall, a fine library, a collection of hundreds of long-playing classical records, and ancient Peruvian artifacts of museum quality. We enjoyed studying fine pieces of tapestry and paintings that adorned the walls while listening to classical music from an elaborate sound system reinforced by the excellent acoustics of the great hall. Because his was a large family with many interests and activities, Señor Valdeavellano sometimes referred to his home as the manicomio (insane asylum). Later, we observed that this was only part of the original building; the remainder was occupied by another family for cost sharing. We were also fascinated by the small factory where Señora Valdeavellano designed and manufactured unique women's fashions. This clothing, based on native Peruvian motifs, was primarily priced for fine stores in Paris and Rome. I purchased for Elaine a small, black purse embossed with stylized bird figures in gold, which continues to serve well on very special evenings.

I had rented a car upon arrival in Lima so that we could visit the city's surroundings independently of the Valdeavellano family. Byron drove it from the airport and for the next two days. Even after driving in the seemingly mad traffic of Madrid, I was intimidated by the drivers in Lima, but then gained confidence to meet the challenge and enjoy the convenience. We spent several days exploring Lima and its marvelous collections of gold art that escaped the Spanish pillage, then proceeded to other historical areas of Peru. We flew to Cuzco, where the aircraft was depressurized on landing to equal the rarified atmosphere of that high-altitude city, then followed advice to acclimate to the altitude by resting upon arrival at our hotel, the Virrey.

On being awakened from a nap by band music, I threw open the wooden shutters of the balcony to look an elephant directly in the eye. Partially blinded by bright sunlight, I first thought it a hallucination, perhaps caused by the altitude change. No, it was real; a circus procession passing the hotel was circling the main square before presenting a tented performance outside the city. We made a two-day visit to incredible Machu-Picchu, located high above the Urubamba River gorge. Byron then joined us for tours of Ollantaytambo, Urubamba, the Pisac market, and other sites.

We next embarked by rail to Puno, a twelve-hour journey traversing two hundred miles in rail cars and equipment of early 1900s vintage, although the locomotive was a diesel. Our seat companion was a Catholic missionary priest who spoke in Quechua to a group of followers at each of the many stops. These assemblages were mostly of native women, who wore brown shawls and the traditional derby hats of that area. Many were accompanied by poorly dressed, somewhat unkempt children, most of whom consistently displayed dripping noses. The priest explained that he listened to appeals for help, presented sermonettes with blessings, and served as messenger and delivery man for letters and small packages. Between stations he fashioned fine art pieces from polished wire, for sale to support his work; his techniques resembled those of my female SAM coworker of 1946.

The train progressed on the altiplano to elevations of ten thousand feet and more, where it was cold and occasionally snowing. In approaching a small stream flowing from high snow-covered mountain tops, we noted a small group of people assembled on one bank near the water. As the train crossed the bridge, a nude young woman emerged from her bath in the icy stream, rushed to the shore, and was quickly covered with a blanket held by the other women. Such are the rigors of the altiplano and the stimulations of travel.

Arriving in Puno on the western shore of Lake Titicaca after sundown, Byron wished to visit a shop where he had previously ordered an Alpaca rug as a gift to me. The taxi took us first through dimly lighted streets and then to an area in total darkness where we found the shop, but it was closed. A passerby alerted the owner of our interest, and she came to meet us. Byron, speaking Spanish, introduced me to her and then immediately left in the taxi to negotiate our crossing of Lake Titicaca. I was to "entertain" the shopkeeper until he returned.

Apprehensive is an understatement of my feeling in the presence of a formidable, middle-aged, female shopkeeper in a room dimly illuminated by a single twenty-five watt lamp hanging from a cord. The woman's dark complexion, brown woolen shawl, and dress blended with the walls stacked in fabric and fur to make her seem unreal. Because her native tongue was Quechua, Spanish was nearly as difficult for her as for me. Communication was limited, and at times her words were not discernible to my ears. Absorption of sounds by the racks of soft woolens and furs added to the unreality of that shop.

Since we were holding the store open after hours, I thought it best to show interest in something or, even better, to make a purchase and was pleased to note a large basket of small, handcrafted items. Peace Corps workers had taught the natives to adapt their knitting of traditional caps for men and children to the manufacture of colorful ornaments for Christmas and other seasons. I examined small knitted birds designed to hang on Christmas tree branches, studying each slowly and carefully to pass the time while occasionally choosing one. When Byron finally arrived, I had acquired enough birds to decorate a large tree. The rug he had ordered was not ready, but it arrived later that year and each winter has occupied a prominent position before our fireplace.

The next morning we set forth on an organized tour to Bolivia. The first leg was by taxi to Copacabana. We passed quaint mission churches, humble dwellings, and pre-Incaic towers silhouetted against billowing white clouds hovering over the distant, eternally snow-covered Andes. The roads were dry, very dusty, and even at that high altitude windows were open for cooling. Seated in the back-center position of an aged automobile, I felt unusually overwhelmed by dust and finally observed a small hole in the dirt-laden roof lining directly overhead. It exuded a puff of dust with each impact between the car and the multitude of rocks and ruts of the road. Insertion of a tightly twisted tissue in the hole made the remainder of the drive more bearable.

In Copacabana, Byron and I insisted on climbing a five-hundred-foot hill to gain a cliff-top view of Lake Titicaca. I arrived there breathing heavily from climbing above twelve thousand feet elevation and disappointed with the view, for it was little better than from the shore. However, the monuments along the path to the top were interesting because each seemed to have religious significance. Apparently, the visitor was to achieve a new status in personal inspiration from the experience of climbing and studying the inscriptions. I hope that achieving heaven is not as formidable as climbing hurriedly at that altitude.

During the crossing of beautiful Lake Titicaca on a hydrofoil to Desaguadero, we were fascinated by the isles attributed to the sun and moon gods and by the floating native homes and boats made of tortora reeds. On the bus to La Paz two U.S. tourists seated behind us were puzzled by the many native pedestrians seen both on the road and crossing fields. One said, "These people are walking as though going home, but we have not seen a house anywhere. Where are they going?" The tourists had not recognized that the many small, adobe, thatched-roof buildings that dotted the treeless landscape were, indeed, their houses.

In La Paz we were shocked to learn that a military coup had occurred on the previous day. Fortunately, it was bloodless and the government had changed quickly after only a few shots were fired. Except for newspaper headlines no visible evidence of a revolution was apparent; the bullet scars we observed on the walls of government buildings were from previous events. We had invited the young woman escort of our tour group to be our guest for dinner, if she would show us a few more points of interest and select a typical native restaurant. She chose a steak house, secured a good table near a window, and ordered grilled pork for four. We were served samples of every part, no exceptions, of the male pig. Despite some inhibitions concerning food choices, we found the pork to be well-prepared and served with excellent condiments.

Good fortune continued to prevail. During the previous summer, Byron had met a guest professor of the Universidad Mayor de San Andrés while studying Spanish at the University of Michigan. He had urged Byron to telephone if ever in La Paz. When Byron made contact, we were invited to accompany the professor and his wife on Sunday to visit a small fair near Cochabamba. It was a quaint village of low, adobe structures and narrow, winding streets that was hardly noticeable on the windswept plain. Our group of five, and a lone professional photographer, were the only Caucasians present at that native festival.

The parade units, comprised of many dance groups in varied and colorful costumes, conducted themselves in unusual and sometimes alarming ways. For example, a group of young men wearing fringed jackets whirled and cracked long rawhide whips. Positioned on a rock and sheltered under an overhanging roof to avoid the very hot sun, I attracted their attention. These men, most with flask in one hand and whip in the other, targeted the ledge on which I stood. They seemed delighted by my fright but were remarkably accurate in striking the rock to dislodge a shower of dust and sand while skillfully and mercifully sparing my feet. An hour later the hot sun was obscured by clouds, and soon light snow fell on us as we walked among groups of farmers offering to sell cattle, sheep, and garden produce.

An important purpose of this Sunday excursion was to visit a young Peace Corps couple who were teaching the natives to improve their lives under the guidance of our host. We traveled increasingly narrow roads, finally to hold tightly to the Land Rover seats as it bounced over fields of potato ridges to a modest adobe cabin topped by a galvanized iron roof. The young man was teaching the natives to grow potatoes from improved varieties and to use better farming and marketing practices. His wife provided daycare and training for small children. They had built the adobe house, using their personal resources for the roof and decorative materials; the Peace Corps allowance did not cover such luxuries. Except for the roof, the house exterior was typical for the area and similar to those that had gone unrecognized by the tourists on the bus. However, it was a bit larger, and the interior was divided into two rooms instead of the traditional single space in an attempt to demonstrate to the native people the advantages of privacy. This fine young couple provided most welcome and refreshing afternoon tea with freshly baked delicacies.

We were most favorably impressed with these Peace Corps workers; both were college graduates, but had only brief training for this work. They had carefully studied the needs of the community and reportedly blended well with the Bolivian farmers, although they traveled by motorcycle instead of on foot. This young woman repeatedly commented that Elaine resembled her mother. I wondered if the resemblance was real or whether the distance and isolation had been a factor, but Elaine, as usual, related well to the young woman. Other Peace Corps members we had met in Cuzco were equally impressive in their dedication and in the effectiveness of their missions. These young Americans and their actions were a tribute to President John F. Kennedy, who inaugurated the program.

The atmosphere turned cool and clear as we drove over open land from the potato fields toward the Royal Range of the Andes and through herds of seemingly wild llamas and the more heavily furred alpacas. We finally reached a mountain pass and radar station at eighteen thousand feet elevation. Although having been at altitudes above ten thousand feet for several days, we moved very slowly in the rarefied air of this new height. Even the Land Rover struggled for oxygen when mounting many steep grades. At one stop I ventured forth to photograph a herd of alpacas but returned soon to rest in the vehicle seat. Elaine managed only to emerge once for a photograph made with the Rover and mountains in the background.

A couple of days later we traveled by taxi to see the Tiahuanaco pre-Incaic ruins on the way to the airport for return to Lima. We were stranded at the terminal for five hours without food or knowledge of the cause and extent of delay. Elaine had exhausted her supply of Coramine pills purchased to alleviate altitude sickness. She spent the time lying flat on a hard gurney in an otherwise unequipped room marked "First Aid." On arrival of the Peruvian Airliner, we learned that the crew had encountered delays in returning from London via Lima and had reached their time-limit for piloting. We waited uncomfortably as they slept. (Later, at home, our friend and physician, Dr. Paul Spray, identified Peru's over-the-counter Coramine as a powerful heart stimulant available only by prescription in the United States. Paul said, "Elaine, we only give that medicine to patients who are dying," to which Elaine retorted, "That's right, I was expiring." Neither doctor nor patient convinced the other about the use of Coramine in the Andean highlands.)

Back in Lima, Fausto Valdeavalleno Jr. invited me to discuss nuclear technologies with his colleagues at the Bureau of Mines and Energy. The meeting was to include engineers from the private utility, Empresas Eléctricas Associadas, that supplied the Lima area with electric power. Those assembled were to pose questions concerning nuclear safety, economics, and technologies to which I might respond. The specific issues raised for both pressurized and boiling light-water reactors suggested that they had a serious attraction to nuclear energy for Peru. They seemed quite familiar with information in the literature but had little nuclear training and no direct experience.

A break for box lunches afforded more informal discussion after which I provided a brief tutorial on advanced reactor types. The conversation then drifted to other energy systems. These engineers, of course, were quite conscious that a small change in oil prices had precipitated the revolution and rise of the military dictatorship in their country. Thus, they had given much thought to energy resources, capabilities, and systems and were quite aware of Peru's need for energy in its eastern provinces. Transport of their oil of the northern Pacific Coast across the mountains was extremely expensive, as was the maintenance of power lines in that severe mountain climate. We engaged in extended discussion about the importance of oil in the world energy picture. They startled me with a prediction that the price of crude oil, then about one dollar per barrel, would reach five dollars within five years. This was October 17, 1969!

As the discussion wound down in mid-afternoon, they asked if I would meet with the chairman of the Peruvian Atomic Energy Commission at his request; they would provide a driver to take me there. I accepted the invitation and found their AEC office located beyond one of the most impoverished barrios of Lima. It was a bit frightening, even with a Bureau driver, to pass through an area where, at the time, a North American would have been ill-advised to travel alone. On arrival at a square, multistory office building of stark outline, I was escorted past guards to the top floor. The chairman, General Rubio, a heavy-set military figure in a braided and ribbon-decorated uniform, was cordial. We moved from his office to an adjacent conference room to be seated in large, leather-upholstered chairs. The walls were adorned with framed photographs of scenery and facilities in Peru but with nothing that seemed related to nuclear or other energy technology. We first engaged in pleasantries as I talked about our enjoyment of the visit to Peru, and he discussed his hopes and expectations for solving the country's economic problems. As we chatted, I remembered reading about this man in newspaper reports. Less than a year earlier, General Rubio had been a leader in the coup that had delayed Byron's arrival.

The general seemed interested in the possibilities for nuclear energy, but I had difficulty in phrasing questions and observations for someone, even a chairman, so new to the technology. He then asked if I would review two documents they had received and report my observations to the Bureau of Mines and Energy. I agreed to examine these informal proposals for nuclear energy installations in Peru, if time permitted before our departure. Also at his request I gave him a short briefing on the discussions at the Bureau. An awkward silence followed when I asked if he had questions. I attempted to revive the dialogue by asking, "General Rubio, I am embarrassed but cannot recall if there is a nuclear reactor in Peru. I know, of course, that there is no power reactor here, but do you have a small research reactor of the swimming pool or Triga types?"

The silence intensified with obvious distress on the part of the Chairman of Peru's Atomic Energy Commission. Finally, he replied, "I think that we have a reactor in Arequipa." It seemed best not to pursue further questions or comment, and I proposed to depart saying, "General Rubio, you must be very busy, and I appreciate your having taken time to meet with me." We closed with more pleasantries, and I was relieved to rejoin the Bureau's driver for return to the Valdeavano's house. (A week later I found in the ORNL library that Peru had no reactor, only a cobalt irradiation source; but indeed, it was in Arequipa.)

I read the reactor proposals during the next day; one was from Motor Columbus, a Swiss firm, and the other a modified doctoral thesis from the University of California at Los Angeles. The former proposal, which called for a 300 MW(e) station combining electric power generation with desalting of sea water, was technically sound and perhaps economically feasible with foreign assistance if located in a desert region of the west coast. The other, which proposed a 1MW(e) plant apparently adapted from a military design by North American Aviation, seemed totally impractical. However, the paper did include an interesting analysis of the costs for transporting fossil fuel and building transmission lines to service remote locations in mountainous Peru.

I was uncomfortable about providing a review that would be unfavorable to an American compatriot and conferred with Thomas Roesch, attaché for Commercial Affairs at the U.S. Embassy. He saw no problem in providing objective comments. I found both the entrance and exit of the Embassy uncomfortable, for despite the military dictatorship, it was the only facility we saw in Peru that was protected by guards with automatic weapons. I discussed my observations concerning the proposals with Fausto Valdeavellano Jr. and gave him written comments for the Bureau. It also seemed appropriate to suggest that nuclear energy was inadvisable in Peru until a much greater infrastructure in nuclear technology had been developed, and that expansion of other energy systems should be considered first.

In appreciation for my time and effort Mr. Calmet of the Bureau and Mr. Svojsik of Empresas took Elaine, Byron, and me on Saturday to visit the Huinco hydroelectric power station. Even though located at about five thousand feet elevation in the Andes' foothills, the plant operated with four thousand feet of hydraulic head. (For comparison, Hoover Dam in Nevada is 726 feet high and provides a maximum of approximately 600 feet of head.) The water from Peruvian mountain lakes is carried through mostly unlined conduits tunneled in granite and impacts eight-foot diameter Pelton wheels. The high-pressure water, directed through nozzles to strike shaped buckets on the periphery of each drive wheel, provides power for a 30 megawatt electric generator. The row of ten units was providing 260,000 MW(e) of energy to the Lima area at the time of our visit. Bucket erosion is a major problem because small amounts of sand impacting the bucket surfaces at high velocity can quickly destroy them; the Swiss manufacturer of the wheels had developed remarkably hardened surfaces to prolong bucket life. Lakes that store and provide the water are interconnected from snow-fed locations very high in the Andes. This water traverses two smaller power plants below Huinco, then provides the domestic and industrial water needs of Lima.

We had lunch at a fine restaurant, the Granja Azul, located on a projection of the mountain located above the garúa (a persistent gray mist) that blankets Lima for six months of the year. This hostelry has served for decades as an escape for residents of Lima from the garúa of the coast. Our hosts described the country's problems with maintaining electrical transmission lines in the Peruvian coastal desert. During the sunny six months before the cold Humboldt current produces fog, insulators become coated with dust. Then moisture from the garúa causes short circuits and system outages. Washing the insulators is periodically necessary, for even dense fog or a brief rain shower on dust-laden insulators will produce power outages and, on occasion, extensive damage.

We were reluctant to leave this fascinating country, but only a few days of vacation time remained. We also wanted to visit Bogotá briefly and spend a short time in Caracas where Elaine would renew acquaintances made during her month in Venezuela a year earlier. She had been the foods and nutrition expert on a women's team that visited under the Partners of the Americas Program. Having prepared by obtaining nutrition books and leaflets written in Spanish, she had been quite in demand for further information provided through translators. I enjoyed meeting several of her friends, especially Dora Palacios. We stayed with Bill and Sue Ross, friends formerly of Oak Ridge.

Bill had invited me to speak at a seminar of colleagues at the Instituto Venezolano de Investigaciones Científicas, where he worked as an assignee from the International Atomic Energy Agency. I was impressed with the quality of the staff and their work but disappointed in their mission. The program centered extensively around a 30-MW(t) test reactor supplied by the General Electric Company with U.S. government assistance, but included other advanced studies in basic research. The senior staff, largely comprised of European scientists, produced results applicable in the United States and Europe but of little value to Venezuela. The country needed research in agriculture and in the chemical properties and uses of their petroleum, aluminum, and iron resources. However, they were proud of the research facility, and I found no diplomatic way of suggesting that a different program was needed.

———•◦•———

After a wonderful month in South America, our first vacation of that length, Elaine and I returned to Oak Ridge, and were soon busy catching up on work and other activities. We presented Byron a Christmas present of funds to extend his return trip by way of Chile, Argentina, Uruguay, Brazil, and Venezuela. Rotarians in Lima volunteered to contact clubs in the major cities of his tour, and he was met and escorted by Rotarians at each stop. In return he was pleased, when his schedule permitted, to speak at their club meetings. Byron spent Christmas Day in Uruguay with Luigi Bassano and the New Year's Holiday with Lillian Harbo and her family in Brazil. Both had been AFS assignees in Oak Ridge while Byron was in high school. Our family continues to be grateful to AFS and the Rotary Foundation for their excellent student exchange programs.

A week after returning to Oak Ridge, I flew to AEC Headquarters in Washington, and among other business reported the Peruvian projection that oil would cost five dollars per barrel in five years. The word was received with skepticism and derision. The general attitude was, "Those Peruvians cannot possibly understand future international oil markets." I corresponded with Mr. Svosjik for some time after the Peruvian visit but never did learn the source of their projection. Svosjik, as a foreigner, finally was required to leave, as the military dictatorship nationalized the utilities, and I lost contact with him. Thus, it was not possible to determine clearly whether the Peruvian projection was serendipitous or based on insight derived from Peru's association with the Organization of Petroleum Exporting Countries (OPEC), which I personally suspect.

Impelled by the OPEC embargo, the price of oil reached eighteen dollars per barrel five years after 1969. Subsequently, many programs and attitudes changed across the nation and the world. For a few years the United States had a national policy of "Energy Independence," a proposed plan to eliminate the heavy importation of oil. Unfortunately, it was never implemented realistically, even though research in energy production technologies increased significantly. ORNL, as an energy research laboratory, was impacted favorably by the perception of an energy crisis that followed the OPEC action. In addition my Laboratory role was altered, partially because of this more realistic understanding and projection of energy cost and resources.

The Peach Bottom and Fort St. Vrain reactors were designed by General Atomics, who also provided the fuel and other components. The utilities Philadelphia Electric and the Public Service Company of Colorado, respectively for Peach Botton and Fort St. Vrain, operated the reactors to provide electric power on their grids. These reactors, together with the gas-cooled reactors of Europe, provide a background of experience for considering the future potential of the GCR in energy production. The 42 MW(e) Peach Bottom reactor, first operated in 1967, was an early demonstration. After initial operation difficulties were overcome, including the steam generator repair mentioned earlier, the reactor operated well for seven years to complete its mission. The several HTGRs of Europe also operated successfully but, except for the German THTR-300, were not large prototype units. THTR-300 was a pebble-bed reactor that started operation in 1985 and was closed down a few years later, primarily for lack of a continued supply of fuel.

In contrast Fort St. Vrain began operating in 1976 and had a checkered operating experience that cast an unfortunate and unjustified shadow on the HTGR concept. All of the significant problems stemmed from one set of components, the helium gas circulators and the control system for their shaft seals. These units were of a new design using untried water-lubricated bearings and shaft seals and were driven by water turbines to provide the rotational energy. This provided a compact drive and seal assembly, which was advantageous for the gas-circulator installation and operation for the reactor design, where the circulators were installed in deep penetrations through the bottom of the thick-wall, prestressed-concrete pressure vessel. However, the shaft-seal requirements were particularly difficult, because maintaining the helium at 700 psi pressure was necessary to prevent the higher-pressure water from entering the reactor under all operating and shutdown conditions. The design and some of the manufacturing was done by General Atomics, which had little experience with this type of equipment.

This clever arrangement of bearings, seals, and water-turbine wheel-drive seemed to fit the Fort St. Vrain design. The complex control system for maintaining balanced pressures at the seals could have served well if tested thoroughly to reveal and solve the seemingly inevitable problems for any new design of an energy system. Unfortunately, the testing consisted of only about one hundred hours of operation in a test stand provided by the Public Service Company of Colorado. Even that limited testing might have been effective if it had been more prototypic and included a wide range of operations to test the equipment through all normal modes and potential abnormalities.

I urged the General Atomics staff to do more testing, but their confidence in the design was unassailable. The utility also was not interested

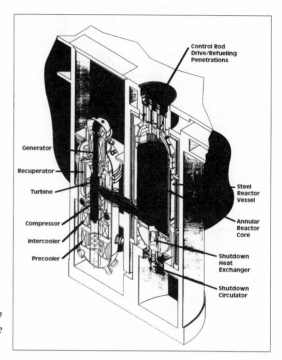

Gas Turbine-Modular High Temperature Reactor. Figure courtesy of General Atomics.

in further testing of their equipment and seemed assured by their purchase contract that General Atomics would be required to solve all problems. I appealed to Milton Shaw of the AEC but to no avail; money for such testing was short, and AEC considered the issue an industrial rather than governmental concern.

With the gas circulators installed in their cavities and the reactor under startup testing, the seals lost their pressure balance and introduced water into the reactor chamber. Since graphite is porous and helium must be very dry and clean to avoid corrosion of metal structures, the process of restoring the reactor to operational status was costly and difficult. The corrective process required removal, repair, and reinstallation of the gas circulators to provide heat from gas compression to dry the graphite. This not only delayed the startup, but similar failures occurred after the reactor was in operation. Even so, the repeated introduction of water led only to minor corrosion of other components.

In other respects the reactor operated well: it maintained a very clean system from which operator exposure to radiation was far less than for LWR systems, and the fuel, pressure vessel, and heat-exchange equipment met their performance requirements. Primarily because of the long shut-down times repeatedly required to remove, repair, and replace circulators,

Fort St. Vrain did not demonstrate the long period of trouble-free operation necessary to convince the utility industry of HTGR viability.

After my active participation in HTGR development ended, the design evolved to modular concepts in which a large installation for generating electric power employs several smaller units, each using a helium turbine to drive an electrical generator. This arrangement offers both high efficiency and exceptionally safe nuclear operation. At ORNL we had examined gas-turbine technology several times during an earlier period, but its viability was considered uncertain because of stringent heat-exchanger requirements. Now, new materials and much development and experience with heat-exchange equipment make the modular helium reactor with a gas turbine attractive for its high efficiency and reliability. Its nuclear safety is based on two fundamental features: (1) the excellent capability of coated-particle fuel to confine radioactive fission products at their source, even for the higher temperatures of severe accident conditions (each coated particle being a miniature pressure vessel of great strength) and (2) with an underground installation, the temperatures possible during a total loss of forced-convection cooling are limited to an acceptable level by conduction-heat loss through the containment vessels to the earth.

Safety achieved in a gas turbine-modular high temperature reactor (GT-MHR) through retention of radioactive materials in this reactor's fuel

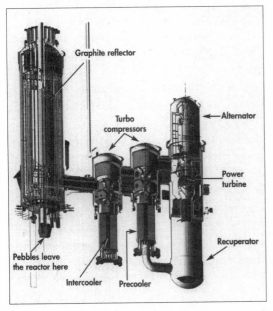

PBMR reactor system, isometric view. Figure courtesy of PBMR project.

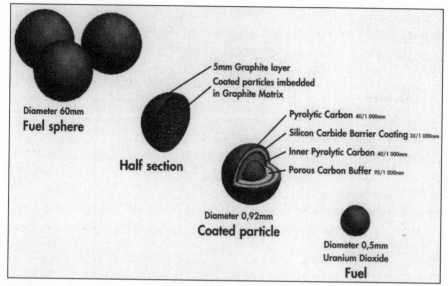

PBMR fuel sphere in cross section with detail of coated particle. Figure courtesy of PBMR project.

particles would avoid the costly systems otherwise necessary solely to maintain fuel element cooling in the event of failure of the primary heat transfer system, as with the emergency cooling systems and multiple containments of current commercial nuclear power plants. Such GT-MHR units in sizes of 100 to 200 MW(e) could be built in factories, thus minimizing field construction and gaining benefits in cost, quality, and safety. Protection of the public would be assured, and projected electricity-generating costs for the GT-MHR appear competitive with those for present large, 1,000 MW(e), light-water-cooled reactors. Multiple GT-MHRs could be grouped to meet power needs, with units added as demand increases.

Full confidence in the GT-MHR is dependent on demonstration of good performance of the coated particles under the high temperatures of 1,600 degrees Celsius (2,900 Fahrenheit), associated with the most severe accident conditions. The safety capabilities of coated-fuel particles near these temperatures have been demonstrated by several laboratories, and the data indicate that the required integrity can be achieved. Repeated favorable tests would be necessary to assure safety, and this fuel research and testing should be pursued to obtain the needed confidence.

I envision the GT-MHR as suitable for use in countries where the extreme and experienced care required for both construction and operation of other reactor types is not available. It could fulfill my dream of safe and abundant nuclear power for use anywhere in the world. Even with presently demonstrated fuel capability, the GT-MHR reactor appears economically attractive, but additional engineered safety features may be required.

The South African government also has an extensive program to develop small HTGR gas-turbine units to produce electricity using designs based on the German pebble-bed reactor concept. These 110 MW(e) units would be grouped in sets of ten modules for efficiency in operation and maintenance. This design aims to limit the maximum fuel temperature, in loss-of-coolant accidents, to temperatures previously demonstrated in severe loss-of-cooling tests of German-made coated particle fuel. For either the U.S. or South African designs, manufacture and qualification of the fuel particles must always be by fully qualified and licensed companies.

Since GT-MHR and PBMR reactors operate at higher temperatures than are possible with light-water reactors and thus achieve higher thermal efficiencies, they are more conserving of nuclear fuel. Their ultra-safety features also make them potentially acceptable for locations near cities or factories where waste heat could be used to achieve an even higher total energy efficiency and further conservation of fuel. Use of these reactors could extend nuclear fuels sufficiently to

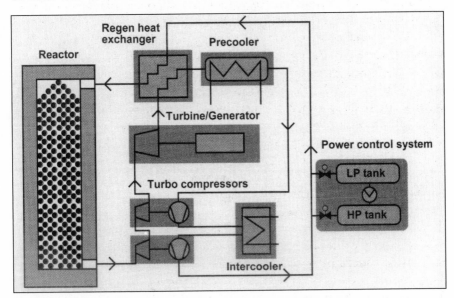

PBMR schematic flow diagram. Figure courtesy of PBMR project.

delay the need for breeder reactors for an extended time in the United States, but perhaps not beyond a one-hundred-year horizon. This class of reactors—HTGR, GT-MHR and PBMR—also can operate with thorium fuels, but reprocessing would be required to recover uranium 233 from spent fuel. Thorium is converted by neutrons and radioactive decay to uranium 233 and would become the fuel. Because thorium ores are abundant, this option would extend nuclear fuel supplies well beyond a one-hundred-year time frame.

Unfortunately, the U.S. Congress and the Department of Energy have not yet seen fit to fund the relatively modest program of development and testing needed to prove or disprove the integrity of coated-particle fuel at accident temperatures for the GT-MHR. Should this program be approved, the larger cost for a demonstration reactor would be justified to utilize the concept's outstanding safety, efficiency, and economy. Fortunately, work continues on variations of helium-cooled HTGRs in Japan, China, South Africa, and in Russia with U.S. cooperation. In the Russian program, enriched uranium from weapons can be used as reactor fuel to destroy the weapons materials. Although that program is encouraging, it may not lead to exploitation of the most attractive features of the HTGR. The United States might regain its former position as leader in nuclear reactor technology by starting with a renewed effort in coated-particle fuel development and testing, followed by a helium gas-turbine reactor demonstration.

The high temperature gas-cooled reactors have interested me since ORNL first chose to work on a pebble-type graphite structure for nuclear fuel. I was particularly interested in the safety features, the versatility in siting, and the adaptability of this reactor to a broad range of energy uses, such as process heat for manufacturing. By the late 1960s, with the success of the Dragon, Peach Bottom, and AVR reactors, it appeared that this class of reactors, often referred to as the high temperature gas-cooled reactor (HTGR), would be successful. The future of the HTGR seemed assured when General Atomics entered the commercial reactor market with the Fort St. Vrain Reactor and several electric utilities contracted for HTGR purchases. Unfortunately, the erratic operation of Fort St. Vrain, unproven costs of the large HTGR, and a sudden national recognition that utilities had over-extended their orders for new generating plants brought the HTGR market to a close. Many orders for light-water-cooled reactors also were canceled at that time. Market expansion had been based on a projection of the nearly linear growth of electric power experienced from its beginnings late in the nineteenth century. By the 1970s households and factories had been electrified, air-conditioning widely installed, and the oil crisis forced recognition of the importance of energy conservation. Electric utilities were even cautioning against construction of new power-generating plants.

Today, amid projections of inadequate energy supplies and growing concern over global warming and increased use of fossil fuels, a need for clean generation of electric power and energy use in new applications seems to be emerging. The small GT-HTR and PBMR could fit those needs admirably. Promoters propose with credibility that these smaller reactors, with gas-turbines and manufactured in factories with minimal site construction, can be installed at competitive costs when compared with other energy sources. The potential usefulness of these smaller power units seems unlimited in countries or locations with inadequate electrical distribution grids or with small indigenous energy resources.

This chapter has primarily described aspects of international cooperation in gas-cooled reactor technology and anecdotes connected with travel often related to nuclear interests. It thus has omitted mention of important technical work in the effective development of fuels, materials, and designs of reactors and their components, because such descriptions would violate the nontechnical nature of this writing. The ORNL contributions in fuel development, graphite technology, fission-product behavior, and concrete pressure-vessel development during this period could well provide the content for a technical reference work or a series of monographs.

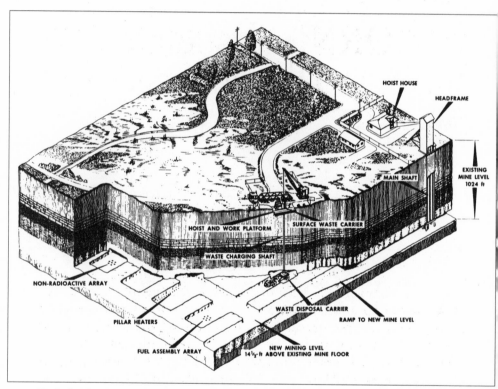

High-level waste repository in a salt bed. ORNL drawing by J. G. Blomeke.

ORNL Associate Director

A lvin Weinberg, director of ORNL, called me to his office late one morning in mid-March of 1970. Anticipating an inter-esting request or suggestion concerning a nuclear reactor program, I arrived with notebook in hand. The message was star-tling: H. G. MacPherson, the deputy director, was leaving for a professorship at the University of Tennessee, and Floyd Culler, the associate director for Nuclear and Engineering Technologies, would take his place. Surprisingly, Weinberg then asked me to accept the position of associate director, vacated by Culler. The post of program director had suited me well, but this was a promotion of significance. My request for time to think the matter over was met with, "You can have a few minutes, but I wish to make the announcement after lunch to avoid rumors." I accepted with appre-ciation, much apprehension, and a resolve to serve well. The bulk of this chapter is given to a summary of the seminal events and major developments in which ORNL participated during my years as associate director.

With an appointment in Oak Ridge early that afternoon for a purpose I have long since forgotten, I lunched alone at a small restaurant to contemplate this heavy responsibility. The May 1 starting date and the complexity of the job demanded careful planning for a smooth transition. In late April I was to chair an International Gas-Cooled Reactor meeting of three hundred attendees from ten countries. This would limit preparation time for the AD position; I had no time to waste or to worry, which was just as well.

Associate directors are members of ORNL planning and administrative committees and often serve as representatives to national organizations and governmental units. Within the Laboratory, I was to have responsibility for six divisions and ten programs. These encompassed most of ORNL's energy-related research that would be vital, as ORNL moved toward and became known as an energy laboratory. The divisions and their directors, respectively, were Chemical Technology, Don Ferguson; General Engineering, Mike Bender; Instrumentation and Controls, Cas Borkowski; Metals and Ceramics, John Frye; Reactor Chemistry, Warren Grimes; and the Reactor Division, Sam Beall. The last division centered around development of nuclear-reactor technology and would continue as the home base for Irradiation Engineering and the Gas-Cooled Reactor Program. Except for Reactor Chemistry, these were large divisions that employed about one thousand engineers, scientists, and technicians; in addition the Instrumentation and Controls Division had about a hundred craftsmen. I could not possibly come to know all of these people before assuming responsibility for their work and well-being. I resolved at least to walk through each laboratory unit as early as possible and succeeded fairly well in doing so. We chose Paul Kasten to lead the Gas-Cooled Reactor Program; the forthcoming meeting therefore offered an excellent opportunity for him to meet new colleagues from around the world.

The international conference had two parts: Gas-Cooled Reactors and Molten Salt Reactors. Both also served as annual information meetings for those programs, and many ORNL staff members would attend. The GCR meeting spanned four days starting April 27; the MSR meeting was held on May 1. Murray Rosenthal was responsible for the latter, but Paul Kasten also was a major participant. Both meetings were upbeat, attended by proponents still embracing the nuclear energy euphoria of the 1960s. I closed the GCR meeting with a summary of the status of fuel development and some cautions concerning needed improvements in materials and plant components. My final remarks were complimentary of the excellent international cooperation that had been a hallmark of the program and the meeting. I thought this success might set a pattern for broader international cooperation in nuclear and other fields; perhaps it has, but the results have been less than I had hoped for and anticipated.

According to our family guest book of foreign visitors, Elaine hosted nine of the GCR's foreign visitors, plus Laboratory staff members, for dinners at our home that week. One guest was Günther Ivens of Germany, whose son, Klaus, would arrive in a few weeks to spend the summer with us. This afforded opportunity to make final plans for Klaus's visit. I was concerned that the new responsibilities would prevent my offering Klaus the same excellent hospitality and learning experiences that Ivens had provided Tom.

The summer was to be a very busy time that included an early September vacation trip to St. Louis, Exeter, Nebraska, and Colorado. Tom had chosen to spend his junior year at the University of Colorado, as a change in environment. He had preferred a university in Europe, but that was not possible without losing time toward graduation in engineering. We questioned this substitution and loss of the Langsdorf scholarship that year, but he countered that a year at Colorado would provide learning experiences very different from Washington University and that he could gain some advantages in scheduling courses. Klaus accompanied us on the trip to take Tom to his new college, to see some of the Rocky Mountains, and to spend a few days on the Nebraska farm.

At ORNL I found it desirable to consolidate most of the ten energy development programs into four major units to improve communication and exchange of information. These became the Sodium-Cooled Breeder Reactor work under Bill Harms, the Radioactive Waste Management activity under Al Boch, the Molten Salt Reactor Program under Gene McNeese, and the GCR work under Paul Kasten. Smaller programs were folded into the four. For example, Concrete Pressure Vessel technology with Jim Corum and Patrick Callihan in charge, applied mostly to the GCR and was placed there. Structural materials and some instrumentation development were located in the Sodium-Breeder Program. These program directors provided the day-to-day interface with their Washington counterparts.

Later, we formed a new division, titled Energy, devoted primarily to analysis of the national energy requirements and to future planning.

Sam Beall headed the new division and reported to another associate director. Gordon Fee was then made head of the Reactor
Division and also later served as program director of work
conducted for the newly formed Nuclear Regulatory Commission
(NRC). The Engineering Division was later folded into a unit
serving the weapons and gaseous diffusion plants, but I continued
to provide liaison with the ORNL unit. The Reactor Chemistry
Division primarily supported the Molten Salt Reactor Project. As
that work declined, we moved its personnel and appropriate work
into the Chemical Technology and Chemistry Divisions. These tran-
sitions took place over many months, with further revisions
developing as support from the AEC and its successors was revised
and reorganized.

One AEC program, the Safety of Light Water Reactors (LWR),
became controversial in 1970. This was a period of transition from
government development of nuclear power to that of private
industry. Nuclear reactor stations were being designed and built in
sizes up to 1,000 MW(e) or more, whereas the earlier technology
had derived largely from military reactor programs for propulsion
of submarines and aircraft carriers. The industrial reactors posed
new safety issues: they were larger and land-based, some were
located near populated areas, and they utilized rivers or coastal
waters for cooling. The smaller, portable naval reactors located on
vessels and having an essentially unlimited supply of cooling water
had less severe requirements to assure safety in the event of equipment failure. When a nuclear reactor of 1,000 MW(e) is shut down,
the residual source of thermal energy from radioactive decay is
greater than the full-power thermal operating level for many military units. Thus, LWRs have demanding cooling requirements to
avoid the kind of fuel damage that later occurred during the accident at Three Mile Island.

For these reasons ORNL proposed more extensive tests to determine the effectiveness of existing and proposed emergency cooling
systems. Vendors of reactors had addressed the problem by calculation
and tests, and the National Reactor Testing Station, now the Idaho
National Engineering Laboratory, also conducted a complex safety
program for the AEC. The primary concern at ORNL was for more
basic understanding of cooling-system behavior, including the protection of fuel and its Zircalloy containers for all plausible accident
conditions. Bill Cottrell and his staff who pursued this programmatic

area saw need for many experiments to demonstrate safety margins and to validate better the codes used in their creation.

Milton Shaw, director of the AEC Reactor Division, controlled the funding for all AEC reactor programs, including nuclear safety. I wrote many letters to Shaw that included proposals for funding of LWR safety-related tasks. Some were funded and others were not, with responses that often seemed inadequate. Many of us found this a frustrating time because we believed that this new and major source of electrical-energy generation required more testing of components chosen to assure safety in the event of failure of the normal cooling capabilities.

Sometime in 1970 ORNL received a request for briefings on the status of LWR safety from a group calling themselves "The Union of Concerned Scientists." Their stated goal was to gain an understanding of reactor safety needs and to help in the resolution of concerns. We provided a factual review of the LWR safety status, without reference to ORNL's specific interests or proposals. As a result of concerns expressed by this group and others, the AEC convened a quasi-judicial hearing in 1972 to resolve the issues. The hearings were held under the auspices of one of the Reactor Licensing Boards created by the AEC to review new commercial reactor projects.

As the hearings evolved, the Union of Concerned Scientists was pitted in an adversarial relationship against the major industrial concerns that build LWR reactors. The vendors, General Electric, Westinghouse, Babcock & Wilcox, and Combustion Engineering, were represented by a formidable array of attorneys, the critics by a Chicago lawyer, Myron Cherry. Witnesses were called from the AEC, National Laboratory staffs, university nuclear engineering departments, and other professionals in the field.

I was responsible for coordinating the ORNL effort, including responses to requests from the hearing board. Our principal expert witnesses were Bill Cottrell, Phil Rittenhouse, David Hobson, and George Lawson. These people were called to the stand for extensive grilling by the attorneys from both sides, with interrogation of an individual sometimes extending for several days. Both sides challenged the testimony by Laboratory members because our people could not testify that the status of LWR safety was as poorly developed as Cherry insinuated or as adequate as the industry lawyers argued.

I was called to the stand only once to present the scope of the ORNL work on LWR safety. It was a straightforward story that generated little cross-examination since I testified as a manager and not as an expert witness. Mr. Cherry did not challenge me on the stand but approached me all alone in the middle of the hearing room at the break. In a tone of saccharine sweetness, Cherry suggested that my testimony had been in error and offered opportunity to change the statements when the hearing resumed. I assured him that my testimony was correct. After some discussion he stated, "I have written proof that you perjured yourself, but am willing to call you to the stand to correct your statements to match my information. If you refuse, I can have you sent to prison for perjury." I did not budge, having seen Cherry try to intimidate other witnesses on the stand, but I was surprised that he would go this far. I repeated, "The statements were correct" and added, "Try to send me to jail if you wish." Cherry emphasized the threat of prison, assuring me that he was only trying to be kind.

I reported the incident to the AEC attorneys, who appropriately questioned me to be certain that there was no flaw in my testimony. I asked them to put me on the stand for their questioning to put Cherry's inappropriate approach on record, but they refused. Instead, they confronted Cherry and brought him to me with an apology. His words were proper, but his handshake felt like grasping a dead fish tail; I was tempted to crush his fingers. We could never determine his motive in that challenge but thought it was designed somehow to discredit the overall ORNL testimony. The incident illustrates the hard verbal fighting in which the attorneys engaged, but I never saw the lawyers for the vendor or the AEC stray beyond acceptable ethical bounds.

The total experience was exhausting and exasperating for most witnesses. We observed the power of the adversarial-judicial system to reveal truth, but we were even more impressed in this case by the inappropriateness of the courtroom application. Facts were not in question. What we witnessed was the acrimonious evaluation of informed human judgment, which is always necessary to determine the adequacy and reliability of complex technical systems. The hearing could have been more efficient and effective had it been conducted cooperatively, not in a courtroom, to establish adequate engineering safety factors in areas of uncertainty. That approach could have better illuminated important problems for resolution and defined needs for further analysis, experimentation, and computer-code development.

As conducted, the hearing accumulated some twenty thousand pages of testimony over a period of several months. Much effort would have been required to analyze the mass of testimony and to define specific studies to be undertaken, but I do not recall any organized effort for that purpose. However, both the AEC and the industry soon created larger research programs that more fully defined the margins of safety for light-water reactors. This, in turn, produced some changes in reactor design and substantial verification of the computational codes used to predict and assure reliable performance of safety-related systems. Although most safety factors applied earlier by the industry to compensate for lack of knowledge had been well chosen, the new research produced a more secure basis for the designs of large plants.

The hearing spawned two new industrial organizations, one within the Electric Power Research Institute, which conducted and sponsored research in LWR technology. The other is the Institute of Nuclear Power Operations, formed by the nuclear utility companies. These institutions were created to share experiences more informally and quickly than through formal procedures established by the AEC and now by the Nuclear Regulatory Commission. With improved designs, code validation, and further testing of components and systems, the safety and operational features of LWR plants have been strengthened, and good performance records are now being achieved.

Another ORNL endeavor that has contributed greatly to safety in the design and operation of present and future reactors is the Heavy Section Steel Program (HSST). In 1965 Bill Manly, while serving as chairman of the Advisory Committee on Reactor Safeguards (meaning safety), pointed out that the integrity of thick-wall steel vessels was poorly understood. Light water reactor vessels were being constructed with walls ranging in thickness from six to twelve inches, whereas safety codes for industrial vessels had been validated for walls up to four inches thick. Strains introduced during the initial cooling of thick steel and in fabrication by welding could compromise vessel integrity. Also, light water reactor vessels are subject to neutron bombardment that causes embrittlement of the steel. ORNL was given the lead in a large program to establish the safety margins for these vessels. This too became my responsibility, with Grady Whitman as the manager in charge.

The first step was to cut small sections from the thick steel slabs to determine such relevant properties as ductility, tensile strength, and hardness, each as a function of position in the wall. These properties were also measured following neutron irradiation of the steels in test reactors to the level of exposure for a power-reactor vessel. The initial program culminated in tests of representative steel vessels placed under internal pressure, increasing until failure occurred. The vessels were about a meter in diameter by three meters long with walls scaled at appropriate thickness to represent those of large reactors. The tests grew out of concern that defects in forming or welding or induced during operation might cause failure. Flaws were introduced in sizes larger than those detectable by inspection, both during reactor-vessel manufacture and while in service. Tests were also made at low temperatures that simulate embrittlement of vessels from accumulated neutron bombardment near the end of service life.

The program, continuing today after thirty years, has shown that the vessels have reassuring safety margins when properly inspected and operated. It has also produced restrictions on operating limits and resulted in the development of improved surveillance technologies to assure safety margins for the vessels as constructed and as evaluated periodically during their lifetimes. The HSST Program has benefited greatly from excellent cooperation by vessel manufacturers, instrument companies, reactor vendors, and utility companies. Portions of the work were conducted by industrial laboratories, and some were funded and conducted by industry. This extended research program for pressure vessels reflects the complexity of design requirements for materials and equipment used in constructing, operating, and assuring safety for light water reactors. The findings of this program together with commercial LWR experience have provided the basis for extending licenses well beyond their original approvals. Since capital cost is a major factor in determining the price of nuclear power, life extension of reactors enhances nuclear power's competitiveness as a supplier of electricity.

In the early 1970s I was briefly involved in the field of radioactive waste.   Starting in the 1950s, ORNL initiated serious studies of methods for the safe handling and disposal of radioactive waste. Various geological strata were considered, including shale, salt, and hard rocks such as granite. Large salt beds that are hundreds of feet thick, lying beneath impervious shales about a thousand feet in thickness, seem ideally suited for the purpose. When mine shafts are refilled with salt, the weight of the shale and soil above slowly deforms the salt into a complete seal around the waste containers, forming a very secure enclosure. These uniform geological formations have been stable for two hundred million or more years and would surely be adequately secure over the fifteen thousand years required for radioactive decay. Contact with the high-level waste then would constitute a smaller hazard than that of the original uranium ore located by nature at, or near, the earth's surface. Large land areas underlain with these forma-tions exist in the U.S. Midwest and West, and only a few square miles would be required to accommodate a large nuclear industry if the waste were cooled in a monitored retrievable-storage facility for a few decades before placing it in the salt. Other geological formations such as shale and tuff also seem suitable, but I have not personally studied them in depth.

An abandoned salt mine lying deep under fields near Lyons, Kansas, had been chosen by ORNL and the AEC as the site for testing the behavior of salt deposits under the heating effects of energetic radiation from the waste. Experiments with electrical heaters showed the salt to be suitable with respect to heat transfer, deformation, and the corrosion of appropriate container materials such as copper.

In 1970 the AEC selected the Lyons mine as the site for a demonstration pilot plant for storage of high-level and transuranic radioactive waste. The plan also called for the mine to become a permanent repository if the experiments proved favorable. ORNL was instructed to develop and conduct the project on a crash basis because the AEC and the nuclear industry anticipated a massive expansion of nuclear energy within a few decades. The project was to be completed in three years at an arbitrarily constrained cost of twenty-five million dollars. Al Boch, who had successfully managed the construction of many nuclear facilities at ORNL, was placed in charge, with W. C. McClain and Tom Lomenick assisting. We developed conceptual plans for a plant design including shipping containers and services to deliver and emplace the waste safely. The plan was also to include an environmental assessment to explore the geology and hydrology of the area.

The requested rush schedule soon proved far too short. Even though determining the suitability of the site was a predicate step, necessarily coming first, the three-year schedule required site assessments, detailed design, contractual arrangements, and construction to proceed concurrently. Floyd Culler and I visited Milton Shaw, John Earlwine, and others in AEC Headquarters to seek a longer schedule and a promise of more adequate funding. The requests were rejected, and we were all but thrown out of the offices in response to our vigorous arguments.

The proposed escalation of the study into a project had appropriately brought it to the attention of political authorities in the State of Kansas. The local community of Lyons had been fully informed of the earlier plans and experiments as developed by J. O. Blomeke, W. C. McClain, and others and thus favored the project. Political figures in neighboring counties and at the state level, however, had not been adequately briefed by the AEC, which correctly reserved that aspect as their role. Some Kansans apparently saw this as a useful political issue.

Thus, the project was already controversial when our environmental explorations revealed two potentially serious problems with the site. Beginning in the late 1800s, the region had seen exploration for oil, salt, and, later, for natural gas. We discovered deep bore holes near the mine that had not been properly plugged. Some early holes even contained discarded drilling cables. Although none of the holes penetrated the mine, their proximity raised serious questions about the site's long-term viability. Reopening and properly plugging the holes, particularly those with cables, would be difficult at best.

The second problem emerged with the discovery of a horizontal shaft dug from a nearby mine at some earlier time. The objective had been to join the mines as a safety measure, in case workers in either mine should be trapped by the failure of its single shaft used for entry and exit. Although the mines were never connected, flooding of the second, unattended mine in such close proximity could endanger the waste repository.

These factors resulted in abandonment of the project, with political consequences adverse to development of similar facilities. It was apparent that AEC officials had underestimated public concerns over activities related to development of nuclear power. The political reactions engendered and the associated requirements for assessment and funding for a permanent repository are reflected today in the much delayed and costly Yucca Mountain radioactive-waste repository project in Nevada.

The Kansas project's difficulties may have been exacerbated by the inadequate background of some AEC staff. A meeting in Lawrence, Kansas, with members of the State Geological Survey had centered on the handling and shipping of transuranic (TRU) waste. This material results from transformation of uranium to other heavy elements by neutron exposure in an operating nuclear reactor. The radioactive hazard of these wastes is primarily from their emission of alpha particles. Although alpha radiation is not penetrating and can be stopped by a sheet of paper, materials containing products of alpha-particle decay represent a serious danger if ingested or inhaled.

As we walked from the University of Kansas to a nearby restaurant, the AEC person in charge asked, "Don, what is an alpha particle?" I first thought he was joking or using the question as an opening to make a point, but quickly realized that he really did not know. The dinner conversation became a lesson on some very basic concepts in the physics of radiation and radioactive particles.

A facility of this type was constructed in New Mexico in the 1980s for storing only TRU wastes from U.S. military programs. It was delayed several years by political controversy but is now in service.

For whatever reason, this country has thus far been deprived of bedded salt as a repository for the safe and effective placement and storage of commercial nuclear waste. Although salt beds have received much additional study in development of the Waste Isolation Pilot Plant in New Mexico, no project for placing commercial radioactive waste in these formations is now proposed. Political interests continue to challenge the technology whenever a radioactive waste disposal site is nominated. Such challenges are possible because the geology of any earth site is complex. However, if similar, often ill-founded challenges were applied to other facilities, perhaps many large bridges, buildings, and mineral mines could not be constructed, even where the plans are sound.

Similar arguments have ensued over the disposal of low-level wastes that originate from medical and industrial use of radioisotopes and from the maintenance of nuclear power plants. Although aging facilities of limited capacity continue in use, the several regional repositories authorized by Congress have not yet been opened. The new technology is to pack the waste in metal-lined concrete boxes that in turn are tightly stacked on a concrete pad. When the pad has been filled, the stack is covered with heavy plastic sheeting followed by a thick layer of bentonite or other clay impervious to water. A berm is then formed by a layer of topsoil, properly shaped to drain water and seeded with grasses for stability. The pad's drain is monitored for radioactive leakage, and the assembly is repairable if necessary. Since these wastes decay to harmless levels in about fifty years, this structure is quite adequate.

It seems appropriate to note here that the nuclear industry of today produces about a third of a trillion dollars for the U.S. economy, with only one-fourth of that from nuclear power plants. The highly beneficial medical and economically sound industrial uses dominate and require suitable waste repositories, preferably as regional entities to minimize cost.

Other countries, such as France, where over 80 percent of the generation of electricity is from nuclear plants, have developed systems for radioactive waste storage. Few have the semi-arid regions with extensive salt beds or other suitable disposal alternatives with which the United States is blessed. Most foreign facilities provide centralized interim storage until the waste has cooled and becomes less demanding of a long-term repository. Although no country has been without controversy, we in this country seem to have excelled in both nuclear technology development and controversy.

Nuclear proponents have also been at fault. Funding has never been provided for preparing comprehensive technical documentation of nuclear-waste technology. The work has indeed been reported but in bits and pieces as completed, making it difficult for technicians in other fields to become informed of the designs and their basis in sound technology. Written in technical terms, the existing reports are even less convenient for the news media. Comprehensive documentation with an accompanying translation to lay language might allay the controversy and restore progress in this important field.

Development of sodium-cooled breeder reactors also warrants discussion as an energy source. Work on sodium as a coolant for nuclear reactors, begun with the Aircraft Nuclear Propulsion Project, formed a basis for this reactor concept. Breeder reactors offer potential energy production many times greater than that possible for LWRs using fissionable materials from existing sources. Sodium-cooled breeder reactors operate in a favorable neutron-energy range to produce plutonium from more plentiful uranium 238. The plutonium formed by neutron capture and radioactive decay is a much larger energy source than the original uranium 235 fuel, hence the term "breeder." ORNL never had a major role in this reactor concept; but ORNL's Bill Harms and others provided technical support and ORNL led in developing its fuel-reprocessing system. A new division was created for the program on fuel reprocessing with Bill Burch in charge.

Much breeder reactor development has been done in Europe and Japan, where large demonstration plants have operated well, but the cost of sodium-cooled breeder reactors remains uncompetitive. However, their potential capability remains a potential bulwark against future electrical energy shortages as other resources are depleted or demand increases. At least a modest, long-term effort should be deployed to improve breeder-reactor concepts, their accompanying fuel cycles, and to address the challenges of nuclear proliferation they pose by producing plutonium.

The development of human-controlled robots for these radioactive systems produced designs superior to then state-of-the-art robots and led to a spin-off industry in the city of Oak Ridge. We even gained preeminence over the Japanese in this technology, despite their pioneering and prowess in automated robotics. Charles Weisbin and Joseph Herndon headed the program, while Bill Hamel and Reinhold Mann improved the robots with the addition of sensory powers that provide feedback to the operator's hands for precise and reliable operations. Such development of improved materials, chemical processes, instrumentation, and science in general have all contributed to not only the nuclear field but to other industries as well. As associate director I initially had responsibility for 40 percent of the Laboratory's technical staff. Space does not permit me to recognize all of the talented people who developed programs, produced valuable research findings, and to whom I am heavily indebted, but I can mention a few. Jim Weir followed John Frye on his retirement as head of the Metals and Ceramics Division. M&C was later transferred to another associate director in a reorganization of technology groups that more equally distributes responsibilities. Herb Trammell followed Gordon Fee as head of the Reactor Division, which was renamed Engineering Technology, more properly reflecting its later work and Pete Lotts assumed management of NRC programs. Herb Hill and Fred Mynatt, in turn, replaced Cas Borkowski following his retirement from the Instrumentation and Controls Division.

I also had responsibility for a great many smaller energy-related programs such as materials development for the Galileo mission to Jupiter, development of new alloys to improve energy efficiency in stoves and furnaces, better insulation materials, improved instrumentation for laboratory equipment, and better control systems for improvement of heating and cooling systems, as well as for reactors. The Engineering Technology and Metals and Ceramics Divisions developed heat exchangers for use in the desalination of brackish and sea water. Materials created for nuclear-energy projects have become new alloys now widely used by industry and are applied in household products such as stoves and furnaces.

I was fortunate to acquire the services of my able finance officer, George Warlick, who relieved me of many important, non-technical burdens. Also, Floyd Culler had created the position of technical assistant to carry out many actions and to summarize technical material for effective decision making. Steve Kaplan, John Coobs, David Krause, Bill Roddy, Baden Duggins, and Jim White each ably served with me in approximately two-year assignments. Secretaries Carolyn Coleman,

Lindy Norris, and Lois McGinnis in succession, assisted by Bettye Brown and Carol Holtzclaw, managed the complex work of the office. I worked closely with members of the Oak Ridge Operations Office as they progressed through federal government reorganizations from AEC to ERDA to DOE. David Cope of DOE and his staff assigned at ORNL for energy systems were consistently helpful. Weekly meetings at the Oak Ridge Operations Office of DOE with Joe Lenhard, Dick Egli, and Stan Ahrends, in succession, helped maintain an effective flow of information to and from Washington.

One of the exciting aspects of my position as associate director was to come into contact with other areas of ORNL expertise, such as genetics. Because the genetics of mice resembles that of humans in important ways, their short life cycle and rapid reproduction prove highly useful in research. Studies with large populations of mice demonstrated that low levels of radiation do not cause genetic change. This long-term program showed that nuclear technology could be widely employed; obviously, our society would never accept use of nuclear power if the associated low levels of radiation were shown to cause genetic damage to the human population. This research technology has now extended into the field of chemical hazards as well and is now focused on functional genomics, in which new tools and research patterns of interest to the pharmaceutical industry and medicine have been developed. The research with mice offers promise of important new developments in the prevention and treatment of disease and in correcting hereditary defects.

ORNL has adapted biological research—derived in part from nuclear programs—to other problems, such as pathology for chemical insults. For example, environmental studies first applied in nuclear areas have made important advances in fisheries. Energy-related terrestrial ecology pioneered by ORNL continues to provide new insights into the problems of acid rain. The effect of other pollutants that cause global warming and the reduced ozone layer of the atmosphere are the subjects of other extensive studies. More than ten thousand acres of Laboratory land has been set aside as a National Environmental Research Park that encompasses an entire, highly-instrumented, 270-acre watershed. This hill and valley land has been undisturbed since the Manhattan Project acquired those fields and forest areas, and thus provides valuable information on natural development from abandoned agricultural pursuits. ORNL chemical and chemical-engineering studies that first centered on the processing of nuclear fuels have been adapted to purification of water systems, cleaning contaminated soil, and improving manufacturing methods.

Physics research at ORNL, which focused originally on nuclear systems, has centered on acceleration of heavy elements and their collision with other matter to gain understanding of the basic structure of matter. Solid state physics research has produced improved body-implant devices, for applications such as artificial hip joint surfaces that can now last a lifetime. The science of neutron diffraction, which began at the ORNL graphite reactor and won a Nobel Prize for Clifford Shull, has been pursued at the high-flux isotopes reactor. Neutron diffraction is to be continued and expanded so that it can advance into many areas of physical, chemical, and biological sciences using the Spallation Neutron Source now in design and that will be operated cooperatively with other national laboratories.

ORNL computers are among the most powerful in the world. This capability has practical as well as mathematical significance. For example, in working with industry, ORNL mathematicians and engineers have made it possible to model automobiles in detail sufficient to determine how they respond in accidents to reduce the need for, and to analyze better, costly impact testing. More recently these computing capabilities have been applied effectively to understanding the human genome.

ORNL was the first national laboratory to employ a staff of social scientists to aid in the identification of technological needs, evaluation of environmental impacts, and to assist in obtaining public acceptance of appropriate technologies. Basic studies in chemistry, materials, and instrumentation both support the applied programs for which I was responsible and produce important discoveries in basic science. ORNL development of instrumentation and controls for nuclear reactor systems has been expanded to encompass many fields. A recent focus has been on needs of biological research, extending to the development of a nonintrusive scan of mice (humorously called "cat-scan for mice," in that it is actually patterned on the CAT-scan technology) used in genetic studies. It was challenging, intellectually stimulating, and always demanding to maintain cognizance of and provide leadership in these diverse, important, and often pioneering fields of research. ORNL also conducts educational programs for thousands of students, from elementary school children to training at the level of the post-doctoral degree.

This inadequate listing only illustrates the broad areas of technology that can be developed in a national laboratory to solve

Don Trauger and Alvin Weinberg at the dedication of the graphite reactor as a national nuclear landmark, circa 1992. ORNL photograph.

problems by focusing on important needs. Much of the ORNL preeminence is attributable to Alvin Weinberg's foresight and leadership as its director over twenty-five of the Laboratory's formative years.

Byron's freshman year at Duke University had given us great concern as he joined a campus vigil in the wake of the Martin Luther King assassination, to protest university practices outmoded by the standards of the late 1960s. As other campus protests turned violent, we were apprehensive but tried to be supportive. In retrospect I must agree that these were particularly important and meaningful segments of his education. His peers of the sixties led the nation to new levels of recognition for the need to balance technical and industrial advancement with corresponding social and environmental measures. But that knowledge did not help me sleep any better on the night we knew he was protesting in pouring rain or on the night he and I were on the phone from midnight to 3:00 A.M. discussing a proposed protest in Washington, D.C. (He decided not to go.)

Byron completed his graduation requirements on time despite having spent 1969 in Peru and several weeks of 1970 participating in a political campaign for Tennessee Senator Albert Gore Sr. He also declined a Fulbright Scholarship for study in Chile in order to accept admission to Yale Law School.

As Duke's student graduation speaker, Byron chose the theme "In Perpetual Revolution." He drew on the student revolution of the 1960s and urged continual change for the better. He was also the only person on the stage in a business suit rather than cap and gown. He had joined a student protest over low salary schedules for the nonacademic employees of the university. Students donated to the employees the money saved by not renting the traditional caps and gowns.

At a luncheon following the ceremonies we were seated with Tom Wicker of the *New York Times*, who had received an honorary doctorate that morning. We were impressed by his depth of knowledge in subjects familiar to us, including ORNL and the Tennessee Valley Authority. Others included U Thant, then Secretary General of the United Nations, who also received an honorary degree, and Nancy Hanks, chair of the National Endowment for the Arts, the principal graduation speaker. It was a pleasure and inspiration to meet and discuss national and international issues with these distinguished leaders. Tom Wicker made "In Perpetual Revolution" the theme for his syndicated column the following day, giving credit to Byron.

A year later, May 18, 1972, we received word that Mother Causey had suffered a severe stroke. Elaine immediately drove to Greensboro but arrived too late. Family dedication brought us all together from different points of the compass, as Byron and Tom skipped preparation for some final examinations, and I diverted to Greensboro from business in Washington, D.C. The gathering of friends from the North Carolina Piedmont area gave testimony to this dedicated woman and her work in support of family, church, and social organizations.

Tom's final months at Washington University were a scramble to meet the requirements for two degrees. He was accepted at Harvard, Stanford, and Yale law schools, but rejected all three upon winning a Fulbright Fellowship to study the excellent German land-reclamation practices for strip-mining of coal. He enrolled for this study at the

University of Cologne. In June we drove to St. Louis where Tom received the B.S. degree in electrical engineering and an A.B. degree in economics from Washington University. We had the privilege of meeting and enjoying a brief private discussion with Jesse Jackson, who had been one of the graduation speakers. Tom spent the summer with the ORNL Energy Division as a non-paid guest to learn about U.S. coal-mining and other fossil-energy programs, as preparation for study and work in Germany. Ed Nephew of the ORNL Energy Division, who had studied in Germany and worked at the AVR as an ORNL assignee, was Tom's principal advisor. He received monthly reports from Tom during the fellowship that was extended for a second year.

While at Yale Byron won a Rhodes Scholarship to study Latin American Economics at Oxford University. He chose to study at University College on the advice of three of its alumni—including fellow law student Bill Clinton. With both sons in Europe, we decided to spend Christmas of 1972 in England, followed by a week of the new year in Germany. Since that would be the first Christmas for Elaine's family without Mother Causey, we invited Elaine's father, her sister Eloise, and her Aunt Annice Underwood to join us in England.

We spent the first days in London with plans to visit Bournemouth over Christmas Day. Unfortunately, Elaine and I contracted the London flu, which was epidemic at the time. Despite excellent medical assistance, there was no possibility of making the scheduled trip to the south coast. Knowing that British families often spend Christmas holidays in resort hotels, we gave the prepaid reservations to George Lockett, senior engineer of the Dragon Project. A few days before Christmas I had recovered enough to drive to Oxford, and we traveled there with Elaine in a makeshift, blanketed bed in the back seat of our rented car. Despite my rather extensive travel in the British Isles, I had never before driven a car there; it was a challenge to drive on the left side of the road, starting in the center of London while still weak from the flu. Byron brought his car from Oxford to complete the entourage. Resolved to celebrate Christmas in the English style, we stayed in the Randolph Hotel, a new and memorable experience dampened only by Elaine's weakness from the flu.

The day after arrival in Oxford, Tom drove Dad Causey and others to Herefordshire, the place of origin for his beloved Hereford cattle. The visit was timely because Tom's grandfather had just purchased a fine bull from that region. Back at the Randolph, with Elaine still confined, we had our meals served in the room. While placing the trays outside the door, one of the boys noted what appeared to be a hard-boiled egg left on another guest's tray outside the room door. He grabbed the egg and tossed it into the room to his brother, who caught and sent it back by return aerial trajectory. In the end, instead of catching it, one ducked, and a very soft-boiled egg crashed into a window curtain rod and slowly spread its contents down an exceedingly dusty drapery panel. We spent the next hour restoring the decor as best we could. This incident joined the flooded tub in Nice and the broken pipe fountain of Malaga as vivid memories of gaffes during European travel.

Elaine was well enough to attend Christmas dinner in the hotel's festive main dining room. Everyone was issued a colorful hat or hair ornament, a noise maker, and a package with a small firecracker, properly called a "cracker." The pyrotechnical devices did not require matches; they modestly explode in the middle when their ends are pulled apart. I could visualize the near apoplectic reaction of a U.S. fire marshal observing this scene: firecrackers discharged in a large, crowded room festive with crepe paper decorations and confetti, in a wooden-structured hotel building.

Our party joined the programmed activities as quickly as we could perceive the protocol for each event. Perhaps Roscoe Causey's English heritage was confirmed when Elaine's usually staid and serious father donned a silly paper hat and tooted party noisemakers. (We do not know his family history, but there is a mountain named Causey Pike in the Lake District.) In due course, Father Christmas arrived with gifts for the children and candy for everyone, climaxing this memorable Christmas celebration.

On Boxing Day afternoon, we were joined by George and Mona Lockett, who had driven from Bournemouth for a wonderful two-hour chat. They reported having enjoyed the hotel stay and had paid the bill to credit our account. In addition we had a visit from our close friends the Plails—Ollie, Eileen, and their son, Roger—with whom Byron had corresponded as a high school "pen pal." The boys teased Ollie and me mercilessly for each having urged his son to respond to letters. Both, by then graduate students, insisted that they soon lost interest in the exchange, and it was only the tyrant fathers who caused them to struggle with that correspondence. My perception was that the incentive for continued communication waned when Roger tried, page after page, to explain the game of cricket. I have never understood that game, even when watching, perhaps for having a short attention span for slow action. I recalled with the Plails dinner enjoyed some years earlier at the Barley Mow, a sixteenth-century inn with thatched roof and low-beamed ceilings darkened with fireplace smoke. I told the Plails that after our dinner there an ORNL friend had approached the inn with a room reservation made on my recommendation, only to find it engulfed in flames.

George Lockett reported to me on the excellent progress and status of the technical information of the Dragon Project, and Ollie, still engaged in irradiation test work, talked of progress at Harwell. The British have been leaders in nuclear energy programs and were increasing their production of electrical energy through their Magnox and advanced gas-cooled reactors. Light water-cooled reactors also were under study in England.

The North Carolina contingent returned home later that week as we and the boys flew separately to Berlin. That New Year's Eve is unforgettable. We attended the Berlin Symphony concert, even though our wardrobes did not include appropriately formal attire. The performance was flawless, with Herbert von Karajan conducting in the modern, acoustically balanced, and well-arranged concert hall. We had planned to visit a city park for an advertised fireworks display but were too weary for further venture. At midnight I opened a window of an upper-floor room of our pension on Kurfurstendamm Strasse in hopes of seeing some of the rockets from a distance. We were greeted by deafening sounds that seemed to rival war-zone action. Fireworks were being set off in the street, forcing us to close the window, for rockets and other fiery projectiles were arriving in close proximity; some later struck the glass. The next day the streets were strewn with enough debris from the fireworks to require caution in walking.

From Berlin we flew to Stuttgart, rented a car, and drove through the Black Forest and a bit of Switzerland before proceeding to Tom's base at Cologne. We also enjoyed a two-day visit with the Ivens family. On the way to Jülich, Tom guided us on a driving tour of the German coal strip-mining area and the pleasant landscapes of lakes and land produced by their reclamation program. I had previously seen the region from the air, but its immensity, the size and complexity of mining equipment, and the beauty of the final restoration could only be fully appreciated when viewed from the ground. Although this was strictly a vacation trip, the visits with Herr Ivens, George Lockett, and Ollie Plail and the direct reports of Tom's studies provided an opportunity to learn much about the energy programs of Europe.

Nineteen seventy three was an interesting year for one involved in energy. The OPEC curtailment of oil supplies created a severe shock for a people accustomed to lavish energy use at low cost through the burning of wood, coal, and petroleum resources in successive historical eras. As the price of oil rose to fifteen dollars per barrel and beyond, I was reminded of the 1969 prediction by Peruvian Bureau of Mines and Energy that oil would cost five dollars within five years, and of the derisive response to that prediction by the AEC in Washington.

ORNL had pioneered energy conservation research in the 1960s. That work, suddenly appreciated, had been criticized earlier as unneeded, particularly by utility companies and members of Congress. It had been supported by the AEC but without enthusiasm. With the new emphasis, the ORNL senior staff reexamined, then proposed, expanded programs in nuclear energy, fossil fuels, bioenergy, and energy conservation and were pleased to receive increased U.S. government support. Also in 1973 Alvin Weinberg resigned to form an Energy Institute within the Oak Ridge Institute of Nuclear Studies. Floyd Culler then served as interim ORNL director for a year, and Herman Postma became director when Floyd left to become president of the Electric Power Research Institute in California. This organization, funded by electric utility companies, sponsors research and development in many aspects of energy production, including environmental effects. ORNL often works with EPRI both funding and being funded by the utility organization.

We planned to visit Greece for our 1974 vacation and invited Elaine's father, then nearly eighty years of age, to go with us. As a great admirer of the Greek philosophers, he was pleased to participate. The boys met us in Athens during the spring break from each of their universities. We first stayed at the Plaka Hotel in Athens, where the proprietor assigned us rooms, each with a view of the Acropolis. I sacrificed several hours of sleep to view in rapture that marvelous structure gloriously bathed by the light of a full moon. It was thrilling to visit the scenes of Greek history and to walk the streets of old Corinth where Saint Paul had visited, preached, and counseled. The trip included other historic sites in the Peloponnese and a cruise to several Greek islands. At Santorini (Thira) we mounted horses to ride to the city's quaint stone buildings on the high bluff above the sea.

Descending the long, irregular, stone stairway to the pier, I was fascinated by a blind teenage girl's agility in managing the asymmetrical steps with a cane. As we were seated together on the launch returning to the ship she turned to me asking, "Please sir, describe the view from here for me." Emotion welled in my throat to make speech difficult, as I attempted to portray the beauty of that shining white city atop the sheer, striated, and colorful rock face formed by an earthquake of long ago. Language was inadequate to describe the spectacular beauty of that scene in the enhancing light of the late afternoon sun, but I intensely hoped that my words had helped her to imagine it. The impact of her question, her ambulatory skills, her seeming self-confidence, and her interest in the things she could not see produced admiration and deep and continuing hope for her success.

Returning to Oak Ridge, I found an invitation to attend the 1974 Nebraska Wesleyan University graduation exercises in the accumulated mail. It was accompanied by a letter from President Vance Rogers advising that the faculty and Board of Trustees had chosen to award me an honorary Doctor of Science degree. We flew to Lincoln for the ceremonies of May 18, and spent another day with my father, my brother Bob, and his family.

Gordon Brown sent a letter in the spring to request assistance for his daughter Anne and her friend Geraldine Harris to come to the United States for work that summer. Each had finished a university premedical program, was to enter medical school in the fall, and wished to spend the intervening months in a large U.S. city. Visas for such purposes required a U.S. citizen to guarantee passage for their return home. I prepared the necessary papers for them to work in Chicago. Upon arrival they were not pleased with their choice of cities and moved south for employment as cocktail lounge waitresses in then-popular "Underground Atlanta." These attractive young women, with charming British accents, earned enough in tips to finance a bus tour of the United States, including a visit with us, as well as their air transport home.

A few years ago, while on vacation in England, we visited Anne Brown (now Dr. Anne Beer) in the attractive country home in Devon that she shared with her physician husband and their young son. Both Anne and Mark Beer serve in small hospitals and clinics. We were pleased to learn of the fine service that these young people provide in their country; that function seemed analogous to an earlier medical tradition in this country, in which the doctor saw and treated all members of the family. Although they worked in a national health system, they reported few constraints or problems in their medical practices.

During 1974 ORNL further assessed its programs in response to national attitudes more favorable to energy conservation and the development of new energy systems as alternatives to the direct burning of coal, oil, and natural gas. The energy conservation program grew with relative speed. When AEC guidance and funding led to emphasis on the design of more efficient household equipment, I thought improvement of industrial systems was more appropriate for this high-technology laboratory. However, the program directed by ORNL scientists and engineers and conducted cooperatively with appliance manufacturers produced great advances in the many energy-consuming devices of the home. As a result any furnace, refrigerator, heat pump, cook stove, or hot-water heater purchased today is much more energy efficient while providing equal or better performance than earlier models. Furnaces for home heating now have efficiencies in the 80 to 90 percent range, rather than 20 to 30 percent as in the 1950s when much of the available heat was wasted to the chimney. The heat-pump hot-water heater is another example of that new direction in technology development that was advanced by ORNL.

The energy efficiency program later turned to improving industrial processes, but there the payoff is slower, in part because of longer amortization periods for energy equipment and facilities. New tasks related to improved energy production included the burning of coal in fluidized beds where powdered coal is elutriated by the air for combustion and to coal liquefaction to produce oil. Although technologically attractive, these processes have since languished, primarily because of the low prices for natural gas. Natural gas is a cleaner fuel than coal and is now favored for stationary plants generating electricity. This, unless curbed, will lead to early depletion of the finite proven domestic resources of natural gas, as is already occurring for oil, with accompanying higher prices for this limited natural resource.

In October 1974—in response to the energy crisis and, in part, from criticisms continuing in the aftermath of the Emergency Core Cooling Hearings—the AEC was reorganized into the Energy Research and Development Agency and the Nuclear Regulatory Commission. This required some reorganization of ORNL. That was a busy time at

ORNL Executive Committee in 1984: (left to right) Alex Zucker, Murray Rosenthal, Donald Trauger, Herman Postma, K. W. Sommerfeld, and Chester Richmond. ORNL photograph.

the Laboratory, especially for associate directors, but the new opportunities were addressed with enthusiasm.

The Executive Committee examined the Laboratory's long-term objectives in a series of meetings conducted at the Oak Ridge Associated Universities (ORAU). Program directors and specialists were invited to assist. The ORAU conference room was nearby but sufficiently isolated from ORNL to minimize interferences with these discussions, which at first were free-wheeling, then focused. As a result of these sessions, we decided to expand two areas of Laboratory competence. One was ceramics, a field previously pursued primarily for nuclear fuels. ORNL's newly developed ceramic materials included improvements in hardness, strength, wearability, ductility, and fabricability. A decade later this thrust had resulted in ORNL's becoming a leader in the field, even attracting industrial ceramics companies to locate in the city.

The second new direction was to expand expertise in geology and field hydrology to meet the anticipated environmental requirements associated with expanding population and energy uses. The nation's accumulation of industrial-waste effluents and the storage of radioactive waste also required improved understanding of geological formations and underground hydraulics. These too have become important areas of research producing substantial contributions to our understanding of the geology of the Oak Ridge reservation and many other industrial areas.

In November of 1974 I was to participate in two international conferences in London. I took advantage of airline schedules and fare structures to make prior stops on the Continent to visit European colleagues and laboratories. Brief visits would be made to Saclay, Cadarache, and Marcoule in France and to Karlsruhe and Jülich in Germany. Elaine chose to accompany me to Europe, but to spend more time in London and at Oxford with Byron.

We first flew to Paris to spend a weekend with Michel and Catherine Grenone. Michel, an avid student of historical sites in Paris and Versailles, took us on a tour of the great and beautiful court of Louis XIV. He described in depth the history of many rooms, with more interesting details than are contained in the memorized scripts of most tour guides. Upon leaving the palace, the Grenones announced that we would now collect wood from the Versailles grounds for their fireplace. We proceeded to cut wood and load dead tree branches into the trunk of their car. I instinctively watched for a gendarme to arrest us, but our hosts repeatedly assured that such scavenging was quite permissible for a citizen or guest. We escaped without interference and after dinner that evening enjoyed the warmth of a small blaze in their apartment fireplace. On Monday and Tuesday I met with staff members at Saclay concerning our cooperative research programs, while Elaine flew to London to settle in a small hotel near Hyde Park.

Late Tuesday afternoon I flew to Marseille en route to a scheduled visit at Cadarache. Eager to reach my hotel in Aix en Provence, I attempted to save time by arranging the car rental while waiting for luggage. That was a wrong and costly decision in Marseille. The rental took a little longer than expected, and upon reaching the luggage rack I was confronted by two stern customs agents hovering over my bag. Marseille was then a center for drug smuggling, and they obviously viewed my plan with suspicion. That mistake cost me about forty-five minutes, as they searched my bag and briefcase, opened and sniffed bottles of after-shave, and asked many questions that challenged me to remain calm.

Finally released, and after struggling to understand the controls of an unfamiliar vehicle with a broken interior light, I arrived late at the Paul Cezanne Hotel. Most restaurants were closed by that time, and the one I found left much to be desired. I speak no French, the waiter spoke no English, and consultation with an American student was not helpful. Eventually, I pointed blindly at the menu to order. I then spent a frustrating meal picking small morsels of meat from the breast of a thrush served with feet, head, and eyes intact. The French bread was delicious and filling.

At three in the morning, the thrush took revenge, leaving me deathly ill and deprived of further sleep. The scheduled eight o'clock departure was difficult with breakfast unthinkable, except for a few dry crackers that had narrowly escaped customs confiscation, but I arrived at Cadarache on time. Engineers there were most hospitable, particularly Claude Ringot who had served as my counterpart in the French fuel-irradiation program conducted in the Oak Ridge Research Reactor. We discussed nuclear programs and observed features of French irradiation facilities that he knew in some detail.

Other staff members showed me their large engineering laboratory for development and testing of components for the French reactors in their sodium-cooled breeder program. The building housed massive structures for testing large pumps, heat exchangers, steam generator components, instrumentation and control units, and equipment to measure the interaction of sodium with materials under various conditions of dynamic flow and metal stress. I climbed a ladder nearly twenty feet high to a platform for an overview of this impressive equipment, a physical exertion I quickly regretted and that could have caused a serious fall. Nearing the top of the ladder, I nearly fainted, perhaps an aftermath of the thrush. After briefly kneeling with head lowered, hopefully unnoticed by my host who remained below, I stood to survey this stately array of equipment. It appeared quite adequate to support the French plan to build the 800 MW(e) fast breeder reactor at Creys Malville.

The French had invited me to stay at their guest facility, a large castle converted to provide guest housing at this rather remote location. I had been there before but only for lunch in the banquet hall. Arriving at dusk, I was frustrated to find the castle apparently deserted; even the great hall was dark and locked without apparent prospects for dinner. The main castle structure was open but showed no evidence of life inside. I ventured to enter and, after becoming accustomed to its darkness, climbed the staircase to the second floor and heard distant voices. After ascending to the next level, I saw a faint ray of light across the hall floor from a double doorway. Peering through the crack between doors I saw several men working intensely over engineering drawings; it seemed best not to bother them. Continuing up the staircase, I came to a door that opened to a valley of the roof. After making sure that the door would not lock behind me, I reveled over a spectacular view of the full moon rising above the foothills of the French Alps and the Durante river. Shimmering reflections spread from the many rivulets and ripples of the sprawling stream. I stood spellbound for several minutes absorbing the moonlit spectacle. Fatigue was threatening, and it seemed best to terminate exploration and retreat to the car for rest.

On the way I met a German who had just arrived with news that the staff would be available for room assignments in about ten minutes and that dinner would be served at eight. I walked with him to a locked door of the castle, where we chatted until the proprietor arrived to present me a key to the stables. My room indeed had been a two-horse stall, and although the floor was new, the original exposed and polished plank walls were there, supplemented above by new paneling for privacy not afforded the horses. Even the door hardware was reminiscent of that which our old horse Harry so deftly manipulated with his lip. I set the alarm for seven forty-five and immediately fell asleep in the complete silence of the stable. At eight o'clock, lights and activity were everywhere. Moonlight bathed the castle, dinner was excellent, and I retired early for greatly needed rest.

My drive to Marcoule was uneventful, and the day proved rewarding. Their systems for handling, sealing, and storing high-level radioactive waste were practical, effective, and superior to the systems then available in the United States. The calcined waste from fuel reprocessing that recovers the plutonium and residual uranium was heavily mixed with shards of glass that were melted and cast into cylindrical metal containers. The solidified blocks were stored in deep, metal-lined holes spaced evenly in the concrete floor of a large building. Although not a permanent storage system, it is quite adequate and manageable to hold the waste for a few decades that allow the heat generated by radioactive decay to diminish by conduction to the earth. Thus cooled, the waste becomes less demanding of a permanent repository, an important consideration for France where geological formations seem not as suitable as bedded salt in the United States. After visiting fuel-reprocessing facilities and laboratories, an evening storm threatened. I therefore left a bit early for the drive to Marseille and the flight to Frankfurt. That was fortunate because in providing instructions to reach the major highway, my hosts had failed to note a road to the right that preceded the one intended. I took the first right, the storm struck with fury, darkness fell, and I was soon lost and terribly confused.

A friend at Cadarache had given me a very detailed map of the region for just such necessity. I searched my bag to find the flashlight, but the map was not helpful, for by that time I had no idea where I was. After driving to an intersection with multiple direction signs, the hard rain and darkness made them unreadable. Fortunately, a small, sloping, roadside shoulder allowed me to pitch the headlight beams in such a way that I could read the signs. With the location determined, I soon found the desired highway, though by a route different from that of my faulty directions. Wind and rain buffeted the small Renault all the way to Marseille.

The storm passed, and the flight to Frankfurt was calm. On arrival by airport train to the Frankfurt station, two men fighting on the platform blocked passage to all. It was an interesting fight, between a near-giant who seemingly knew nothing of fisticuffs and a feisty little fellow who was skilled. It seemed that any blow by the larger man would end the fight, but he never connected with his agile adversary. Finally, the police came to take both away, but I had missed my train to Karlsruhe to lose another hour.

The Karlsruhe visit was useful because I could clarify a misunderstanding between the Germans and members of the U.S. NRC staff. After a train ride to Jülich and consultation with Günther Ivens and others of the AVR, I flew to London. There Elaine also reported excitement, having been temporarily displaced twice from her hotels, once by a small fire and another time by a false alarm. Between those events, she and Byron made a pleasant excursion to visit Cambridge University on a day free from his studies.

Several ORNL members participated in the International Conference of the British Nuclear Energy Society, of which Gordon Brown was then president. I was to chair a plenary session, present a paper at an accompanying meeting on the HTGR—including its potential for process-heat applications—and to meet with several counterparts attending from England, France, and Germany. The conference was to be opened by Prince Philip, Duke of Edinburgh, and prior to his arrival we were instructed on proper etiquette in the presence of British royalty. Instructions included when to stand, sit, applaud, and how to meet the duke, who was to be available for personal introductions at the first intermission. The instruction for the introduction was "to maintain a distance of ten to fifteen feet from the duke, queue to meet him if desired, approach only when escorted, and depart promptly when the next person is presented." His opening speech provided a concise but comprehensive picture of the world's energy status and included recommendations for improvement. I thought to myself, *Who wrote that speech for the Duke?* and *Wouldn't it be great if leaders of our country could present such a deep and scholarly discussion of energy related technologies and issues?*

I wanted to meet Prince Philip, but I had to confer with the secretariat on a necessary modification of my session. (We decided to rearrange the order of paper presentations because one of the speakers was expected to plagiarize the work of a later presenter.) But the intermission did provide me the opportunity to ask a conference official if he knew who had written the speech for the duke. He replied indignantly, "The Duke wrote it himself; he writes all of his speeches." In passing the door of the reception room, I met Gordon Brown, who asked, "Don, have you met the Duke?" To my negative reply he said, "Come, let me introduce you." The Browns were known to the royal family through his having been awarded the Order of the British Empire by Queen Elizabeth for leadership in the design and construction of the first nuclear power reactors at Calder Hall in Northern England. Gordon led the way through the crowd to the center of the room, introduced me, and excused himself to take care of duties at the meeting.

For reasons unknown no one was brought to replace me and I had an interesting conversation with this tall, distinguished member of the Mountbatten family, who stood in erect military posture. His penetrating questions on energy systems made it quite clear that the Duke was capable of writing the paper he had given. He was also familiar with ORNL's research and correctly compared it with that of the British Harwell Laboratory. (That was Harwell's nature at the time, but it has now become more commercial.) We continued the conversation until a bell sounded, closing the reception and calling for the next meeting event to begin. During the entire ten-minute conversation we chatted informally like old friends.

My session just before the scheduled late lunch proceeded smoothly until the suspected plagiarizer was introduced. He made no obvious reference to the earlier speaker's work, but the paper was overly long and his presentation sluggish. A three-minute warning was presented subtly, and two minutes later I placed a card before him requesting a one-minute concluding summary. All earlier sessions had ended precisely on time, and I was concerned that the session I headed not delay the crowded program. After the speaker's allotted time had more than passed, I stood to say softly in his ear, "Please summarize now." After two or three additional minutes and perceiving that the audience was losing attention, I made a request for closure into the microphone. He still droned on, following a written text with two remaining pages in evidence. I then took firm grip on his shoulders, turned him from the podium, and escorted the man to his seat. (My notes show that he had spoken half-again beyond his allotted time.) Returning to face the audience that was sitting in surprised silence, I thought, *What have I done?* Then, as I reached the podium to introduce the next speaker, the hall burst into loud applause.

Elaine had returned to London to meet European friends who had come to the meeting. That evening we were invited to a reception at the luxurious nearby home of the duke of Cumberland. The grand staircase, immense paintings, thick carpets, and gold-leaf decor were most impressive. They were especially so to those who lived in the wartime, government-built houses of Oak Ridge, Tennessee. The superb sandwiches and wines also dispelled prevailing negative opinions of English food. We spent the weekend with Byron at Oxford, staying at the nearby Studley Priory, an old British mansion converted to a hotel, before returning home. Tom was no longer in Europe, finally having enrolled in the Harvard Law School after receiving a second-year extension of the Fulbright Fellowship.

Returning to Oak Ridge, I wrote a detailed report to describe findings of the trip. Since I had personally funded the cost for visits on the Continent, that was reported separately for ORNL internal distribution. Floyd Culler was interested in the extent of the French program for breeder reactors and proposed forwarding the full report to the ERDA. A few days later I was summoned to Floyd's office; he had received a telephone call from an ERDA official accusing me of having fabricated the Cadarache portion of the trip. Several senior ERDA staff members had visited Cadarache some three weeks prior to my visit and had seen none of the large equipment I had described. I assured Floyd that I had seen the equipment, both from the floor and the elevated platform. The ERDA official was so skeptical of my report that he had a staff member, stationed in Brussels, sent to Cadarache for verification, the result of which, of course, was that my report was certified as accurate.

Their failure to have seen the equipment was not surprising. When officials visit countries under international agreements, they often only exchange papers, count technical milestones, and prepare schedules for cooperative work. The engineers visited are not interested in sharing extraneous technical information unless the traveler can respond knowledgeably or counter with suggestions that may be helpful. I have participated in such government visits and found them necessary but extremely dull. However, comparing technical findings between laboratories and obtaining confirming information relative to one's work is always helpful, and such was the nature of my visit. The French had been open and cooperative by sharing their research and its findings as well as by extending the visit to their laboratories. Their courtesy was not the result of obligation under the exchange agreement because the United States did not have an equivalent facility.

At ORNL we moved into additional programs demanded by the new (if only passing) national appreciation of the importance of energy. One significant role of national laboratories has been development of improved codes and standards for the design, manufacture, and installation of nuclear equipment. The importance of that role was recognized early because new materials and more demanding requirements were important for safety and operability. National standards had been developed for other industries by professional societies, such as those of mechanical, chemical, and electrical engineers. However, those organizations use mostly volunteer specialists who may confer intermittently, perhaps during annual meetings, and thus progress slowly and deliberately. In response to the rapid expansion of the nuclear field, the AEC had initiated development of standards somewhat beyond that instituted by the American Nuclear Society.

The codes and practices developed and administered by AEC and then ERDA became rather bureaucratic in structure and application— to say that we found them "unpopular" may be an understatement. Even so, they had value and were needed. I was asked in 1974 to assemble a committee to evaluate the situation and make recommendations for improvement. The group included representatives of several national laboratories, the nuclear industry, and regulatory organizations. We met more or less monthly at various sites to utilize local expertise, and in mid-1975 issued a set of recommendations intended to streamline the processes without compromising values. An assessment made a few years later indicated that many of the recommendations had been implemented advantageously.

The first joint meeting of the American Nuclear Society and its counterpart in Europe was held in Paris. I was to participate and meet with several European collaborators during the week. To minimize airfare and hotel costs, I traveled with a tourist group from Texas in a crowded aircraft and stayed in a very small but pleasant hotel. All went well from my perspective except for the discussion at breakfast, which was entirely about Texas; I wondered, did they really intend to *see* Paris? Fortunately, the Texans' travel agent had negotiated an advantageous shopping arrangement with a small French store. That eased my problem of finding gifts not only for Elaine but for the secretaries who had prepared me for the trip and, more importantly, were to prepare my trip reports.

Byron came from Oxford for the weekend following the meeting. On Friday night we walked across Paris to the Left Bank and dined at Le Chat Riton, a small Basque restaurant that Elaine and I had enjoyed on earlier visits. Saturday was spent at the home of Georges Mordchelles-Regnier, director general of the Society Bertin & Co., near Versailles. Mordchelles-Regnier had previously shared data and experience with ORNL on the design and performance of high-temperature thermal insulation for gas-cooled reactors. His firm was engaged in development of the high-speed trains now operating in France and in many energy-related projects. Visits with Mordchelles-Regnier were always interesting and informative. Late that afternoon he took us for a driving tour of a large French zoo, where the animals are free to roam with people confined to their automobiles or buses. We stopped for some time to watch two large male elephants play "King of the Mountain" on a mound apparently constructed for that purpose.

The continued buildup of nuclear weapons stockpiles had become of concern to many people during the 1970s and reinforced my worries of thirty years. Pressures of the Cold War had contributed, along with the corresponding assumptions of multiple targets that required specific models, but the total numbers of weapons in the USSR and United States seemed excessive beyond any legitimate needs. The stockpiles even surpassed projections for numbers and sizes that could destroy all of civilization. Of equal concern was the failure by many to distinguish between weapons capability and nuclear energy systems that produce electrical energy and isotopes for medical and industrial uses.

I first took issue with such confusion in articles appearing in the United Methodist Church publication *esa/Social Action,* and I sent letters to the editor of that publication. That journal was of special interest to me because my earlier petition to the church's General

Conference had helped save the journal from extinction. A publication focused on social concerns had seemed to me appropriate in the program of a large denomination. However, I found distressing the fact that the journal seemed confused on nuclear issues that seemed clear to me. The editors printed my letters, even publishing one as an article and paying ten dollars for its use. Receipt of even that small payment was rewarding since technical journals often charge their authors by the page for printing submitted papers.

———•◦•———

By the mid-1970s, our family's interest in Latin America had become focused on Venezuela. Elaine had traveled there in 1968 with a volunteer team of women working under the Partners of the Americas program, and she and I had stopped in Caracas on the way home from Peru. Elaine was serving as Tennessee chairperson for the Partners' teenage summer exchange program. In 1975 we spent vacation time escorting two dozen Tennessee high school students to Caracas and a like number of Venezuelan students to our state. The Venezuelans were scheduled to travel to Tennessee two weeks after the arrival of our group in Caracas, so we chose to tour Venezuela during the interim. After spending a few days with Dora Palacios, Elaine's counterpart for the program in Venezuela, to be sure that the students were properly situated, we set out to use airline tickets that provided unlimited flights on Avensa, the internal Venezuelan airline. The offer seemed too good to be real, but we soon learned that Avensa's problems made the price high, at best. We made nine flights, five of which had engine malfunction or other aircraft trouble.

We chose to visit Mérida in the Andes, Barquisimeto to meet with other Venezuelan participants of the exchange, Margarita Island, and finally Canima. The last stop was to the 790-foot waterfall of the Caroni river near its confluence with the Orinoco. A highlight on the way was to see Angel Falls from the air. The flight from Caracas's Maiquetía Airport to Angel Falls in an aging two-engine aircraft was smooth, as we gained altitude and crossed mountains. The falls came into view as we descended just beyond the precipice of Angel Falls and flew close to its face. I was thrilled to see this spectacular 3,212-foot waterfall, which had excited my interest since its Euro-American discovery in 1951. The ribbon of water bathed in sunlight seemed ethereal as it assumed slightly waved patterns in response to air currents. We were particularly lucky to see this and the many smaller streams falling from the bluff face or over the crest, because the area is frequently shrouded in clouds or fog.

As the aircraft made its second pass near the face of the cliff, one engine started to miss badly, then stopped altogether. The pilot could periodically coax it to respond erratically, but the engine emitted disturbing sounds, "arr-rr-bang," silence, "arr-rr-bang," as he repeatedly maneuvered the plane across the face of the falls. I was enthralled with the view, but appreciation was mixed with terror, particularly knowing from Harry Gray on a trip to Idaho that this particular aircraft was underpowered at its best. It seemed impossible for the plane to climb over the mountain again. I continued to take pictures, but it was difficult to steady the camera with hands that were tense, if not shaking. The thought came, *If this must be the end, it's nice to have seen Angel Falls.*

I did not know that the alternate, perhaps longer, air route to Canima was "downhill" all the way. Thus the Convair with one engine quiet, landed safely on the dirt strip outside the camp where we were to spend the night and another day. Our thatched-roof hut with concrete walls and floor was near the airstrip, and for most of an hour we could hear the pilot's radio calls: "Maiquetía, Maiquetía, Maiquetía." Finally he contacted the Caracas International Airfield, and a barrage of Spanish followed. We enjoyed a fine dinner in an open-area restaurant, first fascinated by the several colorful and friendly macaws, then annoyed by their intrusions at the table and their perching on our shoulders.

We spent the following day admiring views of the great waterfall from near the camp and touring the Caroni river in a large dugout canoe. That tour and brief hike through the jungle was enhanced by two members of the party who were naturalists with the Smithsonian Institution. They shared stories of their research on the flora and fauna of the Venezuelan jungle. Animal populations are sparse because the flora does not produce much edible fruit. Their mission, jointly sponsored by the Smithsonian Institution and the Venezuelan government, was to consider the merits and dangers of introducing new fruit-bearing forest trees.

While we were touring, a mechanic had arrived in a small plane to work on the ailing engine. Early that evening we hesitantly boarded the aircraft for the return to Caracas. That flight was to be grounded briefly in Barcelona with an engine oil leak; it was soon repaired and we arrived in Caracas without further incident.

The next day I spent a pleasant afternoon with Union Carbide's representative for Latin America in his small but attractive offices in Caracas. His stories of the intrigue and difficulties of international competition with firms that engaged in bribes were disturbing. He described Union Carbide as adhering closely to U.S. State Department and international agreements that prohibit such practices.

Despite the tensions of flying, the two weeks had been pleasant and passed quickly. We then assumed responsibility for escorting the Venezuelan students to Tennessee. Their excitement exceeded even that of the Tennessee kids, but they were somewhat intimidated and thus reasonably disciplined. However, imagine the problem in Atlanta of placing two dozen Spanish-speaking children on five different airplanes departing from dispersed gates of the sprawling old terminal. The maneuver was further complicated by one student's having lost her ticket, requiring us to purchase a new passage to Knoxville. The students were carrying identical bags for notions and we later found that she had placed her ticket in another passenger's bag. Overall, this was a rewarding trip which, despite the short duration, had provided an enhanced understanding of this fascinating South American country.

During the days we spent in the home of Dora Palacios, I assisted her maid by carrying packages from shops and by helping in small tasks in the garden. On reflection I probably learned more from Dora's maid than anyone else on the trip. She knew no English but recognized my mistakes in her language and felt free and obliged to insist that I both say the words correctly and understand their meaning. Members of the Partners all knew English and had been far too polite to correct mistakes when I ventured to speak their language. The maid's lessons were helpful later in the year when I participated in a meeting of the Mexican and American Nuclear Energy Societies in Mexico City. Elaine, accompanied by our friends Jo and Gilliam Horton and Velda and Bill Duttweiler, also joined me for the week in Mexico. I traveled separately, flying directly from Washington, D.C., because of prior business there.

We spent one very pleasant evening with the Salas family in their penthouse apartment in a multi-story building that Mr. Salas had built. They missed their grand home on the edge of Chapultepec Park but by then had found its maintenance requirements beyond their personal energy resources.

The nuclear meeting was interesting both for the technology centered on the ambitious Mexican nuclear power program and for the bureaucratic nonsense of registration and other formalities embedded in the one-party political system of Mexico. Their nuclear program has progressed slowly, and today only one of their proposed Laguna Verde reactors is operating. Thus, their plan to displace fossil fuels with nuclear power in Mexico has only marginally succeeded. The purpose was to save their oil for sale to the United States. We buy their oil, but their nuclear program, like ours, has passed into the doldrums.

Also in 1975, I was elected to the Board of Trustees of the three Holston Conference United Methodist Colleges: Emory and Henry, Hiwassee, and Tennessee Wesleyan (TWC). Each board member was assigned to a Board of Governors for one of the colleges; mine was to TWC. I found board meetings discouraging; much of each session was devoted to tirades concerning student behavior by overly pious, or hypercritical, members. One indignantly reported finding three beer cans in trash receptacles outside the buildings where campus laborers also worked. Another, after lamenting student disrespect for rules and laws, was flustered when her friend, a passenger, quietly noted the radar detector on her Cadillac's dashboard. The board seldom took time to consider future directions, institutional financial security, educational goals, or student needs. Tennessee Wesleyan seemed on a downward spiral for lack of leadership.

In 1977 when I became chairman of the TWC Board of Governors, the college seemed in such dire financial straits that I feared having to preside over its demise. However, stability was achieved through a close working relationship with the newly elected president, George Naff, and through support from the Athens community and many friends of the college. Slowly, by the grace of attrition, the board developed into a force for improvement.

After a furlough I returned to the colleges' Board of Trustees and again served as chairman of TWC's Board of Governors. Later, the Holston Conference Colleges Board dissolved in favor of each school's forming its own Board of Trustees. A strong coordinating committee was created to assure continued cooperation among these United

Methodist Colleges. I chaired the TWC Board of Trustees and served on the Coordinating Committee. Even though these roles were time-consuming, I enjoyed my association with the many fine people involved, the opportunity to gain new management experience, and the sense of having made a contribution.

Elaine continued many volunteer roles, including teaching money management to homemakers, conducting gourmet food classes, teaching high school seniors in Sunday school, and teaching Girl Scouts and other organizations food preparation and homemaking skills. She also remained active in student exchange programs, and was the first woman chair of our local church administrative board.

By 1975 all members of the family were back within the same national borders. Even though Tom had spent the previous two years studying the German language and within the German system of higher education, he experienced little difficulty as he settled into his law studies at Harvard. We enjoyed an early spring visit to see the boys and New England. The family also gathered at the Causey farm home for the marriage of Elaine's sister, Eloise, to Howard Butler on June 12, 1976. A simple ceremony to exchange vows, with Elaine as matron of honor, was held before the fireplace under an oil portrait of the two sisters at ages three and five. Late in 1976 we had a disturbing tele-phone call from Byron. His larynx had been injured in a touch football game at Yale; we could barely hear his whisper. Doctors at the Yale Medical School and our friend Dr. Thomas Upchurch, an otolaryn-gologist, feared that the injury might cause permanent impairment. For one aspiring to be an attorney at law, the prospect was most disturbing. Fortunately, nature slowly corrected the damage.

In 1977 ERDA, following extensive international negotiations, was prepared to assist the government of the Shah of Iran in training opera-tors for their proposed nuclear power plants. The task was assigned to the Oak Ridge Associated Universities (ORAU) with technical assistance from ORNL. Philip Johnson, then Director of ORAU, Alex Zucker, and I were to visit Teheran to evaluate their technical needs and provide a plan. The Iranian government, which was to pay for our trip and the training program, had contracted with a German firm to build two 1,200 MW(e) reactors at Busheher, near the Persian Gulf. Iranian negotiations for additional reactors also were underway with French and U.S. firms. The stated goal was to use nuclear plants to produce a large fraction of Iranian electric power and to desalt sea water. The Iranian rationale for building nuclear power plants was similar to that expressed by the Mexicans in the meeting two years earlier: they preferred selling the oil over burning it to generate electricity.

The magnitude of the Iranian energy project seemed staggering for them. The first two nuclear plants would be capable of producing nearly one-half the total Iranian electrical demand. Oil sales provided ample funds, but their technology base for operating nuclear plants and for building the necessary integrated electrical distribution system seemed grossly inadequate. They had no national grid, and few inter-connections existed between local systems. The nuclear experience in the country consisted primarily of one small research reactor and a few medical facilities. Foreign training of reactor operators was essential for safe operation of reactors in their nuclear power program.

Arriving in Teheran in the evening of February 2, 1977, after a twelve-hour flight through eight and a half time zones, I was tired and looked forward to the day of relaxation we had requested. Unexplained circumstances dictated an early start the next morning, and I suffered from jet lag during most of the nine-day visit. After a briefing by A. Etemad, chairman of their Atomic Energy Commission, we entered discussions with R. Khazaneh and his staff of the Esfahan Nuclear Technology Center, then existing largely on paper. We were greatly assisted by Bob Liimatainien, science attaché in the U.S. Embassy. Bob had become fluent in Farsi, was familiar with Iranian customs, and understood the Shah's programs.

The training that we proposed for the Iranians in Oak Ridge was to provide basic knowledge and experience in nuclear technology so that more specific and rigorous training of operators by the reactor vendor could be properly understood and assimilated. The trainees were to arrive with a high-school level educational background, although some would have had additional industrial-type training. They were to receive instruction in classroom lectures at ORAU and gain experience in nuclear technology programs at ORNL, including safety, health physics, and limited on-site reactor experience. Other issues such as language training would require early resolution, though all were to arrive with some facility in English. Some individuals would be accompanied by family members, and housing, recreation, and the other needs of students were to be arranged by ORAU.

In the course of discussions the Iranians frequently asked about the possibilities for Iran to obtain nuclear fuel from the United States as needed for their reactors and about the handling of the spent fuel. These issues were outside the scope of our mission and were sensitive because they involved enriched nuclear fuel; we limited responses to general terms. At one point we were shown drawings for a pilot plant to develop processes for recovery of value from spent nuclear fuel. We were curious and concerned about this facility because it would be capable of providing plutonium and would not be needed in their energy program for a long time. The concern was heightened and the puzzle complicated by the absence of dimensions or scale on the prints. I studied the drawings carefully. If it was to be a small pilot plant, we had little cause for concern over its potential use for producing weapons-grade nuclear materials. But it looked to be much larger than needed, although I could only judge the scale from the apparent size of doorways and the truck access. The observation that nuclear weapons could be a possible objective of the Shah's program was transmitted separately to ERDA, since our official trip report would become available in Iran.

On Friday, their holy day, we were invited to see the site for the planned nuclear technical institute near Esfahan and to visit that beautiful city on the banks of the Zayandeh River. It was a full day with flights to Esfahan at 7:00 A.M. and return at midnight. When we arrived in Esfahan, our host, Mr. Moslehi, surprised us with a breakfast at the elegantly ornate Shah Abass Hotel. Because we had risen early for breakfast in the hotel and had been provided a snack on the aircraft, it was difficult to enjoy this elaborate and delicious meal. We visited the bazaar and were taken to see the Shah's medieval palaces, harem quarters, and the Great Mosque. There the acoustics were remarkable, both in the Mosque and in its courtyard, large enough for some sixty-thousand people. A demonstration by our local host indicated that, indeed, all could hear without electronic assistance. At lunch we met Larry Vondra, employed by an Iranian company and the son of Ben Vondra of ORNL. There we also met a director of the Teheran Musical Center, who invited us to a folk-group ballet that evening.

After lunch we were taken by Mr. Moslehi to the site of the proposed technical center. It then consisted of a meteorological tower and guard shack. We were told that the poorly dressed guard who admitted us had owned the land of the center. Although he had been paid the equivalent of a million dollars for the large tract, he preferred not to leave his land and worked and lived there as a guard. It was a large desert area that lies between two mountains, one of which was said to be a place where the Devil comes to earth; we skirted it cautiously. Having escaped apparently unharmed, we engaged in serious conversation about the facilities to be built there.

This was to be a comprehensive research facility for the development of nuclear technology. It would include the proposed pilot plant (or plant) for processing spent nuclear fuel, and would encompass all technologies necessary for a major nuclear energy program. They projected a staff of one thousand of whom 60 percent would be technically trained; that, too, raised the possibility that the Shah's ambitions might include nuclear weapons.

In the late afternoon, we met Moslehi's wife at a coffee house near a large medieval bridge and park for a pleasant, informal discussion. We tarried on that beautiful veranda so long that no time remained for dinner; we had accepted the ballet invitation. Disappointingly, the ballet troupe proved to be a group of Basque folk dancers, not Iranians, who were touring the Mideast to raise funds in support of their quest for independence from France. Regardless, the colorful costumes were attractive and the dances interesting, although somewhat primitive and hardly ballet. We then proceeded to the airport for a late dinner and the flight to Teheran, but the restaurant was unavailable. Even having eaten four times by mid-day, the very small sandwiches prepared by the bar maid seemed inadequate. Arriving in Teheran after 1:00 A.M., my reinforced jet-lag fatigue challenged my best efforts to be effective in Saturday morning's eight o'clock meeting.

The days in Teheran were never dull, except for the tedious discussions required to bridge cultural barriers. Language was not a major problem because most of our contacts spoke rather good English, but the customs and regulations were difficult to accommodate. Morning rush-hour travel from the hotel to the Institute became trips of near terror. On a fast eight-lane thoroughfare, our Iranian driver constantly changed lanes, even darting into gaps of oncoming traffic to pass a slower vehicle. Phil frequently complained of sore feet; one morning he realized that from his front seat vantage he was instinctively applying excessive foot pressure on the floorboard. In lighter traffic I was quite comfortable riding in Iranian autos with their drivers but was frightened in U.S. Embassy cars. Those Chevrolet Impalas were the largest vehicles on the streets, which were obtrusive in Iranian traffic and limited the margins for error.

Growing tension with Iran had led the U.S. Embassy to fortify their cars. The doors had heavy, steel-panel inserts, and the windows were covered on the inside with bulletproof, clear plastic. On cold, damp evenings the windows fogged between the panes so that no one, including the driver, could see out except through the Lexon windshield. The automatic weapon of the Iranian escort in the front seat only added to my apprehension. One evening our driver inadvertently lightly bumped the rear of a taxi. There was no damage, but the taxi operator complained vociferously. He might not have been so demonstrative had he seen the firepower in our front seat. I could not understand the embassy logic in using those large Impalas that would always attract attention, when General Motors marketed a small Chevrolet in Iran that blended unobtrusively with other traffic.

One day we were invited to the U.S. Embassy for lunch, and for some reason I walked there alone from the Iranian Energy Department offices. Upon approaching the embassy buildings, glances from the Iranian people made me increasingly uncomfortable. On other occasions when we visited sites of interest or I simply walked to stores or for exercise in the evening, I felt no discomfort even though readily identifiable as an American. Shopkeepers seeking sales were particularly friendly. However, the body language I observed near the embassy was different and seemed hostile.

The high wall surrounding the embassy grounds was forbidding and contrasted with the relatively ineffective gate at the entrance. I was to recall that entrance three years later as it was shown on television during the terrorist occupation. A brief tour of the facility also made more vivid the later descriptions of the trauma experienced by embassy staff and of the aborted rescue attempt. During informal dinner discussions our hosts and embassy staff members advised us of the Iranian people's increasing displeasure with the Shah and his harsh rule. The dissatisfaction with the U.S. derived in part from the CIA's having been involved in overthrowing the previous elected government that was presumed to be favorable to the USSR and a Cold War threat.

From a Western point of view the Iranian attitude was difficult to understand, for there were was much evidence of economic expansion. I once tried unsuccessfully to remember having been at any point in Teheran where new construction could not be seen. Even so, it seemed that a major political change was inevitable, but knowledgeable people felt sure that such an event was at least a decade away. Our negotiating sessions continued on friendly terms, and a workable plan was developed for the training of Iranian technicians.

I had been invited by the U.K. atomic energy staff to visit Risley for discussions concerning nuclear engineering projects and stopped there on the return from Iran. I was pleased to see their fine facilities, which had been greatly enlarged since my previous visit, rivaling those I had seen in Cadarache. They were focused on their advanced gas-cooled reactor and breeder reactor programs, but Britain was beginning to think more seriously about American and French LWR designs.

———•◦•———

In the spring of 1977 Elaine decided to run for a position on the Oak Ridge City Council. Her opponent was a well-regarded incumbent, and the campaign was quiet and without the rancor of today's elections. Elaine was well-known through her many volunteer activities and won with a comfortable margin. The next week the *Oak Ridger* carried the following story in a column entitled "Ridge Runners" by the editor, Dick Smyser:

> Weeks that were—These have been two rather spectacular weeks for the Don Traugers of 510 Delaware Ave.
>
> Tuesday, June 7, Elaine Trauger was elected to the Oak Ridge City Council from the Elm Grove District. That same day, son Byron was graduated from the Yale Law School.
>
> On Friday, June 10, Don Trauger was awarded an honorary Doctor of Science at Tennessee Wesleyan at Athens. Don, ORNL scientist, is chairman of the executive committee for the Board of the Holston Conference of the Methodist Church Colleges and a member of the Tennessee Wesleyan Board of Trustees.
>
> Then to complete the succession of family accomplishments, son Tom graduated from Harvard Law School today.—RDS

In August we joined with Byron and Tom to participate in "Old Timers Day," an annual event of my brother's family held at the House on the Corner. On these occasions Bob directed the firing of steam engines and the starting of early-model kerosene or gasoline-fueled farm tractors from his collection. Older men from a wide community volunteered to help and to renew their acquaintances with these historic machines. People came from as far as neighboring states to these events that were partially financed by donations. On this day, with rain falling all morning, the attendance was about half of the two thousand expected. I took photographs and presented Bob with poster enlargements of the more active demonstrations such as shelling the kernels from ears of corn and sawing logs into lumber. The volunteers also demonstrated threshing techniques ranging from the ancient flail, all the way to near-modern equipment and included one hand-cranked threshing machine of a type supposedly used by George Washington at Mount Vernon.

The aged gasoline tractors were typically difficult to start, even by strenuously pulling long-handle cranks to turn the engine. One steam threshing engine suddenly released steam at its side with a penetrating screech that overwhelmed the steady roar of the twenty or more gasoline engines. Those standing nearby looked frantically for a valve to close or other action that might be taken to stem the release. When the elderly operator of that engine returned, he calmly removed a large wrench from the tool box to strike one side of the machine strategically three times; the release stopped immediately. Expert knowledge has no replacement.

Returning to Tennessee, Byron took a position as judicial clerk to Judge Frank Wilson of the United States District Court in Chattanooga. He was pleased with an appointment in Tennessee but was more excited over working with Judge Wilson, whose scholarly decisions laced with common sense had made him a nationally respected jurist. Wilson, for example, was the judge who had successfully sentenced union leader Jimmy Hoffa to prison following a dramatic trial. Byron was quite surprised to learn that Judge Wilson had been our attorney and close friend before his appointment to the bench. Following two years of that experience and training, Byron was appointed an assistant U.S. attorney for the Eastern District of Tennessee.

Tom spent three weeks of that summer touring the mountain areas of Colorado with Chuck Knowles, childhood friend, former neighbor, and newly trained veterinarian. To finance the trip in part, Tom and Chuck sought fill-in work for Tom as a pianist in hotel lobbies, bars, and nightclubs. Chuck served as Tom's agent. In October Tom joined the law firm of Spiegel and McDiarmid, then located in the Watergate Building of Washington, D.C. His choices had included a large firm in Seattle with the attraction of its proximity to western mountains and open spaces. The decision was determined by the emphasis of Spiegel and McDiarmid on consumer and social issues. We were pleased with the positions chosen by our sons and noted a substantial improvement in our finances, after having paid tuition and living costs for three years at the Harvard and Yale law schools.

A few years earlier our Swiss friends Alfred and Kungolt Schurch and their two sons had spent some days with us in our home. They gave us a small oil painting of an area in the Black Forest, where they owned an old farmhouse used as a summer retreat. It was described as typical of traditional German farms, arranged so that house and barn form a single large structure. In the course of discussion they invited us to visit Kusnacht and travel with them to their Black Forest house, the girlhood home of Kungolt. The prospect of learning first-hand about a farm structure known only through books was exciting, and I cautioned Kungolt that she should use care in extending the invitation. If she really wanted us to visit, I definitely would find a way, and we did in 1978.

Our vacation trip to Europe that year centered around visits with both the Schurch and McKown families. We joined the Schurchs in Zurich and traveled from their home in Kusnacht via picturesque Stein am Rhein, the thundering Rhein Falls, and Waldhut to the farm at Kussaberg, Germany. It was, as portrayed, a house three stories high coupled with a barn of two levels, having common exterior walls and roof. The stone and exposed-beam construction presented an attractive but imposing facade. A farmer, who tilled the land and tended the building, lived with his family in the first floor of the house. The Schurchs maintained ample quarters on the second floor, and we occupied one of the guest rooms and lounge on the third. The two levels we visited were furnished in beautifully preserved and polished antique furniture appropriate to the setting.

The history of the farmhouse has been traced to the seventeenth century, and although known to have burned twice, it has been rebuilt within the original walls. As dinner was being prepared, Alf invited me to accompany him to the wine cellar. I was struck by its semblance to similar structures in Spain that are known to date from Roman times. Alf observed that the Romans had occupied this mountaintop location for several centuries, but the cellar structure had not been dated. The ruin of a medieval castle stands on a promontory beyond the small village of Kussaberg, only a quarter mile from the farmhouse. The logic of that location for a Roman outpost and later for the castle was evident as I stood on remaining walls of the castle to view the wide panorama of the Rhine valley with Swiss Alps forming the horizon. German elite troops had used the castle structures and nearby cliffs for training during World War II. The site no doubt had seen similar exercises in the times of Germanic kings and Roman centurions.

Our quarters in the old house were very comfortable. The walls, even at that elevation, were nearly three feet thick so that the windows were set in deep wells insulated by shutters, glass, and heavy drapes. When all were closed, darkness was total and the quiet even disturbing; we soon adjusted, placed a small lighted lamp in the lounge, and slept well under featherbed comforters. Breakfast at tables on the lawn one morning was a lavish spread under majestic trees with a background of geranium blossoms and other flowers in a nearby garden.

Although the barn, where the cows occupied comfortable stalls, shared a wall with the house, they were so well isolated that no stable odors reached the home. The barn loft is accessed by a dirt ramp so that racks of hay were pulled by horses or a tractor through the large door for unloading. A conversation I had with the milkmaid as she washed the dairy equipment in front of the house was pleasant but halting, limited by my inadequacy in her language. The fields were well tended and the Schurch garden luxurious with vegetables as well as flowers. Our days on that mountaintop farm produced long-remembered visions of beauty and history.

We next visited Hank and Helen McKown in Vienna, where as an ORNL staff member, Hank was assisting the International Atomic Energy Agency with the development of improved mass-spectrometer equipment to monitor the world's atmosphere for evidence of nuclear weapons tests. I visited his laboratory at Seibersdorf and arranged to meet with members of the IAEA staff in Vienna through Leonard Bennett, who had previously been at ORNL. On the weekend the McKowns took us to the eastern part of Austria and the boundary with Hungary. I stood at a border barrier that could be crossed as easily as a farm fence and peered through binoculars at a guard tower, to see the well-armed guard watching me through his high-magnification glass. It was a chilling experience that made the Cold War seem more real than even the view from within the technological activities it spawned.

Our trip concluded with a tour of Yugoslavia, starting in the historic, well-preserved and picturesque walled-city of Dubrovnic and continued with a drive along the coast to Split. Occasional sojourns inland provided some insight to this complex country. The most notable excursion was to Mostar, where we marveled at the well-preserved medieval bridge and evidences of the long Turkish occupation. I enjoyed several conversations with natives who spoke German as a second language and hence at my speed of comprehension. The Yugoslavs we met were friendly people, who often called greetings from windows and balconies. Having experienced such pleasant and helpful people, I have struggled to imagine the hatreds and atrocities reported during the recent civil war. It was equally distressing to read about the destruction of so many historic sites, particularly of the remarkable bridge at Mostar.

On March 29, 1979, the nuclear reactor accident at the Three Mile Island electric power plant near Harrisburg, Pennsylvania, brought many changes in nuclear policy and programs. I happened to be scheduled to speak two days later in Akron, Ohio, at a United Methodist Church forum titled "Energy Futures and the Concerned Layman." My topic was "Nuclear Energy." The audience was politely quiet during the presentation, but I perceived correctly that they were not listening, only waiting to ask questions about the accident. I was delivering that speech during the height of news commentators' concerns over the hydrogen bubble in the reactor vessel and at a time when little was known about the true nature of the accident. I, like many others, had spent the previous two days in a concerted effort to learn what had happened and to provide help. A team sent from ORNL gave badly needed assistance to stabilize the radioactive iodine solution that had escaped to a poorly contained area. This action prevented an accident phase that might have produced an iodine release, which would have made necessary the limited household evacuations ordered as precautionary. The background of only two days' experience did not help much in fielding questions for nearly two hours following conclusion of the Akron meeting. I did my best to assure the audience that the containment systems would hold and that the distance from the site to Akron was sufficient to protect them from any eventuality.

Evaluation of the Three Mile Island accident brought about major improvements within the nuclear industry. Much of the effort was led by the Electric Power Research Institute under Floyd Culler and by national laboratories of the U.S. Department of Energy.

Several safety-related programs that ORNL had proposed earlier were then funded to provide greater understanding of the complex LWR nuclear plants. The utility industry founded the Institute of Nuclear Power Operations to self-monitor their designs and practices, and the World Organization of Nuclear Operators was formed to share experience in training and operation.

The Three Mile Island accident and the accompanying sensational news coverage also increased public misunderstanding of nuclear power. For example, the television cameras focused on the water vapor emitting from the cooling tower of the second reactor on the site, which continued to operate properly. The impression was that this enormous tower and the column rising from it must be something to fear, but the image was a perfectly normal component of power plants whether powered by coal, natural gas, or nuclear energy. An erroneous prediction that hydrogen would accumulate or was accumulating in the reactor vessel was also disturbing. Actually, the reactor was so well protected that had the operators not been confused through the poor location of instruments and inadequate procedures, or had they done nothing at all, the automated reactor safety systems would have protected the reactor. They then could have leisurely repaired the defective valve that initiated the event.

During the next few years, I was frequently called upon to speak about the accident to church, civic, and professional groups. As the nature and extent of the accident unfolded through excellent teamwork of industry and government, the presentations became easier with the discourse less speculative and more factual. No one had been injured and many lessons were learned, including a new level of realization by the industry: nuclear energy requires precise procedures and a stringent discipline to protect the investment; systems for ensuring public safety are effective and reliable.

The Iranian trainees arrived in Oak Ridge in 1979 but were soon stranded here by the revolution in their country, which immediately terminated the construction of nuclear plants and the operator-training program. The ORINS then had a new problem: how to facilitate the return of these people to Iran? The trainees had little money, and both communication and financial transactions between Iran and the United

States were severely constrained. Finally, a solution was found and they returned. Reports received later indicated that the trainees succeeded in rejoining Iranian society without political difficulty. We had enjoyed entertaining the entire group at our house soon after their arrival, had come to know a few individuals fairly well, and were pleased that they were safely home.

I was selected as chairman of the American Nuclear Society National Planning Committee while also serving on the boards of the Oak Ridge Hospital and of the local chapter of the American Red Cross. In my work for the Red Cross I negotiated an agreement with Medic, a private supplier of donated blood, to eliminate duplication in the operation of blood banks. By a happy coincidence the Medic Board representative in those negotiations was Clinton Campbell, whose wife was the daughter of Pat and Mimi Murphy, who had participated in our cattle venture some twenty years earlier.

Elaine proved to be a highly active member of the City Council and was active in the Tennessee Municipal League, Tennessee Women in Government, and the National League of Cities (NLC). In those capacities she frequently became involved in correcting misinformation concerning the safety of nuclear energy and in promoting the mission and accomplishments of the Oak Ridge facilities. In Oak Ridge she was a stalwart supporter of public schools and was courageous in the face of controversy. She opposed the building of a local airport because of its excessive cost in the hilly terrain and stressed that a projected direct road from Oak Ridge to the Knoxville Airport would largely fill the need. Her proposal to allow use of the City Senior Center by other groups when not in use by seniors distressed many members. Elaine also insisted on tight conformance by builders and others to zoning laws and city ordinances. As a result of these controversial issues, she was opposed in the 1981 election by many groups and nearly lost reelection to Bob Keil, a friend and popular labor leader.

We failed to prepare Christmas greetings to friends and family in 1980 and, properly embarrassed, decided in January to maintain communication by sending cards and messages for Groundhog Day. The card showed a sketch, made for us by Diane Hochanadel, of a groundhog erect on his haunches with a paw shading its eyes. It was titled "Happy Groundhog Day" and carried a note: "I'm not looking for my shadow—I'm still looking for the Traugers's Christmas Card!" Somewhat carried away with the groundhog theme, we chose to use it for a party on Groundhog Day, which fell on a weekend that year. The invitation used the same sketch but extended an invitation to a Groundhog Day dinner at the "Traugers's Burrow," 510 Delaware Avenue, Oak Ridge. "Burrow" fit because the formal entrance to that hillside house was partially below ground level.

As we planned the party, more and more friends came to mind, and we sent 125 invitations expecting perhaps a two-thirds acceptance. Ingenuity was stretched when, to our delight, 112 people accepted and came. Fortunately, the boys had taken the furniture from our largest bedroom to their new housing. By borrowing tables for that room and rearranging furniture in other bedrooms and the den, simultaneous seating for forty was available. With the invitees scheduled for two extended time periods, the plan was feasible.

The boys came home for the weekend and Tom entertained with piano music in the living room. Guests partook of drinks and snacks in the adjoining dining room and Byron served as maitre d'. The menu, except for the pork dish labeled "Ground Hog," was of foods that groundhogs sometimes eat. This included various vegetables and fruits and culminated with persimmon pudding for dessert. Friends were appointed to greet guests at the formal entry on this night of torrential rain and to assist with placement of outer apparel in the bedrooms not used for seating. Others served drinks and provided table settings from a bathroom converted to an improvised butler's pantry. It was an evening we remember for the enjoyment, the effort required, and the many requests for a repeat performance. Some things are best done once.

During the year 1981 we vacationed in the Carolinas to visit facilities and friends at the Savannah River Nuclear Laboratory, at a Westinghouse nuclear-fuel fabricating plant, and at the Electric Power Research Institute's Nondestructive Test Facility at Charlotte. The jaunt also included a visit to our Moore County, North Carolina, tree farm to evaluate plantings and the vitality of trees. We also coupled a business trip to San Diego with a pleasant weekend in Yosemite National Park. Later, I met several times with Catholic bishops and other clergy to plan an ecumenical pavilion for the world's fair scheduled in Knoxville for 1982. It was a rewarding cooperative venture that resulted in a small theater with excellent projection of sound and pictures depicting church-sponsored programs of care and nurture. Elaine was appointed to the Board of Trustees of the United Methodist Holston Conference Foundation, which holds and invests funds contributed for church programs and building projects.

In December we flew to Nebraska to spend Christmas with my family. It was the first Christmas Day I had spent there since 1941. I deeply regret the annual deferral of a return home for Christmas, even though I visited the farm almost yearly. Even this occasion was seriously marred when my father suffered a sudden but temporary illness; I spent most of my time with him in a hospital room.

Nineteen eighty-two was a year of two major family events. Pop Trauger died on Saint Patrick's day. Suffering from a light stroke, he insisted on walking into the hospital for needed treatment and care. A second hemorrhage later rendered him totally incapacitated, and death shortly followed. He had passed his ninety-fifth birthday on the previous September 22. I saw his body lying in a casket in the same place, and prepared by the same funeral home, where I last saw my grandfather. Those similar images blend remarkably. In contrast the second event was a wedding. Byron and Nancy-Ann Min were married at a church in Rockwood, Tennessee, on April 3. The rehearsal dinner of the prior evening and breakfast of the wedding day were both at our home. Guests at these two meals included five Rhodes Scholars, friends that Byron and Nancy-Ann had come to know during their respective terms at Oxford. The couple took residence in Boston so that Nancy-Ann could finish law school at Harvard while Byron worked in a major law firm. They later moved to Nashville, Nancy-Ann as a judicial clerk and Byron to form a new law firm. The marriage lasted only four years.

Elaine and I spent many pleasant summer evenings of 1982 at the world's fairgrounds in Knoxville, having dinner at restaurants featuring ethnic foods then visiting a pavilion. Those evenings were interesting and enjoyable but not so prolonged as to interfere with normal activities. The theme of the fair was "energy" and was based primarily on the multitude and quality of energy programs in Oak Ridge. Thus, it was distressing to find almost no mention of Oak Ridge. Only the small pavilion sponsored by the American Nuclear Society made reference to the accomplishments of our city. In chance encounters we often told fair visitors about several exhibits in Oak Ridge that had been prepared for the occasion, and we later received comments that they excelled those of the fair.

Those Oak Ridge facilities—the Museum of Science and Energy, the Children's Museum, the graphite reactor, and several overlook points for viewing the massive structures of the Manhattan Project— continue to be enjoyed by visitors. I find encouraging the ever-growing signs of acceptance of Oak Ridge by citizens of neighboring cities and counties. ORNL is now recognized as the largest employer for each of several surrounding counties, including Knox. An increasing number of persons are now being employed by firms in work derived from new discoveries at ORNL.

Late in the summer we decided to visit China as an autumn vacation and had one of the hostesses in the China Pavilion prepare calling cards written in Chinese characters with suitable greetings for encounters there. During that intensely interesting tour we saw the country before major changes had evolved to modify dress and customs of the Mao regime. We visited Beijing, several major cities, and Guilin. I like Chinese food and enjoyed each menu. As we studied the customs and foods of Guilin, I read that bamboo rat was a specialty. On one occasion, I jokingly identified a mysterious morsel as rat. In deference to the obvious sensitivities of companions, the jest was not repeated. The limestone mountains, with nearly vertical walls and conical tops, seemed ethereal in the mists above the Lijiang River. Before seeing these mountains, I had thought their portrayal in Chinese art the product of fanciful imagination. During our return from China we stopped in Japan to visit our long-term friends the Inoues and the Takeuchis.

I had written to Kiyoshi Inoue of our desire to see him and his family. He responded with an invitation to visit the Tokai nuclear facilities, but I declined with appreciation; we were to be on vacation. He then suggested that we first stay at Mito, a city near the Tokai nuclear research facilities and his place of employment at the Power Reactor and Nuclear Fuel Development Corporation (PNC). I agreed to meet with officials and staff members of the PNC and Tokai on the evening of our arrival, after a Japanese dinner at the Hotel Keisei. Although tired after traveling from Hong Kong, I could not refuse this invitation, and the discussions proved informative and valuable. Japanese nuclear-energy programs were beginning their march toward world leadership.

On the following day we were to visit the Nikko National Park as guests of Kiyoshi and PNC. We were ready for departure at six o'clock on that holiday morning; early travel was necessary to avoid impossibly crowded traffic on the roads. To our surprise and delight, Kiyoshi and a driver arrived in a company limousine with a prepared breakfast, tea, and coffee aboard so that we could make a nonstop dash to the park. The early morning precautions were valid; we progressed very slowly over the last few kilometers. Had the start been an hour later, I doubt we could have arrived in daylight. Having limousine service in the park was wonderful. After hiking on trails through spectacular mountains and enjoying the brilliant fall colors of deciduous trees, we would find our transportation waiting at a road crossing to take us to the next point of interest. At the end of the day we traded car for train. Kiyoshi took us to Tokyo for dinner at a very old restaurant of the traditional Japanese style. The next morning I met with other officials of PNC at an extended breakfast to discuss the possibilities for a cooperative program with ORNL in nuclear fuel reprocessing. As a result of this meeting and many negotiations by others, an effective program evolved that also resulted in Japanese support for ORNL research totaling tens of millions of dollars.

The next day Kiyoshi, his wife, and sons met us at our hotel. Mitsuko was wearing a particularly beautiful dress of her own design and construction, an example of her dressmaking hobby and small business. The boys, who at ages two and four had enjoyed visiting our house by running and chasing up and down the two staircases, were now young men. One was in medical school and the other an attorney. They and their mother had learned English, and we greatly enjoyed and appreciated their sharing a long afternoon break with us.

Another highlight in Tokyo was an invitation by Hiroshi Furusaki, Board chairman of Nagasaki Wesleyan College, for lunch at a rooftop French Cafe of the towering Imperial Hotel. (The school had been spared from the Nagasaki nuclear weapon by its location outside the city.) A missionary graduate of Tennessee Wesleyan College had founded the Nagasaki school in 1875. Upon reaching their one hundredth anniversary, this institution replaced its Japanese title with "Wesleyan" to recognize the founder of Methodism. As part of that action a collaboration was established with Tennessee Wesleyan College that includes an exchange of students and faculty. Mr. Furusaki was funding many Japanese students for a yearlong study in Tennessee and was interested in extending the program to other Methodist schools. That became a major topic of our luncheon discussion. We suggested Nebraska Wesleyan University and Elaine's alma mater, Greensboro College. NWU already had a large program with Japan, but this luncheon led to the formation of a student exchange with GC that continues today.

From Tokyo we took the bullet train to Nagoya. The smooth and quiet ride with the countryside whizzing by was a marked contrast to the U.S. trains of wartime travel. On arrival we were met by Kunio Takeuchi, who took us to his parents' home where we spent several days with Byron's hosts when he was an AFS student in 1966. Their house was an interesting mixture of traditional Japanese and Western styles. The Buddhist shrine room with tatami floor and ornate altar contrasted with the modern kitchen. We enjoyed walking in the small, attractive garden with its carefully raked stone surface, strategically spaced large rocks, shaped trees, shrubs, and flowers. The parents spoke no English, and we knew no Japanese beyond simple greetings. After Kunio left for his nearby home, communication was challenging and slow but pleasant. Mr. Takeuchi and I persisted in a labored conveyance of ideas and information word by word, each armed with dictionaries that translated both ways. Elaine and Mrs. Takeuchi may have been as effective and certainly more animated with gestures, smiles, bowing, and pointing.

The Takeuchis wished to talk with Byron by telephone and one evening placed a call to him in Boston, where the time was early morning. I first spoke briefly to give him time to awaken and to recall Japanese words and language structure. It was touching to see the excitement exhibited by his "Japanese mother," kneeling on the floor by the telephone stand, awaiting his words. Of such is the personal bonding established through student exchange programs like AFS. Fortunately, Byron mustered enough Japanese words for a few minutes of conversation, thus greatly enhancing the pleasure of this visit for both families.

The next stop was at the beautiful and historic city of Nara. There we exchanged hosting roles with the Hiromu Abe family, first with steak at our Kintetsu Station Hotel and the next evening with a traditional Japanese dinner in their home. Mrs. Abe's sister, Shigeko Fujimura, then of Oak Ridge, had insisted that we visit her family while in Japan. The Abe's arranged for us to stay in the Buddhist guest house, Kegon-ryo, near the Nara Park. It was a small facility of traditional Japanese style, austere in appearance, but offering superb hospitality. At the entrance we changed from shoes to slippers. I walked awkwardly in their largest slippers, their heels probing my metatarsal arches. The proprietor and his wife spoke no English, and after introductions by Dr. Abe, Elaine and I were ushered to our two-room suite.

As we surveyed the accommodations, wearing softer slippers, the hostess arrived with tea and delicacies to be served on our living room table. I failed miserably in trying to find a way to fold long legs and place large feet in an appropriate configuration at this attractive table about nine inches high. Nevertheless, the cakes were delectable and the tea refreshing. After our hostess departed, we were struck by the ascetic nature of the quarters. There was no dresser, closet or chairs, only cushions on the floor and a few hooks on the wall, presumably for the robes of monks. The simple decor of exposed beams, plastered walls, French-style windows, and tatami floor was attractive; the beautifully woven and patterned sliding screen, which separated the rooms, particularly held our attention. The windows were well protected from exterior view by a remarkable growth of bamboo. After reading brochures about Nara by light from the floor lamp, appropriately named since it was only eighteen inches tall, I slept well on the futon.

The only English language encountered at the guest house was a hand-lettered sign at the end of the main hallway with an arrow pointing to the left. It read, "toilet—male or female," clearly a case of false advertising; because the facility had no door, no screening of urinals, and lavatories for common use, the sign should have read, "male and female." The three toilets, two of the Eastern floor design and one Western, were in stalls partially screened. The Western-design toilet was complete with stick-figure drawings on its entrance to show in detail how the facility was to be used, respectively, by male and female guests. Traditional Japanese bath facilities also were quite open to those passing by in the hall. Since we were the only guests at the time, our Western inhibitions were not challenged. Breakfast also was a cultural guessing game at which we were provided raw eggs, very hot rice, seaweed, shrimp, and some unidentified vegetables at a Western-style table. After some thought and experimentation, guided by Elaine's expertise in food preparation, we managed to transform these ingredients into edible portions to the apparent satisfaction of our hosts.

Professor Abe had arranged a delightful walking tour of the Nara Park, conducted by his wife, Neriko, who spoke no English. The well-planned tour, with the names of stops written in Romaji on a map of the park, included the Nara National Museum of Art, where he was curator. The museum featured a special exhibit of Chinese artifacts with a very long line of people waiting outside, but our map identified a private entrance for us. In touring the park, I was especially fascinated by a large, well-preserved treasure house of the eighth century. Its log structure is supported by many large, free-standing, wooden posts, each resting on a flat rock base. This remarkable post design with ends shaped to be stable, allows rocking to and fro with ground movement to prevent transmission of seismic-induced earth movement to the building. The structure has stood for more than a millennium in that area of intense seismic activity. It was said to house some of the most rare and unusual artifacts in the world but was not open to us; in fact there was no stair to reach the floor level, perhaps eight or ten feet above the ground.

Ordering the luncheon we hosted was made simple by a display of food, modeled in plastic, in a cabinet outside the door of the park restaurant. Following the entree course, I wished to add a particular dessert displayed in the cabinet, but by that time I had forgotten its name. Gestures and sketches were to no avail; finally, I arose, motioning for the waitress to follow. As we approached the door, the manager joined us with a great flurry of Japanese words excited, apparently, by my actions. He seemed to think that I was trying to leave without paying and/or to kidnap the waitress. When we reached the display case and my mission was clearly understood, there followed another flurry of Japanese; this time in apologetic tones, with much bowing. The dessert was excellent. After leaving Nara, we visited briefly with Shigeko's parents at the Osaka airport before flying to Narita and home.

———————

I returned to Japan in August of 1983 as guest of the conference to participate in an international meeting on "The Future of Nuclear Energy." This was the third annual meeting sponsored jointly by the governments of Germany, Japan, and the United States. (Previous sessions had been held in Honolulu and Bonn.) Principal attendees from Japan included Takaki Yasuda, minister of the Science and Technology Agency, two members of Parliament and presidents of industrial companies. Germany sent Heinz Riesinhuber, MdB, federal minister of Research and Technology, Joseph Bugl, MdB, German Bundestag (Bugl had been my friend for several years through the HTGR Program with Jülich, Germany) and their secretary of defense. The U.S. was represented by Thomas Roberts, commissioner of NRC, Danny Boggs, assistant to President Reagan, and minor members of the Department of Energy. I was pleased to be included and was welcomed as the only U.S. representative with a significant background in nuclear energy.

The meeting emphasized the value of international cooperation, particularly between the three countries of the conference, and the value of the International Atomic Energy Agency. The discussion sessions also considered proliferation of nuclear weapons and availability of energy resources adequate to meet the long-term needs of the world.

The amenities of the old Hotel Nara and the conference plan provided excellent opportunity for informal discussion of issues and ideas. This location on the edge of the Nara Park would have been tempting for sightseeing, but the unbearably hot and humid August weather and the remarkably talented participants and interesting presentations kept all inside. The enthusiasm displayed in presentations and discussions suggested that the future for nuclear energy was promising and on the verge of a major expansion, but that has not proven true. Production of electricity by nuclear power has expanded only in France, Japan, and some other countries of the Far East.

Following the conference, Commissioner Roberts and I traveled to Kyoto and to visit the Kansai Electric Company's nuclear plants, Ohi and Mihama, on Wakasa Bay of Japan's north coast. One plant was in operation and the other in a maintenance shutdown, providing a good overview of their procedures and practices. The intensity of operator attention, cleanliness of equipment, and orderliness of maintenance operations seemed outstanding. They were more orderly than similar operations one typically sees in the United States. Tom Roberts commented that the facilities were so clean that one could eat lunch from the floors.

I also visited the Mitsubishi nuclear-equipment manufacturing plant in Kobe to see their shops and the training facility for teams who conduct scheduled shutdown maintenance and refueling for nuclear plants they have built. Realistic training facilities included full-scale mockups of a reactor vessel and a steam generator, arranged with accesses restricted as in a nuclear plant containment vessel. Trainees accomplished each action while working under a strict time limit, as they would in a reactor where restrictions are imposed to minimize radiation exposure. In discussions as we walked through the shop, the superintendent's knowledge concerning details of manufacturing equipment and quality control was impressive. I also visited the PNC and Tokai Mura facilities near Mito, in response to a repeat of the 1982 invitation by Kiyoshi Inoue, and participated as a principal speaker at a meeting at the Hilton Hotel in Tokyo on the technology and advantages of the HTGR.

Discussions with PNC representatives included the proposed joint effort in the reprocessing of nuclear fuel, which had been discussed in

Don Trauger (left) and Thomas Roberts of NRC (right), 1983. Photograph by the Kansai, Japan, Nuclear Power Plant.

Tokyo during my 1982 visit. The objective was to separate the valuable plutonium and enriched uranium remaining in spent fuel from the radioactive waste. The effort's primary feature was to conduct all of the work remotely in sealed shielding cells with electronic controls so that operators would have no opportunity to divert material with a potential for weapons. Constant monitoring by plant management, by independent agencies such as the Nuclear Regulatory Commission, and even by the International Atomic Energy Agency would be possible. Spent fuel would enter the plant as fuel elements that could be counted, and the valuable (and weapons usable) materials would exit as finished fuel elements ready for use in a nuclear reactor. These also could be counted to ensure that no diversion had occurred. In the arrangement ORNL was to provide most of the development of equipment to be operated remotely, Hanford would build a demonstration pilot plant, and the Japanese would build the first unit of commercial size. Once demonstrated, this type of facility design could be used for other chemical and biological processes that have inherent dangers.

As the United States lost interest in nuclear fuel reprocessing, ORNL funding was decreased and the Japanese made up some of the deficit for several years so that the necessary development could be completed. The plan was for ORNL and the Hanford Engineering Works to develop the equipment and the technology for remote operation, drawing heavily on their previous fuel-reprocessing programs. In return DOE would receive operating information from the first Japanese plant. In fact all information was to be shared by the parties involved. Fortunately, the Japanese have proceeded with construction of a plant, now nearing completion, but the United States has totally withdrawn from this program.

Support is not available even for the one or two staff members needed to monitor and receive information from the Japanese plant's construction and operation. This seems most unfortunate; the design is based on years of experience in fuel reprocessing at our own national laboratories, and much of the work was done specifically for that plant. The United States contributed its share, then dropped the ball when we would have benefitted at minimal additional cost. Many countries continue to build and use fuel-reprocessing plants that have less effective features for the prevention of clandestine fissional material diversion to weapons. The United States should have adequate information available through experienced staff to be properly prepared through negotiation and international action to prevent proliferation of nuclear weapons materials. Nuclear-fuel reprocessing will also be necessary if breeder reactors are used in the future to utilize fully our nuclear ore resources.

Oak Ridge experienced a time of transition when Union Carbide announced its intention not to extend the contract for operating the Department of Energy facilities beyond 1984. The company had served well and, accordingly, contracts of nominally five year extensions had been made beginning with K-25 in 1944. Many companies and institutions expressed interest, and screening narrowed the field to three from which the Martin Marietta Corporation was chosen. The new contractor was obligated to keep all personnel in the transfer, except for those at or above my level, and chose wisely to place Fred Mynatt, who had served several years as head of the Instrumentation and Controls Division, in my position. I was nearly sixty-five, the age at which one was expected to step down anyway, so the move was acceptable. In fact Fred was one of the people that I had recommended as well-qualified for the responsibility.

It was distressing, however, to receive a letter from Martin Marietta falsely stating that the change was made because I had expressed no further interest in the job. A few more months might have made the transition smoother, but that is conjecture. Because of the letter, I considered other options but chose to accept the proposal from Herman Postma, the Laboratory director, to serve as his assistant. This made possible continuation of a comprehensive program, previously started with Jim White, named the Nuclear Power Options Viability Study. The purpose was to review the status of nuclear power and make projections to determine what could be accomplished by the time period 2000 to 2010. The assumptions were that a need for additional electric power would then be evident and that nuclear energy would be acceptable, if properly installed and economically competitive. Thus, the study focused on projected capabilities of the several nuclear systems then being developed by industry and government. The scope focused on the U.S. market and manufacturers, although some foreign concepts and influences were included. Participation was obtained from other national laboratories, nuclear-reactor manufacturers, and nuclear-electric utilities. Having time to pursue technical studies proved a delightful change from the incessant pressures of management.

Don Trauger interviewed concerning the ORNL fiftieth anniversary recognition activities, 1992. ORNL photograph.

ELEVEN

Special Projects in History and Energy

M y years as senior advisor to the Oak Ridge National Laboratory proved interesting beyond expectation and frustrating only in their limitations. New and unusual tasks flowed from the innovative mind of Laboratory Director Herman Postma, and with time to reflect, I too identified challenging tasks and also enjoyed more technical work. The Nuclear Power Options Viability Study (NPOVS), designed to evaluate possibilities for establishing new or improved nuclear energy plants for electricity and for process-heat production in the time frame 2000–2010, had been initiated a year or so earlier. To ensure credibility, assistance was sought from a diverse group of industry participants and other experts. Many other tasks were of interest, but the NPOVS was a prime focus.

I had much to do but progress seemed slow; several hundred people were no longer available for assignment on my projects. I had two choices: do the work alone or make the task enticing to others. Fortunately, the NPOVS was of interest to many technologists. For example, Bob Braid and Robin Cantor joined the NPOVS to examine the social aspects of the nuclear power industry. Since nuclear energy had become controversial and politically sensitive, social evaluation was important. Howard Bowers and Jerry Delene analyzed economic factors, and Jim White, Ray Booth, Dave Moses, and others assisted with reactor-vendor contacts and in technical reviews. The task was first supported from discretionary funds available to the Laboratory director, but soon it was funded directly by the Department of Energy. This became a comprehensive study of the major U.S. programs in the development of nuclear reactors and fuel cycling, both in the private sector and in government. I am proud of the NPOVS reports and appreciative of the assistance by the many persons from ORNL, other national laboratories, utility companies, universities, and the nuclear industry. NPOVS included an extensive bibliography available in computer retrievable form prepared by Jackie Sims, my able secretary of this period.

Among my new tasks was a role in developing the Consortium of Research Institutions, founded to coordinate research efforts in areas of interest common to ORNL, the University of Tennessee, and the Tennessee Valley Authority. Don Eastman, assistant to the UT chancellor, John Stewart, then a vice president of TVA, and I served as a triad to encourage and facilitate this cooperation. Later, Homer Fisher, UT senior vice president, replaced Eastman, and Brown Wright joined Stewart to serve as secretary for the consortium.

The combined technical resources of these institutions are immense and rival the best and largest of the world. UT focuses on basic studies with some applied tasks, the ORNL components are somewhat more applied, and TVA offers a "real world" laboratory for practical evaluation or implementation of energy-related technologies. An initial step was to plan periodic meetings for presentations by staff members in areas of common interest to the three institutions. Senior managers participating regularly included Lamar Alexander, president of the university, Marvin Runyon, Board chairman of TVA, Herman Postma, director of ORNL, Kenneth Jarmolow, president of Martin Marietta Energy Systems, and Joe Lenhard representing the Department of Energy.

Consortium meetings focused on programs in energy-related research, since that was the major common interest. In addition to broadening and coordinating programs, the consortium succeeded in attracting new high-technology organizations to the area. By assigning task leadership to the most qualified members of each institution, the consortium offered sponsors better service than any of the three institutions could provide separately. The programs became so well coordinated that a single staff member could present a comprehensive report on research at the three institutions. The consortium continued in this vein until I retired in 1993. During that period Clyde Hopkins replaced Kenneth Jarmolow in 1988, and Al Trivelpiece replaced Herman Postma in 1989.

Tom Trauger and Jana Belsky married on May 26, 1984, at the Four Ways Restaurant in the old Fraser Mansion of Washington, D.C. We had come to know Jana well as she accompanied Tom on visits to the North Carolina farm, and she was warmly welcomed into the family. The wedding also introduced two families, the Belskys and Traugers, who have enjoyed meetings in Washington and New York on subsequent occasions.

As the NPOVS work neared completion, the DOE sponsor suggested that we compare our findings with those of similar studies known to be under way in Europe. We had included work in Sweden, where a bold new design for a light water reactor had been proposed and brought to our attention, but we did not know much about other European studies. My DOE counterpart requested that I plan a travel itinerary to present portions of our work and to learn of their studies and findings. Colleagues in Germany were enthusiastic about the exchange and responded by arranging a comprehensive array of appointments, even to include meetings with members of appropriate committee members of the Bundestag and their staffs. The United Kingdom Central Electric Generating Board and other European groups offered cooperation. Each understood that not all of our findings would be shared because of U.S. national competitive interests.

The trip was planned well in advance and submitted to DOE for formal approval, but weeks passed with no response. For convenience and a saving in travel time and cost, the European trip was to start from Chicago after I had participated in a session on the uses and conservation of helium at the annual national meeting of the American Chemical Society. My presentation was on the use of helium as a coolant for nuclear reactors. Since DOE headquarters had requested the trip, I was advised to pack for Europe, expecting a late approval. (I had previously experienced late approval for government travel, once having been paged at the Knoxville airport to confirm an overseas trip, necessitating a sprint to the gate to make the flight.) After finishing my presentation, I still had no word, even though urgently requested. I canceled the trip and carried the heavy suitcase home to unpack that evening. The travel request had remained on the desk of DOE's manager of nuclear programs and apparently died there.

I was embarrassed and apologetic in phoning colleagues in each country to cancel appointments; one German official angrily responded that he would never again agree to a visit by a person with DOE credentials. The purpose of the trip had been to learn from European studies that were seldom published, whereas our reports were published in accord with DOE policy. I have been rewarded by word of extensive study and perhaps application of the NPOVS reports through their distribution in Europe and particularly Asia, but I have been saddened to find little perceivable interest or impact in the United States.

A series of investigative activities occupied much time during the late 1980s. The first was initiated following the disastrous USSR nuclear accident at Chernobyl, April 26, 1986. Although the nuclear test reactors then operating at ORNL bore no resemblance to the Russian nuclear power plants, Herman Postma and I decided that all our reactor systems should be scrutinized. I formed a committee of ORNL staff members and others having expertise and experience in all aspects of test-reactor technology, operation, and experimental uses. We systematically evaluated the Oak Ridge Research Reactor, the Bulk Shielding Reactor, the Health Physics Research Reactor, the Tower Shield Reactor, and the High-Flux Isotope and Research Reactor (HFIR).

As the investigation progressed, we first found no problems of an urgent or threatening nature, although we made recommendations for more thorough documentation of design changes, a better understanding of heat removal systems for very unlikely loss of cooling events, and more extensive training of operators. Many operators were nearing retirement, and their replacements would require both more detailed operating procedures and more comprehensive

High-Flux Isotope Reactor, at left, with hot cells for radioactive materials research. ORNL photograph.

indoctrination and training than had been necessary in the past. Many of the senior operators had participated in the design and construction of the ORNL reactors and were thoroughly familiar with their designs, operating characteristics, and controls.

However, upon systematically interviewing retirees and others having past responsibility for reactor safety, we found a potentially serious problem. John Conlin suggested a more careful evaluation of the status of the HFIR pressure vessel, with respect to embrittlement from neutron irradiation. During construction, samples of the vessel material were placed in vessel locations where the most intense neutron exposure would occur and were scheduled to be removed for testing at designated times. Early sample tests had shown no damage, and a set had been declared acceptable on the basis of a later, partial examination. Testing of the most recently removed specimens had been set aside on the premise that earlier samples were thought to be unchanged, but the delay was caused primarily by pressures from other tasks. We insisted that the samples be given a more thorough evaluation.

I ordered a high priority for the pending tests, which showed measurable deterioration. On that basis, the reactor was operating outside allowable limits established conservatively at the time of its design in 1965. It was promptly shut down as required by rules for safe reactor operation. However, data from the Heavy Section Steel Program, if applied to the HFIR vessel, showed that the technical specifications for the HFIR vessel were overly conservative. Thus, the reactor could have continued to operate if the specifications had been revised as the new research data became available. No comparable problems were found in the other reactors, although the intense review had identified numerous areas for improvement in operator training, housekeeping, and preventive maintenance.

The root cause of most problems found was that DOE and its predecessors had funded the reactors and most experiments from one office and had directed their operation and maintenance from another. Thus, year by year, when funds were short for preventive maintenance and training programs, experienced operators were required to find temporary measures by which to operate safely while economizing by deferring longer-term needs. Following the HFIR restart, this was temporarily corrected by placing both functions under one DOE group, but now DOE has reverted to separation of the operational and maintenance funding.

A committee formed by DOE to evaluate all its nuclear reactors functioned concurrently with our review, but reached Oak Ridge after our final report was issued. Even though this was a duplication of effort, independent determinations seemed appropriate to assure the safety of aging reactors. Their observations closely resembled ours, and they found only a few minor items that we had missed. Those findings also did not show need for shutting down any of the reactors except HIFR (which by that time had been shown to be operable). Despite a lack of technical justification, the manager of the DOE Operations Office arbitrarily chose to shut down all of the ORNL reactors.

I found that action very disturbing. Not only was DOE without any finding that should stop the reactors from operating; doing so cost significant scientific loss. For example, the Oak Ridge Research Reactor was scheduled to shut down permanently in six months after completing a set of irradiation experiments to show that test reactors around the world could operate with fuel that is less subject to nuclear weapons proliferation. That ORR work was never finished. The Bulk Shielding Reactor was nearly ready to accept a new facility for irradiating materials with thermally cold neutrons, a powerful technique for understanding and improving the properties of many materials. The Health Physics Reactor also had continuing programs that could best be carried out in its facilities utilizing the broad base of data unique to the machine and its facility. A politically motivated plan was laid to move that reactor to Los Alamos, at great expense and with the loss of services by the experienced staff in place at ORNL. That move, delayed at this writing, may yet take place. Only HFIR has been restarted.

I later consulted with Nuclear Regulatory Commission (NRC) officials who regulate test reactors outside the DOE sphere. It appeared that the ORNL reactors would not have been shut down if they had been regulated by the NRC. Instead, the NRC would have insisted on prompt evaluation of all HFIR surveillance specimens. Since ORNL does research for NRC, we were familiar with their rigorous inspection process of reactors and enforcement of corrective measures. In my judgment their system provides better assurance of safety and a more rigorous, independent, and practical set of regulations than those of DOE.

My second investigative task of this period was to review the gaseous diffusion plants at Oak Ridge, Paducah, Kentucky, and Portsmouth, Ohio, with respect to the environment and the health and safety of plant employees and their neighbors. Again, a committee was formed with John Shoemaker Jr. as deputy and staffed with others who had more recent experience than mine at these plants. The committee included Dave Massey of the Paducah plant, Earl Allred of Portsmouth, Dot Snead of K-25 Maintenance, Don Kellogg, a recent K-25 retiree, and Herb Trammell, director of the ORNL Engineering Technology Division but formerly of K-25. Jackie Sims recorded meetings and prepared reports. Kellogg was perhaps the best qualified member because much of his experience had been in areas critical to our examination, but those technologies were classified and remain so today. Because his security clearance had lapsed, Kellogg sometimes could not participate where his knowledge was most valuable.

In this investigation, as in reviewing the reactors, we visited all areas of the plants and interviewed key personnel, both individually and in groups. For me it was nostalgic to traverse the streets and visit buildings at K-25 where I had worked more than thirty years earlier. Although I had retained security clearance and visited K-25 many times for specific missions, it was exciting to have broad interests again in gaseous diffusion.

We found no problems comparable to the HFIR surveillance deficiency, but we made numerous recommendations. In follow-up meetings with the management of each facility some months later, we found that our recommendations had received early attention and action. Even so, additional investigations, audits, and reviews by DOE and others of its contractors continued to a degree that seriously interfered with completion of recommended corrective measures and staff effectiveness.

———•◆•———

On Elaine's and my fortieth wedding anniversary, September 2, 1985, Byron and Tom presented a note stating that their gift would be an all-expense-paid-trip "to where it all began," New York City. They and their wives joined us there on a weekend in October. The itinerary included restaurants, theater, music, the Metropolitan Museum of Art, and sightseeing with "reruns" of some favorite excursions from our previous, low-budget, New York life, such as riding the Staten Island Ferry and visiting the Cloisters. Some events were prearranged and others left to our option.

The weekend was magnificent with only one unpleasant moment that occurred when we were leaving the St. Regis Hotel (where we enjoyed a suite) for the Saturday evening theater. The doorman had difficulty finding cabs, and with a long line waiting, I chose to go around the corner from whence the cabs were arriving. Hailing the first one, I forgot how quickly one must act in New York, and someone beat me to its door. The next time I was more alert but heard footsteps from behind me closing fast. As I quickly reached and opened the door, a large man arrived to confront me, stating firmly, "I am taking this cab." I replied, "No sir, you are not." He then swung his fist toward my face as I deftly dodged to take the blow on my left elbow. With an apparently sore hand, he stepped back as I moved toward him to regain my position at the cab door, simultaneously feeling for my wallet on the chance that this was a pickpocket scheme; the wallet was there. I squared away as though to fight, presenting a determined front, but boxing skills were fifty years out of practice. Had he advanced, my next move would have been to hold onto the cab door and kick my heavy boot into his knee. It was a potentially dangerous confrontation. I had often been concerned that when faced with such a threat I might lose control and resist irrationally, out of sheer panic. Fortunately, for both of us, he suddenly turned and ran away. I entered the cab as nonchalantly as possible. The cab driver seemed to consider this a common occurrence.

We turned the corner to pick up the others of our party and proceed to the show. I spent some time massaging a sore arm and shoulder, while listening to Lily Tomlin's superb monologue *The Search for Signs of Intelligent Life in the Universe*. The injury was slight and did not interfere with sleep or the activities on Sunday.

On our last day in New York, Jana had the use of her father's car, and we toured the city from the Cloisters to the Battery, after attending services at Riverside Church. The celebration of that wedding anniversary could hardly have been better. On Monday morning we discovered that the airline had oversold seats, delaying us to a later flight. In recompense, they issued Elaine and me free tickets to any U.S. destination, as well as for the next flight to Knoxville. We later made a pleasant long-weekend visit to Phoenix and Sonoma, Arizona, using the bonus tickets.

Not long after, we attended my fiftieth high school reunion, a pleasant occasion, though it was challenging to remember names and properly greet the seventeen of thirty-six graduates who were present. Friendships were quickly reestablished even though I had seen very few class members since leaving high school. Elaine and I later spent a very busy weekend at the farm in North Carolina to assist Mr. Causey in entertaining members of the Eastern Section of the American Hereford Association. They had come to review the outstanding facilities, management, and superb herd of this ninety-two-year-old's Hereford cattle farm.

In Oak Ridge Elaine's City Council activities intensified as she both did extensive homework for meetings and continually pursued goals for city improvement. She also worked in related organizations of city officials on state and national levels. Since most of her meetings were on weekends, I sometimes accompanied her to places like San Antonio, Texas, and Santa Barbara, California, for National League of Cities events. It was great to discover the luxurious life afforded an accompanying spouse at a conference. I was invited to all of the entertainment events and had the day free to read or do sightseeing with no conference work or responsibilities.

We attended another meeting of the National League of Cities in Charleston, South Carolina. From there we added a side trip to inspect timber and plan for planting trees on our farm in Moore County, North Carolina. While there, we discovered that an oil company had trespassed to drill holes and make seismic measurements for oil. Although the damage to trees and soil was slight, it was disappointing to encounter such irresponsibility. They left no trace as to identity and had not been observed in that remote location, but it may have been one of several companies to whom we had refused to sell options for their exploitation of mineral resources.

During that period I worked through my church to take an active role in shaping public awareness of the threat from strategic weapons, while defending the benefits of nuclear technology for purposes of benefit to humanity. I was pleased to have a small part in refining the document prepared by the United Methodist Council of Bishops addressing nuclear proliferation issues, *In Defense of Creation*. My contribution to the document was to clarify some of the wording that did not properly separate the beneficial uses of nuclear energy from the weapons issues. The treatise was written in 1986, when the Union of Soviet Socialist Republics and the United States both maintained huge arsenals of nuclear weapons, many of which were on missiles that could be targeted for delivery to any place in the world. The quantities exceeded by far the estimated numbers that, if exploded, could make life untenable. Such projections are dire even without considering the concept of a nuclear winter created by blanking the sun with atmospheric debris from a major nuclear conflict.

I include here segments from the overview and summary of *In Defense of Creation*. It begins:

> We write in defense of creation. We do so because the creation itself is under attack. Air and water, trees and fruits and flowers, birds and fish and cattle, all children and youth, women and men live under the darkening shadows of threatening nuclear winter. We call The United Methodist Church to more faithful witness and action in the face of this worsening nuclear crisis. It is a crisis that threatens to assault not only the whole human family but planet earth itself, even while the arms race itself cruelly destroys millions of lives in conventional wars, repressive violence, and massive poverty.
>
> We seek the fullest and fairest possible discussion not only of the convictions that we have tried to state clearly but also of alternative and critical views. We pray that our churches may become redemptive models of peaceable diversity even as they struggle for reconciliation and unity in Christ.

The passage closes with an appeal for ecumenism:

Christian unity in all the fullness of baptism and Eucharist and common life throughout all the earth—is crucial to peacemaking. We call upon United Methodists everywhere to pray regularly for our Christian sisters and brothers in the Soviet Union and every other land, to Russian religious life and thought, and to support ecumenical exchanges with churches in the Soviet Union. We especially celebrate the voice that the World Council of Churches gives to the world's poor and most abused peoples, whose partnership in peacemaking is the plain imperative of the gospel of Jesus Christ.

I would extend the admonishment by the United Methodist Bishops to engage all peoples and all religions. This message, so expanded and made non-specific with respect to religion, also seems appropriate for the United Nations. Although tensions of the past decades have diminished, admonitions about war remain appropriate. However, this may be futile unless the causes of war are addressed. Territorial expansion has been the apparent goal of many wars, but an underlying cause has often been to access natural resources. For example, the coal deposits of the Alsace-Lorraine and Manchuria contributed to major conflicts in Europe and Asia. The recent Gulf War, though touching on aspects of territorial and human rights, was basically a conflict over oil reserves. Regardless of the rising and falling of international political tensions, we would do well to understand that the relationship between the perceived availability of energy resources, on the one hand, and a potential for war, on the other, has always been a constant in modern world history.

Having participated in the review of the bishops' letter, I was invited to join Gordon Fee, then manager of the Y-12 plant that manufactured nuclear weapons components, and Alan Geyer of the Washington Theological Seminary in interpreting the document at the annual meeting of the Holston Conference of the church. (Professor Geyer had served as secretary in writing the bishops' letter.) This led to my receiving invitations for several similar presentations at schools and churches, even to presenting the eleven o'clock "sermon" on the subject for the Gatlinburg, Tennessee, United Methodist congregation.

First United Methodist Church, Oak Ridge, had previously sponsored seminars for clergy of the Holston Conference to enhance their understanding of current technology. The meetings were strengthened

Don Trauger with bishops of the United Methodist Church at the ORNL High Temperature Materials Lab, 1987. ORNL photograph.

by technical staffs of ORNL and other DOE facilities. In 1987 the First Church's Science and Theology Committee chose to use the bishops' letter as the subject for a conference also titled "In Defense of Creation." It was to inform both clergy and lay persons about the issues of nuclear weapons. Bishops of the church were invited, and fifteen came to participate. Bishops Dale White and David J. Lawson made presentations from the perspective of the church, and L. G. Christophorou of the ORNL senior staff spoke on the subject, "What We As Engineers and Scientists Expect from Church Leaders." Alvin Weinberg presented, "An Approach to Peace from a Science and Technology Perspective." Persons from several churches and technical disciplines served on panels.

The meeting was open to the public and approximately three hundred people attended. It was followed two years later by a second conference on Arms Reduction. Senator Albert Gore Jr. and Alan Geyer were key speakers on that weekend. That event, dedicated to William G. Pollard, also was well attended. Pollard's brief appearance was his last in public before succumbing to cancer a few weeks later. The Science and Theology Committee continues to be active and has since sponsored conferences on the human genome, the environment, and medical ethics at approximately two-year intervals. I have been privileged to participate in the planning and conduct of these interesting symposiums.

In 1987 I was invited to a meeting on "Small and Medium Power Reactors" in Lausanne, Switzerland. The conference offered to pay travel and lodging expenses if I would chair a session and present a paper, "Safety and Licensing for Small and Medium Power Reactors." I asked Elaine to go, suggesting that we take a second week as a vacation in Switzerland. She thought it too expensive for the time available but accepted my offer: "The Swiss are paying my way and I will pay yours, so there is no cost." The money, of course, came from our joint bank account, but I wanted to have her company.

Representatives of a dozen countries presented mixed messages, optimism over the prospects for adoption of smaller and safer reactor designs, and pessimism over the continuing decline in the nuclear industry. Following the meeting we moved to Interlocken and enjoyed the convenience and savings of Swiss rail passes for five one-day trips to visit mountain vistas towering above beautiful farms and villages. The last evening, spent with Alf and Kungolt Schurch in Zurich, capped a fine and rewarding visit to a favorite country.

I spent much of 1988 serving on a task force on Nuclear Production Reactors for production of tritium and plutonium for weapons in a task of the Energy Research Advisory Board of the Department of Energy. The purpose was to make recommendations to DOE concerning the production of tritium for the manufacture and maintenance of nuclear weapons. The requirement for maintenance arises from the steady decay of tritium to helium. An earlier review for DOE had determined that the Savannah River plants used to make tritium would require either extensive modification or replacement. Options included new or refurbished heavy-water plants, modified light-water power reactors, nuclear accelerators, and small high-temperature gas-cooled reactors, with my role focusing on the last. Four teams consisting of knowledgeable people and each team representing one concept, functioned well in a somewhat competitive mode.

The effort resulted in a decision by DOE to pursue two options, a small HTGR and a new heavy-water reactor. However, with the end of the Cold War and dismantlement of weapons offering an extension of the tritium inventory, pursuit of both options was placed on hold and remains in that status. In the interest of budget constraints it now appears more likely that tritium will be made in an existing TVA reactor. I preferred the HTGR option both for its merits and because the project would place its remaining development costs in a military budget, thereby securing its continued funding. Early development of the LWR had been accelerated through the funding of Navy and Army reactors.

We were pleased and excited on May 6, 1988, with news from Tom and Jana of Hallie Causey Trauger's birth, our first grandchild! Two weeks later we made a very special trip to Washington, D.C. to meet the new arrival. It was a great experience to again hold a tiny baby and to participate in some of her care. I was privileged one evening to be alone with Hallie as she lay in her basket while her parents and Elaine were otherwise occupied. I sat looking at the serenely sleeping little girl in her basket and pondered what she might encounter in the later period of her statistically projected life of over seventy years. The house in the Woodley Park section of Washington was suitably heated by natural gas to protect us from that cool spell of late spring. Would she enjoy such convenient comfort in her later years when age again demands more warmth? Natural gas reserves are projected to be depleted well before that time. Would our society shoulder the necessary responsibility and effort to meet her needs? Should it not also have concern for the grandchildren of Hallie and her contemporaries?

The year 1988 brought triumph and trauma to Elaine. She continued to function effectively on the city council while suffering a prolonged viral infection during the first months of the year. Despite the illness, she engaged a meeting in Oak Ridge on the technologies of waste management for the Energy, Environment and Natural Resources Committee (EENRC) of the National League of Cities. Senior staff members of ORNL responded to her invitation with a comprehensive briefing that encompassed most aspects of this vital issue. Participants told her that misconceptions about energy and the environment had been corrected, and one reported later that, based on the information received, he had been able to save his city a substantial expenditure.

By midsummer Elaine became afflicted with chest pains. As these progressed and were coupled with increased pain in her damaged left arm, concern for a potential heart problem brought us from a Rotary meeting to her doctor. His judgment, coupled with an electrocardiogram test, was that her heart was sound, but to be sure, she was admitted to the hospital for a few days of monitoring and stress tests. This clearly was not heart disease. Nevertheless, as the summer progressed, the pain persisted, and, in sequence, inflammation of the chest lining and several neural disorders were studied; extended treatment was pursued in response to several alternative diagnoses, to no avail. Each diagnosis was complicated by variations in the intensity and distribution of pain, but it was consistently within the frontal chest region. In early September she felt able to attend a meeting of the EENRC to be held in Gillette, Wyoming.

Elaine wanted me to accompany her to the meeting but I resisted, having seen Gillette during forays into eastern Wyoming following Nebraska family visits to the Black Hills of South Dakota. She then proposed that we follow the meeting with an additional week of vacation in the Grand Teton National Park. I could not resist a visit to my very favorite western mountains, and we were off to Wyoming. Gillette and its surrounding region had changed greatly from my previous impression. Their claim to be the "Energy Capital of the World" had validity when viewed from the perspective of coal domination as a primary energy source; however, we claim the distinction for Oak Ridge in terms of research and development for all energy systems. The committee invited me to attend the meetings to hear presentations on many aspects of energy production, most, understandably, centered on coal as mined, refined, and sold from the nearby deposits of subbituminous coal.

One session dealt with a nuclear issue. A committee member from Colorado had requested time to present concerns over the safety of the Rocky Flats weapons component plant near Denver. In private discussions, I was able to clarify several misunderstandings concerning nuclear technology but carefully avoided the principal issue. I have never been involved with Rocky Flats or, for that matter, specifically with nuclear weapons development. Even my work on the Manhattan District Project was also applicable to nuclear power, not weaponry per se.

Overall, the meeting proved interesting and valuable to me. It included visits to the vast coal strip-mining area where we saw excavations as much as two hundred feet deep from which thick layers of coal were being removed. Coal seams were, respectively, ninety and twenty feet in thickness, one above the other. The size of the mining equipment is well illustrated by the eight-foot diameter tires used on trucks to haul coal out of the mine. I complimented the quality of their roads to the superintendent of the Carter Mining Company, who was briefing the group. His reply was, "When one pays eight thousand dollars for each tire that carries a twenty-ton load, a small road defect can be costly." The land reclamation also was impressive, comparable to that seen in Germany, although this was in a very different climate and for a very different land use. The Wyoming land, as reclaimed and seeded with native grasses, probably supports more wildlife than it did prior to the mining operation. The financial return from mining such thick veins of coal easily supports the costs for restoration.

We also visited research and pilot-plant facilities for removing water and non-combustible impurities to improve the heat content of the coal per unit weight. The objective is to minimize shipping costs and make western coal more competitive in eastern markets. I was impressed with the information presented and prepared a trip report for ORNL. We left Gillette soon after the meeting; Elaine was overly tired, and we both looked forward to a few days rest in the Teton Mountains.

The flight to Denver was a bit rough, hardly restful, and Elaine all but collapsed in the Delta Airlines Crown Room. I violated their no-food rule to bring her a snack. The flight from Denver to Jackson Hole via Rock Springs was one of the roughest I have ever experienced. At Rock Springs the pilot parked the small aircraft in a position that required passengers to walk much farther than usual on that windswept tarmac. In answer to my inquiry, he explained that the high wind velocity forced the unusual orientation to avoid the aircraft's being overturned. We arrived at Jackson Hole in a near-blizzard snow storm. When we finally reached the motel, Elaine was totally exhausted, in severe pain, and could not go down to dinner. As I sat in a nearby small cafe worrying about Elaine while admiring a display of Robert Kerswell paintings, I thought we had made a major mistake in coming farther west. It seemed that an air ambulance might be required for the trip home.

After sleeping late, Elaine was somewhat recovered, but the severe chest pain persisted. The innkeeper suggested a nearby clinic for help, and we saw a doctor who concluded that her problem stemmed from a brain tumor and wanted her admitted immediately to their hospital for examination. She declined, and he prescribed a medication to relieve pain.

Surprisingly, the following week was quite pleasant. Elaine rested most mornings while I hiked trails and indulged my hobby of photography, as warm and sunny days returned to that beautiful place. Each afternoon we visited a new or favorite spot and walked together to see the endangered trumpeter swans on Two Ocean Lake. I recall two particularly good evening meals, in part for the restaurants' names, the White Buffalo and the Blue Lion.

Near the end of the week we took a leisurely stroll along Leigh and String Lakes. In the beauty of those woods and streams with mountains silhouetted, we talked of our love and good fortunes. We verbalized for the first time what each had known throughout the week—that if the doctor's diagnosis was correct, these could be the last quality hours that we would ever have together.

At home we again sought medical help; there was no brain tumor, but our pharmacist, Jim McMahon, on seeing the Wyoming prescription, urged Elaine to stop taking the powerful narcotic immediately. The prescribing doctor probably thought he would make the patient's last days as comfortable as possible. Additional medical tests were made to no avail.

The fiftieth reunion of Elaine's high school class was in mid-October, and she insisted on our going to North Carolina for the weekend event. Elaine, the optimist, after taking the last rationed pill of the strong narcotic to feel well enough to go to the dinner, volunteered to host the next reunion five years later. I listened, wondering if I would fulfill the commitment alone.

In early November, Elaine experienced some respite, and we planned a brief trip to Orlando, Florida, to a Sigma Xi national meeting, since I had been the local president. We had particularly enjoyed the previous year's meeting held on the *Queen Mary*, docked in Long Beach, California. Now, Elaine also wanted to see Disney World. That was to be followed by continuing directly to Washington, D.C., for the American Nuclear Society meeting. I was obligated to attend because plans were to be made there for the 1990 summer meeting, for which I was designated chairman. At the last minute we canceled the Florida trip because of the demands of complicated travel. Just before leaving for Washington, Elaine had an upper body MRI, but we did not learn the results before departing. In Washington, Elaine experienced the worst pain to that time and was essentially confined to our hotel room. After two days, her physician, Dr. David Seay, called to report that the MRI revealed nothing on the front of the chest but that a small irregularity could be seen on the spine.

We immediately rushed home and telephoned Dr. Paul Spray, our orthopedist and close friend. Upon hearing my relay of the medical report, he responded, "Be in my office at eight o'clock tomorrow morning for X rays." The X rays, focused on the spine, showed a defect in the T-8 vertebra. Paul advised that exploratory surgery would be required to obtain biopsy samples and, because of the location, should be performed by a specialist in spinal surgery. He arranged for Dr. Herbert Schwartz of the Vanderbilt University Medical Center (VUMC) in Nashville to examine Elaine and to perform the surgery. Knowing that a probable finding would be cancer, our conversation during the two-and-a-half-hour drive to Nashville was a repeat of the exchanges we shared during the stroll along the lakes of the Teton mountains. On Monday, November 7, 1988, Elaine was admitted to the VUMC.

Although only one day was required for the numerous tests, Elaine endured excruciating pain, and the time seemed endless. The surgery was undertaken on Tuesday afternoon. More than three hours were required to obtain the biopsy specimens, as Dr. Schwartz carefully threaded his tools past nerves and blood vessels that could not be damaged if Elaine was to survive the surgery intact. Byron was a great help with arrangements and stayed with me during the operation. When we learned that Elaine had come through the surgery well and was resting in recovery, we took a long, fast walk to and around Nashville's replica of the Parthenon to relieve tension. As word reached Oak Ridge and flowers began to fill the room, we joked with Elaine, saying that available space could not accommodate both her bed and the flowers and that we would have to move her into the hall. However, there was little real levity because her pain continued with severity, and she had to recover from surgery.

Within hours, we learned that it was cancer, but again days passed before it was determined to be a non-Hodgkins type lymphoma. Telling Elaine of these two diagnoses was the most difficult task I had faced in my life. The doctors had told me that this form of cancer is extremely aggressive and that the only hope for patient survival is to attack it with one of the most aggressive protocols in use at that time—administering severe doses of radiation and chemotherapy.

The plan for treatment was developed by Vanderbilt doctors in cooperation with oncologist Dr. Helen Vodopick of Oak Ridge, who would be in charge of both the chemotherapy and radiography, although Dr. Larry H. Lee would function as the practicing radiologist. We made the trip to Oak Ridge by automobile with Elaine securely packed with pillows. She was actually more comfortable in a sitting position than lying down, so ambulance transport was not appropriate. Byron rode with us in the back seat to help if Elaine had to be moved. Heavily sedated, she fell asleep before leaving the Vanderbilt campus and did not awaken until we reached the Oak Ridge Hospital.

Intensive radiation treatment was scheduled first, to last five weeks. Although her doctors thought that the radiation had killed the identified cancer cells, this type of lymphoma, when located in the spine, is usually secondary to another. No other cancer had been found, but the chemotherapy, even with its accompanying trauma, was judged necessary. At first, the "chemo" caused little distress, but it soon resulted in loss of hair and fingernails and severe debility. Her problems had been amplified by radiation damage to the esophagus; swallowing food and even water was difficult, which led to severe dehydration. A blood clot also complicated recovery.

The treatments and recovery required seven months in hospital beds, alternately in our home and in the hospital. At home Elaine needed constant care and some professional assistance. For the weekday mornings we hired a nursing service and on afternoons when Elaine rested, friends came to sit with her. We are eternally grateful for the many, many people who assisted, showered us with food and flowers, and petitioned for Elaine in their prayers.

Although this was a serious time for Elaine, one morning when the nurse brought cereal to her with an apology for the milk carton, explaining that she could not find a pitcher, she was surprised by Elaine's amusement. At the time her collection of pitchers numbered 250 and now is about 270; several are always on display.

Elaine was first able to dress and come downstairs with assistance to sit in a special chair during the catered rehearsal-dinner party before Byron's marriage to Aleta Arthur. Aleta had come with Byron on many of his nearly biweekly visits during the illness and was easily accepted as a family member. I was so grateful for her expert assistance with the many household tasks that needed attention. We also knew her as a most competent attorney and adjunct professor of law at Vanderbilt.

Aleta chose to have the wedding in Oak Ridge so that Elaine could attend. The date was selected as May 13, 1989, which seemed to allow plenty of time for Elaine's recovery. But we did not anticipate that Elaine's confinement would be so extended. Invitations were sent only to family members and a few of the couple's closest friends. Thus, at this wedding held in First United Methodist Church of Oak Ridge, only Elaine and I, of the forty-six attendees, lived here. It was a beautiful occasion that Elaine was able to attend and enjoy, though three days of constant bed rest were required for her to regain even limited activity. It was to be another month before she could return to a normal bed and resume a few household duties. Slowly but surely Elaine made a full recovery and has been cancer-free ever since, though we approach each bimonthly checkup with some anxiety. Unfortunately, the cancer's injury to the vertebrae has caused frequent back pain, now intensified by arthritis.

The 1990 summer meeting of the American Nuclear Society was held at the Opryland Hotel in Nashville, with responsibility for planning and conducting the meeting assigned to the Oak Ridge-Knoxville Section. I felt honored to be named general chairman; other section members were Norb Grant, Howard Kerr, Trent Primm, and Doug Selby of ORNL, and Marsha Katz and Lillian Mashburn of the University of Tennessee. All served as committee members assisted by Joseph Hamilton of the Vanderbilt University Department of Physics. We selected Alvin Weinberg to be honorary chairman and to give an opening address. We were fortunate also to have Sen. Albert Gore Jr. (later to be vice president), Nuclear Regulatory Commissioner Forrest Remick, and Dr. Leon Partain, director of Nuclear Medicine at Vanderbilt, among other prominent speakers. The meeting attracted a near-record attendance of 1,250 participants.

DOE had urged the nuclear industry to produce guidelines for new and improved light-water reactors. The Electric Power Research Institute was charged with preparing documents that emphasized safety, reduced capital cost, and improved operation. My position as technical advisor, and the only participant from a national laboratory, may have derived from the NPOVS study. At the outset dramatic differences were clearly evident among the many representatives of the American nuclear industry. Some seemed satisfied with the status quo for reactor designs, and others came with more progressive ideas. The series of meetings continued for several years, even after my retirement, when Jim White ably represented ORNL.

That extensive effort has produced documents that propose both large light-water reactors modified from existing designs, and smaller plants offering substantial differences. Both include more passive safety features, and the smaller plants facilitate the manufacture of major components in factories, instead of the extensive on-site construction necessary for large units. I view these designs as offering much-improved plants and only regret that the recommendations do not prescribe a greater degree of standardization to achieve even larger cost savings and safety margins. The standardizing of nuclear plants in France has contributed to their success in providing a large fraction of electrical energy needs from the fissioning of uranium and plutonium.

I next participated in a fun project. The city of Oak Ridge and the Laboratory were approaching their fiftieth year. My role was to chair the ORNL Celebration Committee. First, it was necessary to determine an appropriate date on which the Oak Ridge Community should be declared fifty years of age. We settled on the day when General Groves made the decision to place a major portion of the Manhattan Project in what was to become Oak Ridge. Research showed September 19, 1942, to be the most probable date, which would also allow the celebration to benefit from the beginning of beautiful autumn weather of East Tennessee.

The Oak Ridge city's committee, with which we coordinated plans and effort, focused on a parade and other festivities, as well as on new civic facilities. At ORNL we first looked for improvements in historic buildings that could be enjoyed by employees and the public. The graphite reactor had been designated a National Landmark in 1968 and had been open to the public since the 1982 Knoxville World's Fair. The building needed thorough cleaning, renewal of exhibits that described ORNL's current activities, and improved historical posters. The graphite reactor became a priority focus.

Another project, chosen from many possibilities, was the restoration of a small church building located near the main entrance to ORNL that stood as a reminder of the pre-Oak Ridge era. It had been purchased in late 1942 by the government as part of the land acquisition for the Clinton Laboratories, now ORNL. The congregation had disbanded as members hurriedly searched for available farms, other occupations, and housing. Anticipating that their church building would be destroyed, just as were most of their homes, they used the government's compensation for the building to erect a monument with the inscription that reads, "Forty feet south of this monument stood the New Bethel Baptist Church." The church building remains. It had been used, in succession, as office space during the initial Laboratory construction period, as a facility for neutron studies, which benefited by the building's all-wood construction, and later for equipment storage.

Remarkably, members of the church congregation had stayed in contact and even held annual reunions on the grounds. Chuck Coutant worked with a committee of former church members that included Ralph Magil, Dot Bussell, Buck and Marge Magil, Nancy Van Kimley, Jean Coley May, and Rev. E. M. Sherwood. These and others of that former congregation and community contributed both ideas and artifacts for the museum's displays. We contracted with the American Museum of Science and Energy to design, build, and arrange display cases, with Don Barksdale and Dan Aultman of their staff as designers. It was challenging to convert the church into a museum while retaining its features as a building suitable for occasional worship services.

The aging structure had not been maintained well and required improvement for safety and preservation. The window frames were no longer serviceable and would have been costly to purchase in their nonstandard sizes. Ralph Magil arranged for prisoners at the Brushy Mountain Correctional Facility to build new window frames to replace those that had deteriorated. The original irregular window glass of the

New Bethel Baptist Church and Memorial Monument. ORNL photograph.

1923 construction was reused where possible. Church pews also posed two challenges: first to discover an authentic, rustic design and then to build them. They were patterned after still-existing pews from another church of the 1920s, and Greg and Dan Dahl built a dozen reproductions as Eagle Scout projects. The pews provide authenticity; folding chairs are added when the facility is used for meetings. New front doors, donated by Byron's law firm, were obtained from an old Arkansas hospital, which was forced to modernize with metal doors. The doors fit perfectly, even though they were about fifty years older than the church.

Former church members contributed lumber and labor to replace the pulpit and its platform. They also located the piano that had served the church in the 1930s and had it restored and tuned for the first service held in the renovated building. The church is now available to the congregation for special events, and it has provided the setting for several weddings. Former members with friends and families come to annual reunions and services in numbers greater than the building can accommodate. This church-museum authentically portrays life as it existed in the area before the intrusion of the Manhattan District Project and includes some artifacts of World War II.

Having been drawn to the study of history since Marie Minnick's high school class in American history, I was excited by ORNL's decision to include a written chronology of ORNL as a major feature of the celebration. First envisioning a book, we found that DOE regulations for publication of a historical document would extend publication well beyond both our deadline and our budget. Thus we chose a special issue of the *ORNL Quarterly Review* as the appropriate format. The *Review* staff, however, did not have time to write so comprehensive a document, estimated at three hundred pages. We requested competitive bids from a variety of sources and were pleased when Daniel Schaffer, editor of the University of Tennessee *Forum*, won the bid with an excellent plan in which the university would employ Leland Johnson, a historian and writer, as the principal author and Schaffer would assist with continued planning. Caroline Krause, editor of the *Review*, would be the principal ORNL contact for developing the document. Committee members Ellison Taylor, Waldo Cohn, Ed Aebischer, Mike Wilkinson, and Alex Zucker would collect historical material and contact retirees for anecdotal information.

The work of the Laboratory is well documented in thousands of reports and papers, each written in terse technical language. Our goal was to report the history as accurately as possible in a generally understandable text. We first focused on the early days to capture the recollections of as many pioneers as possible. Assembling the record of and bringing order to the first years was made easier by the perspective of time and by the fact that both the wartime project and the AEC focused on a few nuclear energy projects. ORNL task areas eventually broadened under ERDA and DOE to include essentially all energy sources and their related technologies. More recent projects became either indirectly related to energy systems or properly classified as applied or pure science. The problem of describing this work was further complicated by the DOE practice of allocating research funds in small portions without classification into broader programs. The effects of such micromanagement has made the conduct of research much more fragmented and less efficient, thus also complicating the historical writing.

We had to make hard editorial choices in selecting a representative sample of ORNL's many accomplishments, especially in the basic sciences. Mike Wilkinson, Alex Zucker, and Stan Auerbach were particularly effective in this work. Other committee members who made significant contributions include Deborah Barnes, Charles Coutant, Joanne Gailar, and Bill Alexander, whose prior collection of ORNL artifacts was of special interest. And Ann Calhoun certainly deserves recognition for organizing documentary collections and providing excellent secretarial service. The history was published as combined issues 3 and 4, 1992, of the *ORNL Quarterly Review* and later in book form as *Oak Ridge National Laboratory: The First Fifty Years*, by Leland Johnson and Daniel Schaffer (University of Tennessee Press, 1994).

In November of 1992 ORNL invited retirees and others who helped build and operate the graphite reactor to a luncheon in recognition of that remarkable achievement forty-nine years earlier. Unfortunately, Eugene Wigner's health limitations prevented his participation, but approximately fifty people came to a luncheon at the cafeteria and to a ceremony at the reactor. Retirees regaled their peers and laboratory staff with reminiscences about the hectic days of 1942–1945. The luncheon was but a small tribute to the people who designed and built the first nuclear reactor in the world to produce intense neutron sources, to produce small amounts of plutonium for weapons research, and to make possible advanced nuclear research and nuclear medicine. In recognition of the occasion, a plaque designating the reactor as a National Nuclear Historical Landmark was presented by Robert Long, then president of the American Nuclear

Reunion of early staff at the dedication of the graphite reactor as a national nuclear landmark, 1992. ORNL photograph.

Society, to Murray Rosenthal, ORNL Deputy Director. The graphite reactor had made possible the diagnosis and treatment of cancer and other diseases with radioisotopes, beginning with the first shipment of Carbon 14 to Saint Louis on August 2, 1946. Clifford Shull was later awarded the Nobel Prize for his pioneering work with neutron diffraction at the graphite reactor in the 1940s.

I had several less pleasant but quite important tasks to be conducted in parallel with the projects related to history. These were investigative in nature, often to evaluate supposed or real violations of good environmental and management practices. One task, accomplished with Wilbur (Dub) Shults, was to review a program for the packaging and distribution of tritium, a radioisotope used in medicine and industry. (This was totally independent of the DOE program to provide tritium for weapons.) Earlier, DOE had announced a projected major increase in the price of tritium, thus creating a severe demand among our customers to purchase at then-current prices. The ORNL staff had struggled to meet the demand for tritium and in so doing had taken some steps without all of the required documentation. No releases occurred and no one was injured, but the equipment was found to be in need of improvement and costly maintenance, previously requested of DOE but which they never funded. In the end another national facility was found to have more modern equipment, and the program was transferred there.

Another task was to evaluate the condition of heavily shielded and remotely operated "hot cells" used to dismantle experimental devices irradiated in the High-Flux Isotope Reactor (HFIR), and to examine irradiated material samples of scientific interest. This study extended over several years, both to recommend improvements for the facilities and equipment and to evaluate the potential future workload that might justify such expenditures. Assistants in these extensive studies included Paul Arakawa, Everett Bloom, Charles Devore, Mike Kania, Martin Grossbeck, Alan Krichinsky, Robin Taylor, Ken Thoms, and Lloyd Taylor. The needed cells were cleaned and rebuilt to modern standards in accord with committee recommendations, but the program, though viable, has continued to suffer limitations imposed by the national decline of nuclear research programs.

Reluctantly, I became chairman of the Reactor Operations Review Committee that has provided safety oversight for the ORNL reactors. I preferred serving as advisor to a younger chairman, but the credibility established by my earlier investigative work proved determinative. This committee has served as a model for similar committees at many other institutions having nuclear operations, including the oversight bodies of the Materials Test Reactor and other facilities at the Idaho National Engineering Laboratory. Because the committee staff is rotated, space does not permit mention of all of them, but Mike Harrington, Dave Moses, Howard Kerr, Bob Kryter, and Libby Johnson served as able leaders of subcommittees, one for each of the reactors. Beecher Briggs, an ORNL retiree, also assisted. In addition to conducting annual reviews and special studies, our committee developed a new set of standards and guidelines for its function to assure thoroughness in reviews. Even with the programmatic disappointments associated with the permanent shutdown of all but HFIR, I was again rewarded in reviewing the reactor facilities that had been maintained well by their dedicated and responsive staff members. (Several shut-down reactors also were reviewed in order to be certain that they were properly secured and safe.)

During my later years at ORNL, I developed a new organization called Friends of Oak Ridge National Laboratory. Its purpose is to further the support of good science and engineering with emphasis on programs appropriate to ORNL. This nonprofit organization is independent of the Laboratory and may review ORNL programs.

Celebrating fifty years of service, September 1992: (left to right) Alvin Trivelpiece, ORNL Director; Don Trauger; Clyde Hopkins, President, Martin Marietta Energy Systems; Elaine Trauger. ORNL photograph.

Members are expected to be cognizant of national activities in science and engineering. Where needed, "white papers" may be written to advise congressional and administration members on technical issues. The Friends have programs to promote effective science writing for the public and technical press and to bring prominent technical speakers to Oak Ridge symposiums.

The Friends of ORNL may also take initiative to broaden the interaction between the research needs of industry and Laboratory programmatic capabilities. One activity that originated with the organization is directed toward interaction between pharmaceutical companies and the genomics research at ORNL, utilizing resources centered around the large colony of lab mice. I have continued active participation in the Friends organization, but I soon turned over the leadership to others. At last count it was approaching two hundred members, many of whom are retirees of ORNL.

Other late activities included initial planning for a new type of open house for ORNL termed "Community Day." Its objectives are to inform the public of Laboratory programs, to provide opportunity to see the

Recognition of Don's retirement: (left to right) Byron, Aleta, Elaine, Don, and Tom, 1993. ORNL photograph.

facilities, and to display new developments in exhibits shown at an annual event. Guests, often whole families, had previously been permitted visits only on a few special occasions, and the impression of wartime secrecy remained prevalent among the public, even though most ORNL programs had not carried security classification for decades. I retired before the first event, passing the plans to Tom Row, my successor in this and other activities. Community Day has been a great success.

The year 1992 included memorable family events as well as work. Elaine and I took a vacation trip to Costa Rica with a group of Oak Ridge friends led by Dan Robbins. We had never been bird watchers and were skeptical of our decision to participate once we learned that birding was a focus of the trip. As it turned out, however, that small country is blessed with such a huge assortment of beautiful birds that we declared a new interest in ornithology. I thrilled at seeing the unimaginable quetzal and at both hearing and seeing the three-wattled bell bird.

We later took a vacation trip to the Adirondack Mountains of New York State. After passing age sixty-five, I had resolved to enjoy all annual vacation time, as well as that accumulated in prior years when duties, or perhaps my poor planning, had precluded utilization of many vacation weeks. Although accrued vacation could be converted to money at retirement, I chose to take the days of relaxation, adventure, and travel.

In May we made several trips to Nashville to become acquainted with our new granddaughter, Katherine Aleta Trauger, who had joined Byron and Aleta's family on the date of her birth, April 26, 1992. In June, Nebraska Wesleyan University recognized the class of 1942 at the annual alumni banquet and related events. I was asked to give the banquet address and chose to reminisce, mostly about humorous events during our four years at NWU. I closed my remarks with the following, though slightly abbreviated, words:

> When we were in college, the world was facing one of its greatest threats: World War II was a carnage the U.S. had entered through the surprise attack of Pearl Harbor. We came through our crises rather well. The war was won. We also won the immediate postwar period through the Marshall Plan and the wisdom of many leaders in establishing a better world—a world free of global conflict, now for an unusually long time.

Former secretaries assembled at retirement party: (left to right) Betty Jeffers, Jackie Sims, Lois McGinnis, Anne Calhoun, June Zachary, Don Trauger, Ruth Hale, Kim Peppers, and Carol Johnson. ORNL photograph.

However, the students of today face equally serious challenges. We are troubled by a massive national debt. We are troubled by gross environmental issues. We are troubled by many small conflicts around the world. These problems again seem as overwhelming as they were at the time we graduated from Nebraska Wesleyan.

Today, the college seems poised to offer a foundation of knowledge and skills that exceeds its achievements of the past. It is my hope and wish that future students will rise to their challenge as we did to ours. But they must do better, not creating so many insults to the environment, or frustrations to segments of our people. We are leaving a legacy that troubles many of us very seriously.

We can challenge our successors to respond to these issues. We also must invite them to call on us to help in every way possible. It is a responsibility that each of us should feel as we complete our careers and see ourselves through to whatever remaining years we may be privileged to enjoy.

As Elaine and I returned from spending the Labor Day weekend at the farm in North Carolina and making plans for harvesting and replanting timber, we engaged in serious conversation about our future. She was properly concerned about her increasing arthritic problems. The ever-encountered stairs of the split-level house that we had occupied for over thirty-nine years were becoming a source of pain for her. We agreed that moving to more appropriate accommodations was necessary, and I asked her to see what might be available.

By the following Friday evening Elaine had a list of possibilities, all but one of which I rejected for various reasons. The last possibility was a new structure on a street that neither of us had visited. The house is in a beautiful setting on an inlet of the Melton Hill Lake. It had been built on speculation by Jane Marquiss, a builder we admired. By Saturday evening we had taken an option to purchase the house at 20 Palisades Parkway.

Our sons had chosen to relive the days of their youth with their families during the festivities of the Oak Ridge fiftieth-year celebration. My brother, Bob, his wife, Shirley, and Elaine's sister, Eloise Butler, also came to join in the celebration. We dutifully watched the parade, which was well planned if not spectacular, and visited the ORNL exhibits that I and others had labored to make available.

That afternoon Elaine and I invited our guests to follow our car for a surprise; we had not told them of our plan to move. At the new house Tom noted that I had a door key and asked, "Are you thinking of buying this?" It was a logical question about a retirement house that was larger than the one we were leaving. I replied, "Yes, perhaps it is crazy for old folks like us to buy a large house like this, and if you think that you should have us committed, we will go quietly." Fortunately, they liked the house and approved.

On September 1, 1992, I passed the fiftieth year of continuous employment, and ORNL took note with a small ceremony in the director's office. I was to finish my career at the Laboratory officially on January 31, 1993, my last day of employment. The autumn and early winter were, therefore, a period of intense activity. We were to move in November when minor modifications to the house were completed. It was not possible to accomplish all that I wanted to do that fall, but the Laboratory assisted by approving a consulting contract by which some tasks could be completed later. The remaining work in completing the fifty-year history was so pleasant that I chose to do that without compensation. On Friday, January 29, 1993, I exchanged my employee badge numbered 1014 (badges had been renumbered a few years earlier starting at 1000 in the order of one's hiring date), then the lowest at ORNL, for number 625050, the highest as of that day.

My retirement party occurred on February 3, 1993. It had not been possible to schedule it earlier owing to the press of activities. I had worried for years that a retirement event might leave me embarrassed by my display of emotion; although the feelings were intense, the time was right to retire and there were no problems. Howard Kerr and a committee had chosen appropriately embarrassing reminders from the more than fifty years and were remarkably kind in other presentations. It was great to see the many retirees, friends for decades, who came for the occasion. I was also especially pleased to greet several former secretaries, all of whom had worked hard to meet demanding requirements. They even laughed, perhaps after crying, when I reminded them of the many revisions they had typed for every paper and report I had written. Thus closed an employment career in energy technology spanning more than fifty years.

Clockwise from top left: photographs courtesy of www.arttoday.com, TVA, TVA, and Donald Trauger.

TWELVE

Energy Futures

T his final chapter deviates from an emphasis on the order of past to future by discussing available and pursuable options for maintaining acceptable lifestyles consistent with foreseeable energy resources. I have assumed that current technologies will continue to expand through the next one-hundred-year period but without a new technological revolution comparable to the introduction of steam power, the invention of the internal combustion engine, or the discovery of nuclear energy. The figures of this chapter are from a document entitled *Technology Opportunities to Reduce Greenhouse Gas Emissions, Prepared by National Laboratory Directors for the U.S. Department of Energy,* 1997.

Because the focus of this chapter is on the availability and utilization of energy systems and resources as we progress through this new century, I begin with some trepidation, knowing that today's seemingly sound predictions often look foolish in the unforgiving glare of arrived tomorrows. In this context I often think of an article in an old encyclopedia of my boyhood home that predicted future desert travel. The traditional camel caravan was to be replaced by a vehicle consisting of a metal box carried on large-diameter, wide-rim wheels. This staggering machine of the bold imagination would easily carry goods and people across sand dunes. In the time of that writing, neither air transport nor track-tread vehicles were envisioned as practical. Although camels have largely been replaced, the slow-moving wheeled vehicle carrying an unair-conditioned container to transport goods and passengers looks foolish. In a way this chapter is a search for the camels as well as the unrealistic projections bound up together in our present usage and understanding of energy resources and an effort to look forward with reality.

These projections emphasize the importance of energy in supporting and extending the lifestyle now enjoyed by most people in the United States. Indigenous sources and uses are emphasized, although attention also is given to other countries and regions, particularly those that provide us with petroleum. In terms of energy we are no longer isolated, as was often envisioned in earlier pioneer days. We must focus honestly and realistically on the availability and utilization of energy systems and resources as we progress through this century.

Evaluating subjects for future ORNL missions was part of my endeavor for three decades, at a time when development of energy systems and assessment of potential resources were subjects deemed vital to the United States and to world stability. The availability of water is also critical, a matter fully capable of generating problems comparable to those potentially erupting from issues over energy. Fortunately, our country is blessed with adequate water supplies for some time to come, except in parts of the West and Southwest. However, on a global scale the human population explosion underlies potential tensions concerning the availability of energy, water, and, of course, land. We no longer have large, sparsely populated, and fertile unused areas like the U.S. Midwest of my grandmother's era that attracted pioneers, new communities, and a belief in a limitless future of development. Increasing population density, even in the United States, has created local difficulties in the distribution of water and the use of energy systems.

Avoidance of world conflict over shortages of energy and water resources through technological and political solutions is as important as adequate military defenses and is surely less costly than fighting. Since population and water are outside my area of expertise, they are left for others to address. Energy, the subject of this chapter, can indeed be considered the lifeblood of modern civilization. The availability and utilization of energy have been major factors in shaping the world's nation states and in determining their status.

———•◦•———

The United States has been endowed with a multitude of large energy resources. In order of historical exploitation, these resources have been, first, biological, as wood for fuel and grass for animal energy; then fossil sources, including coal, oil, and natural gas; and, finally, nuclear sources. Utilization of water power also has progressed throughout our country's history: from powering small mills that ground local feed, to commanding factory machinery to spin the cloth for a nation, to energizing massive turbine-electric generators that power our whole way of life by the flow of waters from behind dams that remain some of the largest structures built by any culture. The integration of water storage and power generation into energy systems such as those of the Tennessee Valley Authority also minimizes flood damage and provides for water transportation and recreation. Exploitation of water resources for energy production now seems near its maximum when other demands for water and land use are considered. These U.S. energy resources have enabled building an immense industrial capability reflected in both military strength and in an envied quality of life for most Americans.

Even though the industrial revolution had begun some two hundred years earlier, the buildings and basic energy system of our farm in my youth were not greatly different from those of previous centuries. Although our gasoline engines for threshing, sawing wood, and washing clothes were great conveniences, horse power still tilled the ground and moved products to market. In some farming communities horses even activated treadmills or powered other machines to provide energy.

All of this has changed since World War II. With large and highly mechanized farms and machines powered by petroleum, a single Midwestern farmer can till about eight times as much land as was possible a century earlier. With improved seeds, fertilizers, herbicides, and more efficient machinery, crop yields have increased so that the U.S. produces a disproportionate amount of the world's agricultural products. This both feeds people and contributes positively to our balance of trade with other countries. At the same time, life has been made more comfortable by central heat and air-conditioning, motor vehicle transportation, and economic production of goods and services. Each of these advances has come in some measure through a greater energy expenditure and a corresponding depletion of energy resources.

The effective use and future availability of energy involve very complex relationships among the environment, natural resources, and economic forces and are dependent on the dedication of people, industry, and government to the task. Energy systems and their utilization must change as resources are depleted, environmental consequences are more fully understood, and new technology is developed. The depletion factor is somewhat predictable, but managing environmental consequences and developing new technology depend substantially on the quality and intensity of applied effort. That is one reason why transition from one energy source to another usually requires a long period for development and implementation.

My examination of future energy systems emphasizes technology, but it must be recognized that social and political policies can also affect choices among energy systems. For example, gasoline taxes in Europe encourage the use of smaller private vehicles, discourage new road construction, and fund both the construction and utilization of public transportation systems. Thus far, such measures have not met with favor in the United States. I speculate sparingly on the political and social contexts in which energy technology must function because I lack expertise in those important fields. My personal concern is for the energy futures of our grandchildren and of those whom they may come to love and cherish. If they, their compatriots, and direct successors are to be comfortable, we must look ahead at least one hundred years for the energy systems needed to fill their requirements.

The future prospects are uncertain. It is no great exaggeration to say that the United States consumes its energy resources in inverse proportion to their domestic availability. The resource of which we have the least is precisely the one we use the most—oil. At our current consumption rate, and without imports, our domestic oil reserves would be severely depleted in about a dozen years. More troubling, we have essentially no prospect of finding new large domestic oil deposits, though not through want of searching. In 1973, when ORNL was considering locations for a radioactive waste repository in Kansas, I was surprised to learn that more than forty thousand holes had been bored there in exploration for oil and natural gas. States prominent in oil production have been penetrated even more. Reserves of natural gas will also be depleted well within our posited time frame. Such projections are somewhat dependent

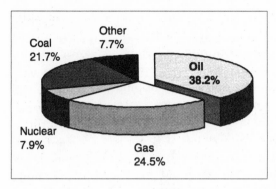

Coal 21.7%

Other 7.7%

Oil 38.2%

Nuclear 7.9%

Gas 24.5%

Fossil fuels account for most of the energy used in the United States.

on pricing; less accessible resources will be exploited, but at great cost to the consumer.

New technologies and improved practices can enable continuation of a comfortable lifestyle as present energy sources become more scarce or expensive, but their implementation will not be a simple undertaking. Aggressive energy conservation measures, use of solar and other renewable energies, and advanced nuclear reactors can contribute. Such development for the longer term is not likely to receive timely support by a highly competitive industry enjoying minimal regulation. A long-term viewpoint seldom seems important to industrial leaders if it extends beyond their tenure. Thus, many recommendations in this chapter imply or explicitly require government financial support for research and development.

Once new technologies are adequately developed, demonstrated, and deployed, the dynamic U.S. industrial complex should function well with more diversified energy sources. The current trend toward cooperation between government laboratories and private industry can accelerate the transition from invention to utilization. As the complexities of new energy systems refine toward practical, real-world use, the importance of innovation, evaluation, careful implementation, and regulation where required can hardly be overstated.

In order of utilization today, United States energy resources are fossil fuels, nuclear-electric, hydroelectric, and nothing else of consequence. For the future, biomass, geothermal, solar electric, solar heating, and wind also have potential. Energy conservation, including enhanced energy efficiency, is economical and attractive for the short term. Its effect can also be long-term, because conserving practices, when built into the social structure, may continue indefinitely. Conservation, however, cannot replace a depleted primary energy source, nor can it provide adequate energy for an increasing population or for new demands. Even with these limitations, energy conservation offers an important option for a more comfortable energy position.

Energy conservation reduces demand; if less energy is needed to accomplish tasks, the resources are effectively greater. Although the oil shortages of the 1970s shifted the national orientation to a more balanced program between resource exploitation and effective energy utilization, a plateau seems to have been reached. As present energy resources are depleted and costs escalate, the incentives for energy conservation will increase. Early implementation of operative efficiency measures is justified because each step both reduces overall costs quickly and helps protect against future eras of energy shortage. However, replacing energy-inefficient appliances, buildings, and industrial equipment before their normal end of life is not often economically attractive. But whether a household freezer or an industrial furnace, energy-dependent systems must be replaced periodically because of deterioration or to meet changing needs. Buildings of the future, for instance, can be far more energy efficient, as well as attractive and comfortable, through application of present and new technological developments.

Conservation of oil should have a high priority as a measure to extend limited reserves. Thus, the automobile, as the greatest user, should be targeted for application of energy savings. Campaigns of the 1970s to promote car-pooling by use of special traffic-limited driving lanes, the 55 mph speed limit, emphasis on smaller vehicles, and substitution of public transportation all first flourished, then faded in importance. Increases in speed limits for major highways to 65 then 70 mph apparently receive public approval, frequently made painfully obvious by how often even the higher limits are exceeded. Increased popularity of larger and specialized vehicles with lower gasoline mileage has further defeated energy conservation goals. Use of long-distance shipment of goods by truck in preference to more energy-efficient rail carriers remains high, although rail freight traffic has increased in recent years.

Urban sprawl that requires expanding automobile travel continues to increase oil consumption; only recently have many local political entities recognized the impact on transportation of our widely distributed housing and industries. This chaotic distribution makes difficult, or perhaps precludes, funding and installation of efficient public transport systems such as those of Europe and Japan. Obviously, zoning regulations designed in part for energy efficiency are a social/government option for long-term energy conservation, but that potential is highly dependent on public awareness and cooperation.

Vehicle technologists are searching for more efficient motive systems, including those using different resources such as natural gas or electricity. Major automobile companies and national laboratories are now linked by a limited liability company, US-Car, to develop more energy efficient vehicles. Prototypic designs feature new engines and fuels, lighter, yet strong, bodies and frames, as well as further streamlining. The goals include highway mileage up to eighty miles per gallon for mid-sized automobiles. Progress has been encouraging. These achievements will also reduce the "greenhouse" effect and global warming. Even so, our travel habits and the spreading of suburban residences, businesses, and factories in wide areas along highways continue to force the use of automobiles instead of public transportation. An intensified search for improved efficiency and new energy sources alone may not be sufficient to win the race with increasing population and personal demands.

In contrast to the public reaction in transportation, energy conservation through recycling of household items such as glass, metals, and paper has received a steadily improving response. That may be surprising since this does not effect a direct saving to the household participant and not always to the business community. However, consistent reuse of containers, aluminum foil, and other products within the household does produce savings. I am reminded of Benjamin Franklin's little rhyme about economics: "Use it up; wear it out; make it do; or do without." It has merit in energy conservation as well.

Industry frequently has objected to the complications of recycling compared to using metallic ores, fresh water, forest trees, and other resources. A few years ago I conducted an informal survey of experience in using recycled materials with company representatives at a meeting of the National League of Cities. Most reported that experience had been favorable, even though they were skeptical at the outset. Industrial action in conservation has been quite effective where it offers an early monetary reward, as with the strengthening and thereby thinning of plastic bags and metallic containers. Improved technologies for the use of plastics and other refuse for energy production seem predictable but will be limited by the logistics of collection and the chemical complexities in furnaces and other equipment. Although recycling remains a small factor in the energy economy, it is growing and offers an increasing potential.

Energy for heating, cooling, and lighting of buildings is important. Insulation products and their use for housing and other buildings have been improved through new technologies and standards derived from research at national laboratories, universities, and industries. Improved insulation together with potential improvements in the devices that provide heating, cooling, and lighting can reduce energy costs substantially when applied in new buildings or by retrofitting older facilities. Some electric utilities offer customers assistance with energy conservation to reduce demand to avoid capital expenditure for new capacity.

The example of our former home in Oak Ridge illustrates some of these factors. The house was solidly built but poorly insulated when constructed under a government contract in 1952. It featured a layer of fiberboard sheathing as wall insulation, single-layer glass windows with aluminum frames, and little attic insulation. Following the sale of Oak Ridge properties to occupants in 1957, we sought measures to improve the energy efficiency. First, an addition built to double the size of the small house was well insulated except for yielding to aesthetics by retaining the design of the window frames. Over the years we replaced the oil furnace with an electric heat pump (with air-conditioning), added insulation wherever practical, and installed double-pane glass in windows. With these improvements, the energy cost per unit area of the house was held nearly constant as fuel and electric prices escalated through the 1970s decade. Precisely determining the ratio of savings over costs is a difficult matter, but I estimate it was about 5:1 over a period of fifteen years. These improvements are now benefiting the present owners and may have helped us sell the house.

Beyond such individual initiatives for reducing energy use in buildings, industry can be expected to develop and employ improved heating, cooling, and ventilation systems. These improvements will derive in considerable measure from research programs such as those of ORNL described in chapter 10. Of course, innovations such as insulation for footings and ground-slab floors cannot be utilized until the building is replaced. A recent document produced cooperatively by five DOE national laboratories indicates that residential homes should last fifty years and commercial buildings one hundred for purposes of estimating and planning their energy requirements. The Building Research Center at ORNL, also mentioned in chapter 10, is providing a database for substantial improvements in building design, insulation, weatherproofing, materials, and mechanical systems. There is much more to be accomplished in the field of energy efficiency through combinations of research and industrial development.

Although overall cost savings should encourage such improvements, lethargy and inconvenience during modification may inhibit or delay desirable actions. The advantages of energy savings need greater publicity to encourage owners of homes, commercial establishments, and industry to investigate the potential for their facilities. The Department of Energy should be applauded for maintaining a technology program in energy efficiency. A new and continuing national emphasis is needed for energy conservation at a level comparable to that which followed the oil price-shock of 1973.

People have difficulty conceiving of depleted oil reserves when gasoline is readily available at acceptable prices in conveniently located service stations. A general misconception about U.S. oil reserves grows from a poor understanding of the Alaska Prudhoe Bay discovery. Those reserves, though substantial, were equivalent to about three years of our total oil consumption. The Arctic National Wildlife Refuge may offer a somewhat larger source, but if drilling does eventuate in that wilderness the results will have little effect on the size of the problem. Perhaps we should leave that fragile environment undisturbed as a last resort in the event of major world events that might justify its exploitation. Some additional deposits will continue to be found, but the oil discovered per foot of drilling has steadily declined in recent years. New technologies, such as slant drilling, chemical additives, and water flooding have increased the recovery from new and existing wells. Even so, more than 50 percent of the crude oil now used in this country is imported; that fraction may rise to 70 percent by the year 2010. This obviously represents a drain on our international balance of trade and a continued military and diplomatic cost for protecting the foreign supplies.

The precariousness of our oil supply is indicated by the need for the U.S. government's costly Strategic Petroleum Reserve. A few years ago I asked the person in charge of the reserve why it was necessary; could we not simply pump the oil more rapidly during an emergency? He replied that most wells were being pumped at an optimum rate and that such increased pumping would cause them to diminish or fail. Actually, the depletion level of many wells so limits the allowable pumping rate that lower oil prices also make them uneconomical to operate. Perhaps equally significant in the long run is the leverage that groups such as the Organization of Petroleum Exporting Countries (OPEC) can exert by limiting the supply or controlling prices to be just below the cost of a newly developed substitute energy source.

Other indigenous petroleum resources can be derived from tar found in sands and those contained in some shale formations. Tar sands offer a large source that is difficult to exploit. Heating the sand is necessary to liquefy the tar to an oil-like substance for extraction, pumping, and conversion to useful liquid fuels. Large amounts of heat energy would be required to warm the deep geological structures in which U.S. tar sands exist. Little experience is available for assessment of costs and environmental impacts of tar extraction and conversion to fuels. However, the cost probably precludes burning the recovered product for use in the extraction process, and, therefore, nuclear energy might be the only practical heat source. Obviously, it would be necessary to compare the advantages of this oil source with the cost of other heat energy used for extraction. Some Canadian deposits are in sufficiently shallow beds so that the oil could be extracted through above-ground operations. Waste products from surface operations, however, may be a greater problem than for in-situ extraction. Either option may have formidable environmental problems that should be carefully evaluated. For these reasons oil from tar sands may have potential for exploitation only late in our hundred-year time frame.

In contrast to tar sands, oil-containing shales have received considerable study since their discovery when early Western pioneers were surprised that rocks placed around their camp fires sometimes caught fire. Initially, the shale's location near the earth's surface in Wyoming and the relative ease of extracting oil from it made the shale an attractive source. Demonstration pilot plants have been built and operated, but the costs have not been competitive. This process has other disadvantages. The resulting debris from above-ground retorts is

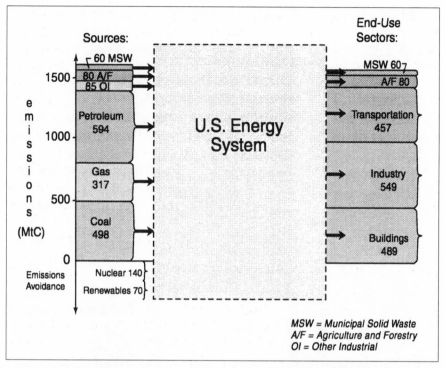

Overview of the sources of carbon emissions in the United States in 1995 (in million tonnes equivalent and including CH4 from MSW, A/F, and OI).

environmentally dangerous and very difficult to stabilize. In-situ recovery, although less complex than for tar sands, is uncertain with respect to control and prevention of poisonous seepage to water systems. The experience to date does not justify reliance on this resource during the time frame of this evaluation, even though oil-containing shales represent a large potential resource.

U.S. coal resources are large and could last for several hundred years at the present rate of use. Coal has its own disadvantages, however. As a solid it is more difficult to transport, purify, and use than natural gas or liquid fuels, and in burning it produces undesirable combustion products. Coal ores are comprised of carbon, hydrogen, sulfur, small fractions of metallic impurities, oxygen, and nitrogen, with the carbon content varying from 80 to 90 percent. Since the hydrogen and oxygen content are found largely as water, the energy from combustion of coal derives primarily from the reaction of carbon with oxygen of the air to form carbon dioxide, which raises concerns for global warming. I recall, even from the high school physics class, that the natural quantity of carbon dioxide in the air allows ultraviolet rays from the sun to pass through, but reflects longer wavelength radiation from the earth to reduce energy escape. This mechanism keeps the planet warm enough for life. (The effect of carbon dioxide gas is analogous to that of glass in the sun's warming of an automobile or greenhouse, hence the terms *greenhouse effect* or *greenhouse gas*. It is interesting to note that on Venus the atmosphere is about 95 percent carbon dioxide, and the surface temperature is 900 degrees Fahrenheit.)

Because earth contains many large and widely distributed deposits of coal, the potential exists for greatly increased levels of carbon dioxide through the burning of coal and, consequently, a dramatic

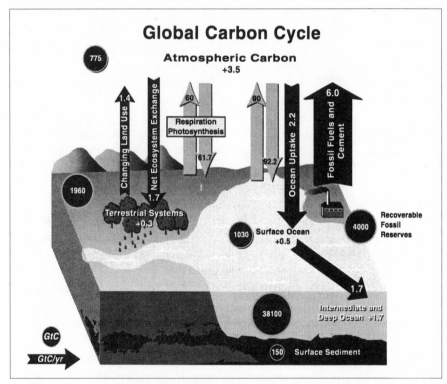

Global carbon cycle. The burning of fossil fuels and changing land use have resulted in human-induced alterations of the global carbon cycle. The solid arrows in this diagram indicate the average magnitude of perturbation in carbon fluxes and the fate of carbon resulting from these activities, averaged for the first half of the 1900s. Courtesy of Technology Opportunities to Reduce U.S. Greenhouse Gas Emissions, copyright 1997 by National Laboratory Directors for the U.S. DOE.

increase in the earth's atmospheric temperatures. Atmospheric carbon dioxide has increased steadily since the middle of the 1800s, when coal first became a major source of energy. Its concentration was 0.029 percent in the early 1900s and is now 0.033 percent. And because other gases such as methane, even more than carbon dioxide, also function in the atmosphere to retain heat, their release must be controlled also.

How then should we view the inherent problem of greenhouse-gas generation and global warming from burning coal? Other factors enter in. For example, the oceans, through absorption, contain large quantities of carbon dioxide at dilute levels that may increase with a higher atmospheric source and serve as an ameliorating effect. In contrast, a small degree of global warming might cause the oceans to release carbon dioxide, as occurs with warm carbonated beverages. Conceivably, this could prove uncontrollable with changes in ocean temperature. Sequestration also occurs on land. Forests consume carbon dioxide in photosynthesis while trees are growing and release it when wood products are burned or decay. Mature forests with much decaying surface debris may produce only a balance between consumption and release. The destruction of major forests for immediate economic benefit or to accommodate food production is an important concern because burning releases their entire inventory of carbon and their absence removes the photosynthetic absorption of carbon dioxide.

The balances between these several large factors remain as major uncertainties in understanding the role of carbon dioxide in the global atmospheric system. An extensive monitoring system is planned by DOE and NASA to measure small changes in carbon concentrations in locations thought to be sources or sinks. Since the warming effect of small annual increments of greenhouse gases is small, controversy swirls between those who would be cautious and those who would burn coal.

All this points to the difficulty of determining the extent to which the increase of greenhouse gases to date has caused a rise in global temperature. Conventional readings of temperatures collected globally are suggestions, but uncertainties in measurement and analysis are also significant. However, recently improved satellite measurements of earth temperatures have detected worldwide warming. Retreat of ice sheets in Arctic regions and local warming of inlets also may relate to or confirm global warming. Small changes in atmospheric temperatures are known to cause instabilities, such as the recent increases in the number of major floods and severe storms.

Coal, with its high production of carbon dioxide per unit of heat produced, is of greatest concern because of its global abundance and widespread use. However, combustion of oil and natural gas, respectively, produce 80 percent and 60 percent as much carbon dioxide per unit of energy content as does coal, so they also are substantial contributors. Projected problems that could result from continued and increased use of fossil fuels include severe climate changes in areas of temperate zone food production. Also, melting of polar ice caps could produce an increase in ocean levels and inundation of coastlines. In some worst-case scenarios, New York City and other low-lying coastal cities could be flooded.

The problem is that the temperature of the earth seems to be delicately balanced. If an increasing temperature and the resulting effects are clearly proven unacceptable with a future society highly dependent on energy sources that produce carbon dioxide, preventing serious or disastrous consequences may prove very difficult indeed; at least a Herculean effort would be required to make rapid changes in global energy production to avoid a substantial impact on world economies. Conversion from one major energy production system to another, if pursued as rational policy, should involve assessment of the environmental and economic impacts and would require major investments of resources, effort, and time. Thus, a prudent course would be to chart methods of limiting the burning of coal, and to a lesser extent the burning of oil and natural gas as soon as practical.

Reducing the need to burn fossil fuels would also alleviate problems with acid rain and other adverse environmental effects, such as those produced from the mineral impurities of coal and the accidents suffered in its mining and transportation. From a global perspective, this objective is complicated by disparities in the economies and resources of nations. Perhaps developed nations can take the lead by changing first to less polluting but more technologically demanding energy systems. Global agreements such as the effort in Rio de Janeiro and the conferences in Kyoto and Brussels may help to lessen the effect of national disparities by the use of provisions such as carbon-release quotas and credits and markets for emission allowances. However, the atmospheric problems are not likely to be solved without changing to more benign systems of energy production.

There are many beneficial steps available in the United States to minimize our negative impact on what is increasingly a global problem. For example, coal from western mines could be transported to eastern markets in trains drawn by electric rather than diesel locomotives. Although this is a relatively minor use of oil, other developed countries power their trains by electricity often produced in nuclear-electric plants. As mentioned earlier, legislative measures such as selected taxes, reduction of auto travel by zoning of residential and industrial areas, and support of public transportation are possible, but I will discuss only technological measures.

The use of hydrogen to convert highly carbonaceous fuels to other, more environmentally friendly forms is a well-studied technology with a number of advantages. Conversion of coal to oil is possible but requires the production of hydrogen and is made difficult by the carcinogens produced at intermediate stages of the conversion. Handling and transporting coal as a liquid would be far more efficient and, in addition, would provide another source of transportation fuel. But the conversion would not reduce carbon dioxide production. Also, such processes must be compared to the advantages of hydrogen itself as a fuel.

Hydrogen is attractive as a direct source of energy for many applications and may become very important if the hydrogen can be produced economically and used safely. However, adoption of hydrogen for general use would require a somewhat difficult infrastructure for safe and economical distribution.

The immense quantities of hydrogen in the water of oceans make it attractive, and since it is renewed when reacted with oxygen to produce heat and again water, there is an unlimited supply. Also, there are no environmental impacts from its direct use as a fuel, except for hazards in the use of this very reactive, even explosive, chemical. Moreover, water is a remarkably stable compound and thus difficult and costly to separate into its components, oxygen and hydrogen. Extraction of hydrogen from cellulose and other components of biomass materials could become attractive. Enzymes have been found that convert these materials successively into sugars and hydrogen. Research in this biological field is at an early stage, the high cost of enzymes is a deterrent, and land-use constraints for biomass as fuel also would apply. However, hydrogen is so attractive as a fuel or fuel component that research and development should be encouraged. Although present processes are not practical except for very advantageous applications, achievement of hydrogen production is a laudable goal. For example, hydrogen used in efficient fuel cells could become an attractive, local source of electricity for homes and businesses and for electric automobiles. Since, however, fundamental invention and extensive development would be required before hydrogen could be used safely and widely as a fuel, this possibility is deemed to lie late in our time frame.

Another potentially helpful technology is sequestration of carbon dioxide, which requires the separation of carbon dioxide from other gases in the large-volume effluents passing up the stacks of typical one thousand megawatt electric-generating plants. Although achievable by known chemical and mechanical means, implementation would necessarily involve increased capital and operating costs. Once separated, several options have been proposed for sequestration of carbon dioxide so that its entry to the atmosphere would be restricted. The carbon dioxide might be disposed of as a gas pumped into the oceans, as a liquid pumped into abandoned natural gas or oil wells, or as a solid, calcium carbonate, by burial. Each has its problems or potential limitations because of the large volumes of material to be handled. Note, for example, that a one thousand megawatt power plant burns about one hundred rail carloads of coal per day.

Each sequestration option also would have limitations that are not readily known today and would require demonstration. Early development of the technologies for separation and emplacement may be justified, but it seems premature to attempt a comprehensive demonstration. Studies to determine the capability of oceans to contain carbon dioxide, although a seemingly difficult undertaking, would have value in better understanding the role of the oceans in the greenhouse cycle. Although a complex and expensive technology to implement, sequestration of carbon dioxide from fossil fuel combustion products needs careful consideration. The greatest incentive for sequestration derives from the large coal resources of the United States. Sequestration's position in this century, therefore, is more a question of need and economics than of invention.

Turning next to natural gas, a 1992 study by the National Petroleum Council, conducted at the request of the Department of Energy, presented a comprehensive assessment of U.S. gas resources. It took into account known reserves and resources, such as gas that resides in "tight sands," even though not economically extractable by then-current methods. (Tight sands are rock-like or sandstone structures that require fracturing to release the trapped gas.) The evaluation also assumed that advanced processes for recovery would be found for other geological formations not then exploited. The report predicted a sixty-five-year supply of natural gas assuming continuation of the 1990 rate of consumption. It recommended that new and expanded uses of natural gas should be employed, including increased burning in plants for electric power generation. Other studies by the Office of Technology Assessment and the Gas Research Institute projected smaller resources lasting approximately fifty years. Recently improved methods for exploration and drilling have validated some of the optimistic projections of those estimates, but these are very brief terms indeed for predictions of severe depletion of a vital natural resource. Price increases will spur exploration and new methods of recovery, but the total domestic supply is limited by what nature has stored, mostly underground.

Even so, increased uses of natural gas for urgent economic or environmental considerations may be justified for the short term, such as to reduce local atmospheric pollution from burning coal until another replacement is possible. However, natural gas is so convenient for home heating, industrial uses, and as a source of chemicals that it should not be burned extensively in stationary, electric power generating plants. To indicate the potential impact of such continued use in the United States: electric energy production from coal and the overall use of natural gas are approximately equal; thus, squandering natural gas instead of coal in power plants to meet future needs for electricity would reduce its depletion time to perhaps thirty years. If natural gas energy were to replace present nuclear-electric power plants, the depletion time would be shortened by an additional six years. Concern over this use should not be dismissed lightly because the suppliers of electrical power chose natural-gas-burning turbines for essentially all new plant additions during the 1990s.

Importation of natural gas is technologically feasible but, for overseas sources, requires refrigerated and pressurized tankers having significant accident risks that may be greater than those for oil transport. Processes for chemical conversion of natural gas to liquid fuels by adding coal, water, and heat have been tested and are projected to achieve production at costs approaching those of conventional oil. Although widespread use of natural gas as liquid oil seems unwise to me, that liquid form may prove practical for transporting the energy of natural gas from Alaskan (and foreign) sources to "lower 48" markets. This could be simpler than transport by refrigerated tankers or by construction and operation of exceedingly long, large diameter pipelines from Alaska. The supply of oil would be extended at the expense of natural gas reserves. This could be quite important to Alaska, but the value of those oil and gas resources on a national scale appears comparable to that of the Prudhoe Bay reserves, thus offering only a modest deferral of oil depletion.

A large source of natural gas is known to be contained in clathrates. These lattice-like structures are found in deep ocean beds, primarily of the Arctic. The gas is largely methane, an excellent source of fuel. However, the clathrates are unstable and disintegrate to release the methane if either the high pressure or intense cold is relieved. Methane is about twenty times more effective than carbon dioxide in greenhouse heating, and, so, the risk appears great for accidentally introducing large quantities of greenhouse gas into the atmosphere from necessary or accidental disturbances in mining. Also, if global warming of the oceans should reach a level to cause release of methane, the effect could reach "runaway" proportions.

The complex operations required for extraction of gas from the clathrate deposits, some of which are dilute in sand, would require much study, careful experimentation, and testing before proceeding to production. The cost of recovery also appears quite high when considering the undersea locations, high pressures, Arctic cold and the measures required to prevent inadvertent disintegration and methane release. This seems unlikely to become a practical source of burnable gas during this hundred-year time frame for evaluation, but the potential magnitude of this resource justifies long-term research.

Electrical-energy delivery systems are monopolies, since it is seldom economical or desirable to extend more than one set of wires to a consumer. However, a transition from regulated monopolies that both produce and distribute power to a more economically competitive system is being considered in the Congress and is expanding within the industry. This requires separating the generation function from the delivery of electricity. The objective is to increase competition among generating companies by offering customers more choices and thus to reduce the cost of electricity to the ultimate users, eliminating or at least minimizing the need for price regulation of energy production. These conditions, coupled with fair and open access to the transmission grid, could facilitate the construction of larger and more centralized units for generation, which in turn might increase the potential for using more complex energy sources such as nuclear. Smaller generating companies have found it difficult to staff and adequately train for more technologically demanding energy sources.

However, under the conditions of deregulation and to the extent that energy suppliers become more competitive, the generating companies will either tend or be forced to choose the lowest-cost source for prime energy. That could discourage or preclude introduction of new energy sources that have long-term benefits but are costly to introduce. Under rate regulation the responsible authority could evaluate the extended value of producing a new energy source and allow inclusion of the required costs in the rate structure. Thus, this new industry structure for generating electricity may discourage adequate consideration of the limited long-term availability of the prime energy sources or their value for other uses. Also, if construction of needed new generating plants is delayed, for environmental, economic, or other reasons, competition would not be effective in cost reduction because of energy shortages. Since natural gas is relatively inexpensive to extract from the earth and to distribute in pipelines, it is likely to be an attractive choice for competitive generation in the near term.

Increased reliance on the competitive marketplace, rather than regulation, to determine what types of electricity generation will be built may be at odds with public policy objectives that take into account the best long-term choices for energy supply. Unless the opportunities, disadvantages, and environmental costs associated

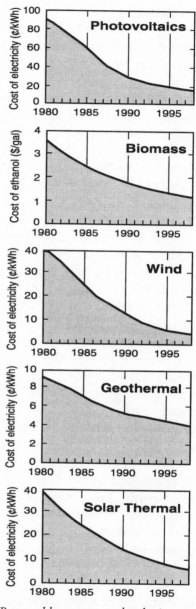

Renewable energy technologies are well along a path of development.

with each fuel source are fully appreciated, experience with markets driven by relatively short-term prices suggests that sound energy practices may not result.

Some have proposed a carbon tax as one method to reduce atmospheric carbon dioxide, which could make natural gas a preferred combustion fuel in relation to coal. Gas derives another advantage from the low capital cost of gas-turbine generators and their small unit sizes that can be incrementally added to meet increased demands. Since essentially all energy sources can produce electrical power and natural gas has advantages for other uses, its application in this field should be limited. Legislation signed by President Carter in the late 1970s restricted the use of natural gas for electricity production but was soon repealed after Carter left office. As mentioned earlier, some increases in natural gas use may be justified to meet special needs, but its more unique applications should be dominant in national energy strategies.

Renewables are attractive energy sources that should be considered in future plans, since by definition they can be replenished. Most energy sources classed as renewables are traceable to the sun and can properly be labeled solar energy. However, the term solar is more commonly applied to systems using the sun's radiation directly to produce heat or electricity, so I will follow that convention. The renewables include hydroelectric systems, solar energy, and biomass as a source of heat, and perhaps biological resources that are convertible to oil or other energy-related products. Geothermal energy from sources within the earth also may be called renewable on the basis that the heat derives at least in part from continuing decay of radioactive elements. Under that rationale, all renewables are dependent on nuclear energy, since the sun's heat also derives from a nuclear fusion reaction.

Hydroelectric energy has been exploited in the United States to nearly its practical limit, but is likely to be sustained at near the present level. Most existing large dams should be replaced at the end of their structural life because the ecological damage caused by the impoundment renders the land unsuitable for other uses. If the dam construction is sufficiently permanent, the impoundment eventually will fill with silt and the dam will lose storage capacity. This permanence also suggests that a careful assessment be made before constructing new hydroelectric projects. Hydropower now produces about 10 percent of U.S. electrical energy generation, having dropped from about 40 percent before World War II, as coal-fired and other plants multiplied.

The aesthetics of free flowing streams in picturesque valleys, natural wildlife habitats, and farm-land uses also must be considered. Some investigators have proposed that increased use of small electric generating plants on river tributaries can produce substantial amounts of electricity. It seems doubtful, however, that those could prove attractive when the costs, the unreliability of small reservoirs during dry periods, and the environmental factors are fully included. Small "run-of-the-river" plants that generate electricity with a minimal dam structure might prove compatible with other stream uses, but even these can disrupt fish movement and change recreational uses.

Biomass as fuel, and solar energy, may be important renewables for the future, even though little experience is available for biomass systems designed solely to produce energy. Of course, the burning of wood constitutes a major energy system for developing societies and is estimated to produce more than 10 percent of world energy in its limited use mostly for heating and cooking. In many cases this is not renewable because the forests are being depleted rather than replanted. Also, because of steep terrain, the nature of some soils and climates, along with the prevailing practice of periodic clear-cutting in many portions of the developing world, restoration is infeasible or impractical. In addition, production of wood or grasses for fuel cannot readily be expanded where additional land is required to produce food for increasing populations. Many questions must be answered before biomass could become a major energy source to power industry and support modern lifestyles. How much land would be needed? Some estimates range up to one-half of available arable land of the United States to equal or replace the energy content of oil now in use; enthusiasts for biomass energy, of course, have projected much smaller areas.

At present, the biomass energy used in the United States primarily produces steam for processing of materials and power to run factories. These units are fueled with scraps from the fabrication of wood products, sugarcane processing, and other waste materials where the industry producing the waste benefits both from the fuel value and avoided disposal costs. There is no established, large-scale biomass industry based on the growing of crops as fuel to produce energy for sale to customers. Studies indicate that biomass could be burned effectively in power plants of economical size, say, about 50 megawatts, fueled with poplar-type trees or saw grass grown within a radius of twenty-five miles. Trees might be harvested on a seven-year cycle, whereas grass would be an annual crop. But other questions must be answered. How much land could be devoted to large energy plantations, either as contiguous acreage or in distributed fields? What would be the best or most feasible pattern for ownership of the land

and the harvesting operation? How long would it take to establish this type of plantation structure to assure the supply? What environmental impact would large areas of these crops produce?

Since we can draw on but little experience with large-scale biomass cultivation for energy production, several other questions are relevant. What overall net efficiency might be achieved when the tally is made for the cost of the energy used for planting, cultivating, harvesting, transporting, and converting to steam, electricity, or other application? What environmental impact would result from periodic clear-cutting that seems to be a necessary practice for biomass to be economically attractive? What processing of the biological materials would be necessary, such as water removal before burning? Alternation of trees and grasses for soil conservation might have environmental merit if the power plant can be designed for both. A small demonstration project sponsored by the DOE is under way in Minnesota to provide an initial data base. In addition, experience will be needed on a sufficiently large scale to evaluate the economics and the environmental effects on native and migratory wildlife populations. Biomass production should not displace land used for foods that have high value for either domestic consumption or for export.

Even so, there are substantial areas of the country that could be adapted to biomass production as an energy resource. Land areas that seem most readily convertible to biomass production are those already in forests, which are environmentally beneficial but not economically optimal. These primarily are in the Southeast, where other agricultural practices of the past have been abandoned, "second growth" forests have evolved, and the annual growing season is long. Biomass as direct fuel could be a significant but not primary energy source and could begin before the middle of our hundred-year time frame. Introduction of biomass energy may be facilitated by combining processed bio-fuel with coal in existing or modified power plants.

The warmth of sunlight on a cool day is comforting and reassuring. Walt Whitman wrote, "Give me the splendid silent sun with all his beams full dazzling." Solar energy systems for production of electricity by photocells, by harnessing atmospheric effects such as wind and ocean currents, and by direct heating of buildings are now practical for some purposes. Passive heating of homes by strategic placement of windows, installation of solar collectors for heating water, and use of Trombe walls seem attractive. (A Trombe wall is an interior wall of material having a high thermal storage capacity that is heated by direct sunlight through glass windows, thus utilizing the greenhouse effect. At night the wall dissipates heat to maintain comfort in the room.) So far, when unusual collectors are required, Americans seem to favor architectural niceties over solar heating benefits. However, architects produce attractive building designs that use window placement and optimum roof overhang extensions to maximize winter heating and minimize summer cooling needs. This is practical for homes and small business buildings but is difficult or impractical for tall central-city buildings.

Solar heat has been demonstrated in small plants designed to produce electricity on a commercial scale. In one design, polished mirrors are arranged in a circular array to focus sunlight on a vessel mounted on a tower. Temperatures high enough to produce viable steam for a turbine-electric generator have been achieved. Such units are limited in size, require special equipment, and are subject to the other limitations of solar systems. Although such towers have been made to work, this appears impractical when mechanisms to follow solar movement and vulnerability to storms are considered. I place this system in the category of the wheeled vehicle in the desert.

Direct conversion of sunlight to electricity by photocell arrays of modest size is an environmentally favorable technology. It matches peak demands for manufacturing plants and areas that require air-conditioning and is increasingly used for remote locations where electric transmission lines would be costly. One successful installation provides power for the Natural Bridges National Monument in Utah. This 50 kW(e) solar power plant and bank of storage batteries covers a little more than one acre of land. In this location at 6,500 feet elevation and with its sunny climate, it provides 90 percent of the energy requirements for the visitors' center and for several nearby families who maintain the monument area. Batteries supply power for two days without sun, after which a backup diesel generator is employed. Petroleum fuels are used for most cooking and heating. That solar installation has been particularly advantageous where a long transmission line would detract from the natural beauty of a pristine area. The installation replaced a large diesel-electric generator that produced noise and pollution. Solar installations need not be this complex to be useful in many countries.

A friend in Bangladesh reports that some remote homes are supplied with photocell units as small as 20 watts. By using batteries and efficient fluorescent bulbs made in China, that size has been adequate for lighting modest huts. Isolated islands of Samoa now effectively use larger centralized solar photo-voltaic units for lighting and communication. Photocells to light highway signs and power other remote or movable installations are now commonplace and effective in the United States and other countries.

Continued development of thin-film solar converters for higher efficiency, and that can be mass produced, offers competitive costs for additional applications. The potential seems greatest for dispersed units of modest size that provide electric power for summer homes or small manufacturing units. In contrast, solar plants mounted on building roofs or walls are environmentally favorable and may be effective for factories and retail establishments. However, central solar units of sizes comparable to present fossil or nuclear plants present very difficult environmental problems. A 500 MW(e) installation of efficient solar collectors, depending on location and terrain, would cover between six and fourteen square miles. Diversion of rainfall and even shading the soil would present difficult environmental situations. Dividing the plant into multiple dispersed units might be more acceptable environmentally but would increase the total area impacted and increase the cost of maintenance, operation, and control.

Sunshine is part time and does not match some load demands in northern areas of concentrated population. Storage of electrical energy appears prohibitively costly where several rainy days can occur in succession. Even if the costs of solar installation are no greater than an alternative source, the capital cost for a reliable power system would be essentially doubled if backup power is provided. Thus solar-electric energy may be expensive for a long time, except for specialized applications, remote locations, or those where the products from electrical-energy use can be produced intermittently and stored. With continued improvement in the efficiency of photocells and other technological developments such as energy storage in superconducting electrical coils, solar-electric may become significant as a substantial energy producer late in our time frame.

The farm windmills of my youth and the picturesque windmills of Holland were used to pump water or to grind flour and animal feed, each of which could be stored. Electricity-producing energy from wind is now nearly competitive with fossil fuels in certain advantageous locations like the California coastal hills that are swept constantly by warm sea breezes. Windmills constructed and operated there, with the help of financial subsidies, have produced valuable experience. This suggests that a larger operation for the manufacture, installation, and operation of windmills would reduce the unit costs to levels competitive with fossil-fuel plants. New windmills having variable pitch blades to accommodate more effectively the variations in wind velocities should improve overall efficiency and extend the regions suitable for wind power.

However, favorable costs for construction and maintenance of installations on U.S. Pacific and European coastal locations may not be achievable for the U.S. Midwest or other regions where average wind velocities are also attractive. There, the weather patterns often produce severe thunder-storms and icing, and thus, construction costs would be higher to accommodate those conditions. Also, winds may not coincide with peak usage, and U.S. interior areas are subject to extended periods of calm. Alternative energy sources would therefore be required at an increased total capital cost. In addition, the aesthetics of large metal towers with widely flailing vanes have not been tested for public acceptance in highly populated areas that require great amounts of electric power.

Even with these limitations, there seem to be many possibilities for expansion of windmill power in the United States, but probably not enough to produce a substantial fraction of our nation's energy requirements. Modern windmills now being installed in some Midwestern and other locations should provide valuable experience for this energy system. If energy prices increase due to shortages or other restrictions in use of fossil energy, windmills could almost immediately be competitive.

Geothermal energy is an intriguing and potentially useful energy source. It has been exploited commercially only where natural steam or hot water springs can be tapped. Although these are limited, hot rock exists at practical depths for exploitation in large geological areas of western states. Deep wells drilled into rock access the thermal energy by injecting and extracting water at pressures of three to four thousand pounds per square inch. The rock is first fractured by high pressures to increase the surface area for transfer of its stored heat to the water. Demonstration units exploiting rock formations about ten thousand feet below the surface suggest that this can be a feasible means for producing electrical energy. However, these geothermal sources produce water or steam temperatures of about 300 degrees Fahrenheit, which compares unfavorably with 900 to 1,100 degrees Fahrenheit for efficient modern fossil-fired steam-electric plants. Practicalities of rock fracture and heat transfer would limit installation to sizes well below those of fossil and nuclear plants.

Special turbine equipment designed for geothermal temperatures must be provided to drive electric generators; at best, the overall efficiency will be low and the turbine size large. Accordingly, the steam condensation equipment also will be proportionately large, and costly cooling systems would be required to dissipate the waste heat. Alternative cycles using condensable vapors may be preferred but would require additional heat-exchange equipment. Much of the hot rock is located in regions of limited water supplies for condensate cooling, further reducing the overall potential. Contamination of recycled water with minerals and resulting corrosion of the system have been longer-term problems for power plants operated from hot springs.

Geological effects of the fractures resulting from a large installation or from distributed long-term cooling of an otherwise stable formation are additional concerns. Some regions favorable with respect to the shallow location of hot rock are susceptible to seismic action; the stability of thermally disturbed formations would therefore require careful evaluation. Drilling costs to reach the required depths for most practical locations also need to be reduced. Hot rock in the eastern United States lies too deep for the process to be economical. Although much testing and evaluation is required, the large areas of hot rock suggest that geothermal sources may offer modest amounts of electrical energy in the future, perhaps late in our time-frame.

Dispersed generating units introduce other problems. In conventional electrical grids, the power emanates from one or a few central generators. In such a system each power line can easily be disconnected for the safety of maintenance workers or for protection during emergencies. But where many sources are contributing power, that safety problem becomes more complex.

———•◦•———

Nuclear energy now supplies about 20 percent of the U.S. electrical energy supply, or 8 percent of total energy consumption. It is commonly estimated that indigenous supplies of uranium available at competitive costs used in Light Water Reactors (LWRs) could supply the present total national electric demand for forty years. LWRs have operated well since the industry matured and have now demonstrated capability to meet major electric energy demands. However, it is only by using more advanced reactors that nuclear technology can become a major, long-term energy source. With breeder reactors, the nuclear capability can be expanded about sixty times to serve a period well beyond our projection. Replacement of fossil fuel use with nuclear energy would dramatically reduce global warming problems, approximately in direct proportion to the substitution.

Unfortunately, social and political concerns about nuclear safety and radioactive waste have discouraged further construction of nuclear plants. With respect to safety the record of existing commercial LWR reactors manufactured and built in the United States, Western Europe, and Japan has been excellent. No nuclear event in these plants has adversely affected the public health. The safety record for nuclear workers has also been favorable compared to safety records in plants dependent on fossil fuels, as well as other kinds of industries; no fatalities in these power plants have been traceable to the nuclear reactions. Newly evolved standardized and smaller LWR plants offer even greater margins for safety. Small plants have a further advantage: more of the equipment can be manufactured and assembled in factories for shipment to the site as an entity. This offers reduced cost, improved quality, and wider margins of safety over that achievable through the difficult field assembly of larger units.

Nuclear power presents a large energy source and demands a highly technological industry. Although prospects for revival of the U.S. nuclear industry are emerging, this capability is substantial, and new technology can be implemented to meet the stated goal of ample energy for the longer term. Although presently applied only to electricity production, it can also extend fossil fuels more directly. Examples are the heating of tar sands and the production of hydrogen in which nuclear power could rationally be used as a means to the end. Nuclear power to desalt sea

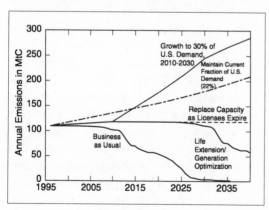

Carbon emissions avoided under various scenarios of nuclear electricity generation.

water also has been proposed to alleviate water shortages in desert coastal areas. All nuclear plants operating today use uranium as fuel. Utilization of thorium would extend nuclear energy capability even further, but new reactor designs and reprocessing of spent fuel would be required. Recovery of uranium from the very small concentrations in sea water provides a possibility for extending the nuclear energy era indefinitely, but the cost and energy required for reclaiming uranium in this way places this source beyond our one-hundred-year time frame.

In order to reach the optimum use of nuclear power as a peacetime energy source, we must develop advanced nuclear reactors different from the ones now in use. Since the 1970s, when General Atomics withdrew its high temperature gas-cooled reactors from the market, only light-water-cooled reactors have been considered for U.S. commercial nuclear plants. Other designs should be pursued.

For example, General Atomics, with national laboratory cooperation, has developed a conceptual design for a modular helium reactor with gas turbine-electric generating capacity (GT-MHR). This new system, described in chapter 9, offers potentially greater safety and operating efficiency than is possible with the LWR. Tests in the small German AVR reactor have shown that this type reactor has the capability to accommodate safely an accidental total loss of the helium coolant. A modest effort for refinement and testing of coated-particle fuel is what is needed for adequate evaluation of GT-MHRs of a commercially viable size, which could be economically competitive with fossil-fuel power plants. Similar nuclear reactors are under study for possible deployment in South Africa, China, Japan, and Russia, which is relying on U.S. assistance.

The design of South Africa's helium-cooled gas-turbine reactor is based on the German technology of the AVR pebble-bed reactor and fuel. The South African concept is to employ up to ten reactors, each of 110 MW(e) capacity, in a complex to be operated from a single control center, and to use common operating and maintenance services. Their plan is to start with an initial unit and add capacity as needed, until the installation reaches the size of modern LWR plants. These units can match growing needs and have the advantage of standardization and shop manufacturing for the small-unit equipment sizes. South Africa's state-owned utility, ESCOM, is to make a decision on the practicality of proceeding in the near future.

The safety features of the GT-MHR and the pebble-bed modular reactor make them attractive for locations near cities, where waste heat could be used to warm or cool public buildings or factories. In addition, they would use uranium ore more efficiently than LWRs and thus extend its utilization. These reactors also seem appropriate for developing countries that lack the infrastructure required to assure safety for other reactor types. Thus, gas-cooled reactor systems have potential to become a major energy source within a few decades.

Since breeder reactors can greatly extend the effective uranium fuel supply, at least a modest, long-term effort should be deployed to improve breeder-reactor concepts and their accompanying fuel cycles. Although much development work has been done and large demonstration plants have operated in Europe and Japan, the breeder reactor remains uncompetitive at present energy prices. However, this capability remains as a bulwark against electrical energy shortages, as other resources are depleted and costs increase correspondingly. Recognition of the need for the breeder-reactor system may occur within this century; the ultimate potential for the breeder reactor extends much further into the future.

Nuclear waste handling and storage are technologically sound operations but continue to be misunderstood and are often criticized by many who are not familiar with the science and engineering involved. This seems most unfortunate since methods for accommodating each facet of radioactive waste technology are well understood by those who have studied the subject thoroughly. Designs for geological storage are well developed, and a facility to store transuranic waste (also suitable in principle for high-level waste) has been constructed in New Mexico. Unfortunately, it remained unused through a decade of intervention and controversy. Engineered interim-storage facilities have been fully qualified and are in use at nuclear power-plant sites. Designs and facilities now exist for storing short-lived radioactive products from industry and medicine, which require decades of storage but not geological disposal. Great need exists for clear and comprehensive technical reports that describe these technologies, followed by publications that provide suitable interpretation for public understanding.

Nuclear waste issues and technology were discussed extensively in chapter 10, but their importance justifies a brief repetition here. Geologically stable bedded salt and shale formations are widely available in the U.S. West and Midwest as suitable sites for storage of high-level waste. These relatively simple geological formations are amenable to evaluation (in contrast to the more complex geology of the controversial Yucca Mountain site). Extrapolation of geological stability of a well-designed salt or shale repository to a time when the waste's radioactivity is less than that of the original uranium ore seems reasonable. The definable stability of such a geological formations warrants high confidence in repositories this type. Although the required fifteen-thousand-years represents a long time in human terms, the geological stability of these salt beds is known to have been free of disturbing events such as earthquakes for the past 200 million years.

Less radioactive wastes from non-fuel reactor components and many medical products, which decay away to safe levels in decades or less, are easily stored in man-made structures that can be observed and repaired if necessary. Those times are comparable to the lives of commercial and residential structures. The problems for disposal of radioactive waste are of perception, not technology.

Nuclear fusion offers potentially inexhaustible energy and less radioactive waste than fission produces. This has provided incentive for a large research effort in the United States and other countries. The extensive study and experimentation to date indicate that a sustained fusion reaction will likely be achieved despite its elusiveness during a half-century of research. It also has produced much "spin-off" technology that may justify the effort. However, fusion, under development since the 1950s, has not reached a stage corresponding to that existing for fission energy in 1945, when practical plants had been envisioned. Note that it has taken nearly fifty years for fission energy to progress and become a supplier of world energy. It is reasonable to expect fusion development and particularly its deployment to require even more time than fission.

The basic difference is that the fusion process produces very energetic neutrons that damage key structural members deep inside the complex and highly radioactive enclosure. Thus, the present concepts for fusion power plants involve maintenance protocols that are costly and time-consuming in the extreme. In contrast, the greatest radiation damage in a fission reactor is to the most exposed structures, which are periodically replaced with the installation of new fuel. The potential of fusion is great, but it probably will not be realized within our time horizon of one hundred years.

———•◦•———

My goal to help develop nuclear power for peacetime purposes, as naively developed in college, is only partially fulfilled. However, an industry that produces approximately 6 percent of the world's energy may be regarded as successful. This important and environmentally friendly energy resource is particularly important when considering long-range energy requirements. I feel privileged to have participated in the development of nuclear technology and nuclear energy for peaceful uses. The world can now utilize the broad benefits of nuclear medicine, radioisotopes for industry, and nuclear energy.

This "new" industry, now estimated to be contributing over three hundred billion dollars annually to the U.S. economy, is important in the nation and the world. Nuclear technology now flourishes in areas such as nuclear medicine to provide many benefits to society, some of which are linked to nuclear-power generation. The economic, health, and social benefits of nuclear technology for industry and medicine are immediate and manifest. Manufacturing plants depend increasingly on nuclear technology for detectors, monitors, and other devices. Sterilization of surgical and other medical equipment by radiation has proven superior over chemical or heat treatment for many applications. Preservation of food by radiation could greatly reduce losses from spoilage and thus expand food supplies, particularly in tropical climates. Elaine's recovery from cancer described in chapter 11 was in large measure made possible by nuclear medicine. In our family's limited sphere, that alone makes a lifetime of nuclear work worthwhile. Nearly every extended family has one or more members who have benefited from the use of radioisotopes and other technologies of the nuclear industry.

The technological change in the world during my lifetime has exceeded that of any prior period of history, and the introduction of nuclear energy surely rivals other momentous occurrences of the past. However, it has come with a large and demanding penalty. The world must effectively control the materials usable for nuclear weapons to ensure that these devices are never again used. This problem is only lightly coupled with the production of nuclear energy, but its importance demands emphasis. Nuclear facilities capable of producing weapons materials must be fully inspectable and protected from catastrophic pitfalls, natural or intended by man. The constraints to prevent use of such weapons of mass destruction must rank among the greatest challenges that civilization has ever known.

Just as pioneers of the American West strove to establish stable communities, all now must help build strong international control of the giant "nuclear genie" that has been released. The beneficial applications of nuclear technology and nuclear energy can be nearly boundless, but their overall use must be within constraints commensurate with the potential for good or evil. I cannot resist another reference to my speech as an undergraduate in college on the benefits and dangers of developing nuclear weapons. That speculative concern is now harsh reality. This issue is of such importance that all members of the planet should feel an obligation and necessity to help in the maintenance of the nuclear weapons constraint so essential to peace. I still cannot conceive a solution that is better than having strong international organizations that can enforce proper uses of the puissant energy of the atom and that also can cope effectively with other devices of mass destruction, including chemical and biological weapons.

In summary, what energy-system choices might be most promising for the long term, starting now? My choice, perhaps influenced by a working lifetime of research, is to pursue improvement in several energy systems. No single option stands out as a preferred choice for all circumstances, but most have merit for some applications. Energy conservation is widely applicable, provides both immediate and extended benefits, and should have the highest priority. Renewables offer lasting sources and security from international crisis and should be employed wherever

Elaine and Don Trauger's fiftieth wedding anniversary gathering on their front porch at 20 Palisades Parkway (left to right): Charles and Lynette Trauger with daughter Nicole in front and son Tyler at far right; Byron and Aleta Trauger with daughter Katherine on Elaine Trauger's lap; Shirley and Bob Trauger behind Don; Lillian and Keith Coley behind Eloise Causey Butler; Jana and Tom Trauger with their daughter Hallie on Don's knee, and Josephine Horton, 1995.

nature and other land uses permit. Nuclear fission is the most abundant of indigenous, non-fossil resources now technologically ready for expanded utilization. The nonrenewables of oil, natural gas, and coal are the easiest to employ for the present, and although these resources are fully developed for immediate utilization in energy production, their impact on the environment is of serious concern. But total dependence on fossil fuels assures the greatest risk for our successors. As a logical consequence research, development, and entrepreneurship are vital to the long-term viability of energy supplies, the economy, and the well-being of our compatriots and successors. These considerations seem even more urgent for those in many other countries now less affluent. Programs of energy development pursued cooperatively among national laboratories, universities, and industry offer practical solutions to alleviate present environmental problems and meet future needs.

The world that evolved during the twentieth century is greatly different from that of my grandmother's childhood. Her home in the dugout was similar to those of inhabitants described by very early historical records. It was only temporary for her, but much of the world's population still lives at such levels. It is now possible for nearly all people to see how the fortunate live, and aspirations so acquired grow into demands—both for survival and for conveniences that help to make life more than just survival. If most human beings are to achieve living standards at all comparable to those of developed countries, the demands on resources, particularly energy, will be of a staggering magnitude. Our planet offers the potential to meet even this challenge, but only if its resources are used at near-optimum efficiency. Continuation of present energy practices can only lead to a decline in quality of life for all, including those who now live comfortably.

We have a heavy responsibility to future generations as we continue to populate the earth. I have read with wonder and amusement of plans to move people to other planets, the moon, and asteroids, even to mine those sources for materials, water, and perhaps energy. No doubt that will someday be feasible, but it surely is easier to populate the Sahara desert, which is well supplied with air and heat and is underlain with some water and oil. Until we see the Sahara widely utilized, space ventures as solutions to earthly problems other than in some areas of research must be viewed with skepticism. We occupy a wonderful planet that struggles to support the life and luxuries that many now enjoy; let us put emphasis on preserving it responsibly and forever for the benefit of all people.

Acknowledgments

The work described in the later chapters of this book was largely conducted by teams functioning in major project activities appropriate to National Laboratories. This interaction often extended to many groups, and I have been assisted by many talented and remarkably cooperative colleagues. Full recognition of all of them would require encumbering the text with an encyclopedic style. I have therefore attempted here to include those who worked most closely with the tasks described, and I apologize to others who have been omitted either by oversight or failure of my aging memory.

The original writing of this book was from memory and was later augmented by researching references, personal papers, and other documents of relevance. To further assure validity in describing the events, each chapter has been subjected to review by one or more persons closely associated with that period of time in my life. Many improvements have been made from corrections and suggestions by John Coobs, Betty Winquest Cooper, Paul Gnadt, Dale Magnuson, David Mickey, Robert Shelton, Caryl Steyer, Bill Tewes, Robert Trauger, Shirley Trauger, and Jim White, but I accept full responsibility for any and all inaccuracies and omissions.

Two persons especially, Marilyn Schuette and Jim Campbell, have assisted throughout the writing of the book. Their encouragement and guidance have reached far beyond what could reasonably be asked of friends and neighbors.

Finally, I am grateful to Bard and Sherrye Young of RedLine Editorial Services for their splendid work on editing this volume and to Andrew B. Miller and his staff at Hillsboro Press for their skillful and timely publication of this book.

Without the encouragement, tolerance, and support of my wife, Elaine Trauger, the text might never have been finished; and the reading with notations for clarification by our sons, Byron and Tom, has been extensive and helpful.

Acronyms

AEC	Atomic Energy Commission	**MSR**	Molten Salt reactor
AGR	Advanced gas-cooled reactor	**MW(e)**	megawatts electric
ANP	Aircraft Nuclear Propulsion	**MW(t)**	megawatts thermal
AVR	Arbeitsgemeinschaft Versuchs Reaktor	**N-A**	Norris-Adler Barrier
EENRC	Energy, Environment, and Natural Resource Committee	**NPOVS**	Nuclear Power Options Viability Study
EGCR	Experimental gas-cooled reactor	**NRC**	Nuclear Regulatory Commission
		NWU	Nebraska Wesleyan University
ETR	Engineering test reactor	**ORAU**	Oak Ridge Associated Universities
GA	General Atomics	**ORR**	Oak Ridge Research Reactor
GC	Greensboro College	**ORINS**	Oak Ridge Institute for Nuclear Studies
GCR	Gas-cooled reactor		
GT-MHR	Gas turbine modular high-temperature gas-cooled reactor	**ORNL**	Oak Ridge National Laboratory
		PBMR	Pebble bed modular reactor
HFIR	High-flux isotopes reactor	**PNC**	Power Reactor and Nuclear Fuel Development Corporation (Japan)
HTGR	High temperature gas-cooled reactor		
IIT	Illinois Institute of Technology	**P&W**	Pratt & Whitney Aircraft Company
kW(e)	kilowatts electric	**SED**	Special Engineering Detachment
LITR	Low intensity test reactor	**TVA**	Tennessee Valley Authority
LWR	Light water reactor	**U.K.**	United Kingdom

Suggested Further Reading

Chapters 1, 2, and 3

Andreas, A. T. *History of the State of Nebraska*. Vol. 1. Chicago: Western Historical.

Barnes, Cass G. *The Sod House*. Lincoln: University of Nebraska Press, 1980.

Creigh, Dorothy Weyer. *Nebraska, American Association for State and Local History*. New York: W. W. Norton, 1977.

Exeter Centennial Book Committee. *They Called It Exeter*. Exeter: Nebraska, 1979.

Gaffney, Wilbur G. *The Fillmore County Story, Geneva Community Grange No. 403*. Geneva, Nebraska, 1968.

Nicoll, Bruce H., and Gilbert H. Savery. *Nebraska: A Pictorial History*. Bicentennial Edition. Lincoln: University of Nebraska Press, 1967.

Chapter 4

Anderson, H. L., E. T. Booth, J. R. Dunning, E. Fermi, G. N. Glasoe, and F. G. Slack. Letters to the Editor, *Physical Review 55* (March 1939): 511–12.

Bohr, Neils, and John Archibald Wheeler. "The Mechanism of Nuclear Fission." *Physical Review 56* (September 1939): 426–50.

Booth, Ethel. *Where Sunflowers Grew*. Lincoln: Nebraska Wesleyan Press, 1962.

Harrington, Jean. "Two Elements for One." *Popular Mechanics*, January, 1941, 1–5, 149A–50A.

Mickey, David H. *Of Sunflowers, Coyotes and Plainsmen, A History of Nebraska*. 2 vols. Lincoln, Nebr.: Wesleyan University Agustums Press, 1992.

Nicoll, Bruce H., and Gilbert H. Savery. *Nebraska: A Pictorial History*. Lincoln: University of Nebraska Press, 1975.

Stewart, G. W. "The Secret of a Teacher: J. C. Jensen." *American Journal of Physics 24*, no. 3 (March 1956): 123–25.

Chapter 5

Booth, Eugene, John Dunning, William L. Laurence, A. O. Nier, and Walter Zinn. *The Beginnings of the Nuclear Age*. Newcomen Society in North America: New York, 1969.

Groueff, Stephane. *Manhattan Project*. Boston: Little, Brown, 1967.

Hewlett, Richard G., and Oscar E. Anderson Jr. *The New World, 1939/1946*. University Park, Penn.: Pennsylvania State University Press, 1962.

Howes, Ruth H., and Caroline L. Herzenberg. *Their Day in the Sun*. Philadelphia: Temple University Press, 1999.

O'Keefe, Bernard J. *Nuclear Hostages*. Boston: Houghton Mifflin, 1983.

Rhodes, Richard. *The Making of the Atomic Bomb*. New York: Simon & Schuster, 1986.

Smyth, Henry DeWolf. *Atomic Energy for Military Purposes*. Princeton, N.J.: Princeton University Press, 1945.

Chapter 6

Federal Writers Project. *New York Panorama*. 1938. Reprint, New York: Pantheon Books, 1984.

Feininger, Andreas. *New York in the Forties*. Mineola, N.Y.: Dover, 1978.

Fosdick, Harry Emerson. *On Being a Real Person*. New York: Harper & Brothers, 1943.

Miller, Robert Moats. *Harry Emerson Fosdick*. New York: Oxford University Press, 1985.

Chapter 7

Johnson, Charles W., and Charles O. Jackson. *City Behind a Fence: Oak Ridge, Tennessee. 1942–1946*. Knoxville: University of Tennessee Press, 1981.

Robinson, George O., Jr. *The Oak Ridge Story*. Kingsport, Tennessee: Southern Publishers, 1950.

Trauger, D. B. "Irradiation Testing of Fuel Materials for EGCR." *Nuclear Science and Engineering 14* (September 1962): 69–82.

Chapter 8

Bussard, R. W., and R. D. Delaver. *Fundamentals of Nuclear Flight*. New York: McGraw Hill, 1965.

Gantz, Kenneth F. ed. *Nuclear Flight*. New York: Duell, Sloane & Pierce, 1960.

Green, Harold, and Alan Rosenthal. *Government of the Atom*. New York: Atherton Press, 1963.

Chapter 9

Landis, J. W., F. W. O'Rourke, C. L. Richard, P. Fortesque, A. J. Goodjohn, J. L. Everett, R. F. Walker, and D. B. Trauger. *Peaceful Uses of Atomic Energy.* Vol. 5. United Nations, New York & International Atomic Energy Agency, Vienna, 1972.

U.S. Dept. of Interior. Bureau of Mines. *A Hundred Years of Helium.* Helium Symposia Proceedings. Information Circular 8417, 1968.

Winnacker, Karl, and Karl Wirtz. *Nuclear Energy in Germany.* La Grange Park, Illinois: American Nuclear Society, 1979.

Chapter 10

Ford, Daniel. *Cult of the Atom.* New York: Simon & Schuster, 1982.

League of Women Voters. *Nuclear Waste Primer.* New York: Lyons & Burford, 1993.

Murray, Raymond L. *Understanding Radioactive Waste.* 2nd. edition. Columbus, Ohio: Battelle Memorial Institute, 1980.

O'Keefe, Bernard J. *Nuclear Hostages.* Boston: Houghton Mifflin, 1983.

Organization for Economic Cooperation and Development. *The Environmental and Ethical Basis of Geological Disposal.* Paris: Bulletin of the Nuclear Energy Agency, 1995.

Weinberg, Alvim M. *The First Nuclear Era.* New York: American Institute of Physics Press, 1994.

Chapter 11

Johnson, Leland, and Daniel Schaffer. *Oak Ridge National Laboratory: The First Fifty Years.* Knoxville: University of Tennessee Press, 1994.

Smyser, Dick. *Oak Ridge 1942–1992.* Oak Ridge: Oak Ridge Community Foundation, 1992.

Trauger, D. B., and J. D. White. "Safety Related Topics from the Nuclear Power Options Viability Study." *Nuclear Safety* 27, no. 4 (1986).

United Methodist Council of Bishops. *In Defense of Creation: The Nuclear Crisis and a Just Peace.* Nashville: Graded Press, 1986.

Chapter 12

Chiles, James. "A Second Wind." *Smithsonian* (March 2000): 50–58.

Cohen, Bernard L. *The Nuclear Energy Option.* New York: Plenum Press, 1990.

Franssen, Herman, John P. Hardt, Jacquelyn K. Davis, Robert J. Hanks, Charles Perry, Robert L. Pfaltzgraff Jr., and Jeffrey Record. *World Energy Supply and International Security.* Institute for Foreign Policy Analysis. Medford, Mass.: Fletcher School of Tufts University, 1983.

Kerr, Richard A. "A Smoking Gun for an Ancient Methane Discharge." *Science* 286 (November 19, 1999): 1465.

Landsberg, Hans H. *Chairman.* Study Group Sponsored by the Ford Foundation and by Resources For the Future. Cambridge, Mass.: Ballinger Publishing, 1979.

Leviticus, Sydney, John I. Antonov, Timothy P. Boyer, and Cathey Stephens. *Science* 287 (March 2000): 24.

Livingston, Robert S., Truman D. Anderson, Theodore M. Bessman, Mitchell Olszwski, Alfred M. Perry, and Colin D. West. *A Desirable Energy Future: A National Perspective.* Philadelphia: Franklin Institute, 1982.

Normile, Dennis. "Ocean Project Drills for Methane Hydrates." *Science* 286 (November 19, 1999): 1456.

Rhodes, Richard, and Denis Beller. "The Need for Nuclear Power." *Foreign Affairs* 79, no. 1 (January/February 2000): 30–44.

Rhodes, Richard. *Nuclear Renewal: Common Sense About Energy.* New York: Penguin Books, 1993.

South, David W. "The Importance of Nuclear Power in Emissions Avoidance." *Nuclear News* (November 1999): 26–32.

Stone, Richard and Phil Szuromi, eds."Powering the Next Century." *Science* 285 (July 1999): 677–711.

Suess, Erwin, Gerhard Bohrmann, Jens Greinert, and Erwin Lausch. "Flammable Ice." *Scientific American* 281 (5): 76–83 (1999).

U.S. Congress. Office of Technology Assessment. *Decisions for a Decade, OTA-SET-490.* Washington, D.C.: GPO, May 1991.

U.S. Department of Commerce. *Nuclear Energy Research and Development Agenda.* Springfield, Va.: National Technical Information Service, no. UCRL-ID-129209, vol. 2 (December 1997).

U.S. Department of Commerce. *Scenarios of U.S. Carbon Reductions.* Springfield, Va.: National Technical Information Service, no. LBNL-40533.

Index

Italicized page numbers refer to references in captions. Acronym list can be found on page 425.

428

429

431

432